Law, Policy and Climate Change

Focusing on systemic risks caused by climate change, this book examines how these risks can be effectively regulated to ensure resilience and avoid catastrophe.

Systemic risks are risks that threaten the systems upon which society depends, including ecosystems, social systems, financial systems and systems of infrastructure. Such risks are typically characterised by inherent complexity, profound uncertainty and overwhelming ambiguity. In combination, these features pose significant regulatory challenges for policy- and law-makers. Examining how different types of systemic risks caused by climate change are being regulated in four different jurisdictions – the EU, the UK, the US and Australia – this book identifies deficiencies associated with regulating systemic risks using a traditional approach, based on a linear relationship between risk and regulation, which is widely used to regulate risk. The book advances a regulatory approach that is, instead, founded on the concept of "risk governance". This involves a structured yet flexible, holistic, interdisciplinary and inclusive basis for responding to systemic risks; and it is, this book argues, a more effective basis for regulating systemic risks given their uncertainty, complexity and ambiguity.

This book will appeal to academics, policy- and law-makers and practitioners working at the intersection of law and policy in the areas of regulation, risk management and climate change.

Dariel De Sousa is Director at Dart Legal & Consulting, Melbourne, Australia. She was formerly a Counsellor in the Legal Division of the World Trade Organization and a Director of Compliance & Enforcement at the Australian Energy Regulator.

Law, Policy and Climate Change

The Regulation of Systemic Risks

Dariel De Sousa

Routledge
Taylor & Francis Group
a GlassHouse book

First published 2023
by Routledge
4 Park Square, Milton Park, Abingdon, Oxon OX14 4RN

and by Routledge
605 Third Avenue, New York, NY 10017

A GlassHouse book

Routledge is an imprint of the Taylor & Francis Group, an informa business

British Library Cataloguing-in-Publication Data
A catalogue record for this book is available from the British Library

Library of Congress Cataloging-in-Publication Data
A catalog record has been requested for this book

ISBN: 978-1-032-18214-8 (hbk)
ISBN: 978-1-032-18215-5 (pbk)
ISBN: 978-1-003-25346-4 (ebk)

DOI: 10.4324/9781003253464

Typeset in Bembo
by Taylor & Francis Books

For Mum & Dad and Scott, Abby & Marley

Contents

Illustrations

Figures

Tables

List of Acronyms and Abbreviations

Adaptation Strategy	EU Strategy on Adaptation to Climate Change
AEP	Annual exceedance probability
AS 3959	Australian Standard 3959–2018 Construction of buildings in bushfire-prone areas
ASCI	Areas of Special Conservation Interest
BAL	Bushfire Attack Level
BCA	Building Code of Australia
Bern Convention	1979 Bern Convention on the Conservation of European Wildlife and Natural Habitats
Bill	Health and Care Bill 2021 (UK)
Biodiversity Strategy	EU Biodiversity Strategy for 2030
Birds Directive	1979 Wild Birds Directive (EU)
BMO	Bushfire Management Overlay (Vic)
BPA	Bushfire Prone Area (Vic)
Brexit	The withdrawal of the United Kingdom from the European Union
CBD	Convention on Biological Diversity
CCC	UK Climate Change Committee
CFRAC	US Climate-Related Financial Risk Advisory Committee
Climate Change Act	Climate Change Act 2008 (UK)
COAG	Coalition of Australian Governments
COP26	2021 UNFCCC Conference of Parties
COVID-19	SARS-CoV-2
CSIRO	Commonwealth Scientific and Industrial Research Organisation
Dodd-Frank Act	Dodd-Frank Wall Street Reform and Consumer Protection Act (US)
EU	European Union
FSOC	US Financial Stability Oversight Council
FSOC Climate Risk Report	FSOC Report on Climate-Related Financial Risk

Global Financial Crisis	The 2008 crisis affecting the global financial system
GM	Genetically modified
Habitats Directive	1992 Habitats Directive (EU)
HCA Report	Third Health and Care Adaptation Report (UK)
ICS	Integrated Care Systems
IPCC	Intergovernmental Panel on Climate Change
IRGC	International Risk Governance Council
IRGC Framework	Risk Governance Framework developed by the International Risk Governance Council
NATO	North Atlantic Treaty Organization
NCC	National Construction Code
NHS	National Health Service
NHS Act	National Health Service Act 2006 (UK)
OECD	Organisation for Economic Co-operation and Development
OFR	US Office of Financial Research
Order	US Executive Order on Climate-Related Financial Risk
SAC	Special Areas of Conservation
SCI	Sites of Community Importance
Secretary	UK Secretary of State for Health and Social Care
SPA	Special Protection Areas
SPPF	State Planning Policy Framework (Vic)
SRC	US Systemic Risk Committee
UK	United Kingdom
UKHSA	UK Health Security Agency
UN	United Nations
UNDP	United Nations Development Programme
UNFCCC	United Nations Framework Convention on Climate Change
US	United States
VBRC	Victorian Bushfires Royal Commission (2009)
VPPs	Victoria Planning Provisions
White Paper	2021 White Paper entitled "Integration and Innovation: Working together to improve health and social care for all"
WHO	World Health Organization

Preface

This book sits at the intersection of areas about which I am passionate and have had the good fortune of providing advice for much of my career – namely, regulation, risk and governance. It is based upon my PhD, which started to germinate while working as a consultant in a law firm, mostly advising government clients.

David Rennick, former CEO of that firm, is a visionary leader who understands the importance of and need for thought leadership in the legal sector. David encouraged me to think about big picture issues that did not immediately translate into revenue for the firm. That encouragement resulted in a large piece of work undertaken for the Australian federal government about how regulation was being used to manage climate change risks to infrastructure. I had the great privilege of working with John Thwaites, former Deputy Premier of Victoria and former Minister for Climate Change on that project, whose passion and concern about climate change was infectious and ultimately led to me taking on my PhD.

Undertaking the research for the PhD and this book has been a profoundly humbling experience. I have been astounded by the wisdom and insight of so many of the academics to whom I have been exposed. Their contributions have highlighted for me the importance of academic endeavour in helping to solve some of society's most complex and challenging problems. And I am in awe of the efforts made by a vast swathe of institutions, governments, independent bodies and individuals to tackle what is proving to be the most serious existential threat faced by humanity to date – namely, climate change.

I am indebted to my outstanding PhD supervisors – Professor Graeme Hodge and Emeritus Professor Arie Freiberg. Their combined intellect, astute observations and sage guidance were critically important for my PhD and, consequently, also the book. I am equally grateful for the insights shared by colleagues over the years, which have helped to shape my views about various regulatory matters, particularly Greg Campbell, Peter Adams and Michele Dawkins. There are many others who have provided intellectual and moral support along the way, including Jayashree Watal, Matthew Kennedy and Cedric Saldhana.

This book would not have been possible without the faith, love and support provided by my family – specifically, my ever-encouraging mother Yvonne, reassuring siblings, and saintly husband Scott who has abounding patience and tolerance. Throughout the process of writing my PhD and this book, my thoughts have been with my late father Lionel who was truly inspiring in his commitment to education, his love of knowledge and his unwavering support for me. It is in his honour that I dedicate this book in the hope that it provides a similar source of inspiration for my two precious little miracles – Abby and Marley – who arrived in the midst of this amazing endeavour and are a constant reminder of why addressing climate change is so important.

Chapter 1

Introduction

Climate change is widely considered to be one of the most significant threats to the future well-being of humanity. Indeed, it has the potential to threaten the very existence of mankind. Following release in 2021 of a report by the UN body responsible for assessing the science related to climate change[1] – namely, the Intergovernmental Panel on Climate Change (IPCC) – the UN Secretary-General warned that the "alarm bells are deafening", that "greenhouse-gas emissions from fossil-fuel burning and deforestation are choking our planet and putting billions of people at immediate risk," and labelled the IPCC report as "code red for humanity".[2]

The threat to humanity arises not only from the physical hazards associated with climate change – like rising temperatures, increasing sea levels, changes in rainfall patterns and extreme weather events[3] – but also from the impact of these hazards on systems upon which society depends,[4] such as

1 IPCC, *Climate Change 2021: The Physical Science Basis. Contribution of Working Group I to the Sixth Assessment Report of the Intergovernmental Panel on Climate Change* (Cambridge University Press, 2021).

2 Secretary-General Press Release SG/SM/20847, dated 9 August 2021 accessible at: https://www.un.org/press/en/2021/sgsm20847.doc.htm. When addressing the UN Security Council, well-known naturalist, Sir David Attenborough, labelled climate change as "the biggest threat to security that modern humans have ever faced" and stated that "[i]f we continue on our current path, we will face the collapse of everything that gives us security", such as food production, access to fresh water, habitable ambient temperature and ocean food chains. Security Council Press Release SC/14445 dated 23 February 2021 accessible at: https://www.un.org/press/en/2021/sc14445.doc.htm.

3 These hazards are outlined in a chapter entitled "Weather and climate extreme events in a changing climate" in IPCC, n. 1 above.

4 Various climate change impacts on systems are described in IPCC, *Climate Change 2014: Impacts, Adaptation, and Vulnerability. Part A: Global and Sectoral Aspects. Contribution of Working Group II to the Fifth Assessment Report of the Intergovernmental Panel on Climate Change* (Cambridge University Press, 2014). An embargo on citation of the report containing the contribution of Working Group II to the IPCC's Sixth Assessment Report was in place at the time of publication and, therefore, has not been referred to in this book.

DOI: 10.4324/9781003253464-1

ecosystems,[5] systems to deliver health services,[6] financial systems,[7] and systems of infrastructure, including critical infrastructure.[8] In short, climate change produces "systemic risks", which are risks that unexpected and unlikely interactions could threaten the survival of natural and human systems.[9] At worst, systemic risk can result in the collapse of an entire system.[10]

A system can be defined in general terms as "a set of elements or parts that are coherently organised and interconnected in a pattern or structure that produces a characteristic set of behaviours, often classified as its 'function' or 'purpose'".[11] Systems exist in a broad variety of different contexts, including natural systems (the micro-organisms, insects, animals, plants, land and water that exist in a particular local environment comprising an ecosystem), technical or engineered systems (consisting of the physical components that work together to perform a useful function) and socio-economic systems (comprising networks of people, institutions, culture and practices that enable or support the provision of services). The scale of a system may vary, ranging from extensive global systems (such as, the global financial system) to much smaller systems (for instance, a local transportation system). Systemic risks may arise in each of these contexts regardless of the type of system, its particular components, scale and context.

Systemic risks are typically characterised by inherent complexity, profound uncertainty and, at times, overwhelming ambiguity. Complexity arises from non-linear cause-effect relationships between a system and the context in which the system exists and operates.[12] Uncertainty is linked to the lack of knowledge or understanding about the behaviour of the system when stressed.[13] Complexity

5 Ibid. See, especially, Chapter 4 – Terrestrial and inland water systems, Chapter 5 – Coastal systems and low-lying areas, and Chapter 6 – Ocean systems.
6 WHO, *COP26 Special Report on Climate Change and Health: The Health Argument for Climate Action* (World Health Organization, 2021). See, especially, Chapter 4 – Build health resilience to climate risks.
7 Financial Stability Board, *The Implications of Climate Change for Financial Stability* (Financial Stability Board, 2020).
8 IPCC, n. 4 above. See, especially, Chapter 8 – Urban areas and Chapter 9 – Rural areas.
9 M.A. Centeno, M. Nag, T.S. Patterson, A. Shaver, & A.J. Windawi, "The emergence of global systemic risk" (2015) 41(1) *Annual Review of Sociology*, pp. 65–85, at 68.
10 J.L. Carboni & H.B. Milward, "Governance, privatization, and systemic risk in the disarticulated state" (2012) 72(s1) *Public Administration Review*, pp. S36–44, at S37.
11 A. Corbett, "A systems approach to regulatory excellence", in C. Coglianese (ed.), *Achieving Regulatory Excellence* (Brookings Institution Press, 2017), pp. 255–6 citing D. Meadows, *Thinking in Systems: A Primer* (Chelsea Green, 2008).
12 IRGC, *IRGC Guidelines for the Governance of Systemic Risks* (International Risk Governance Center, 2018), p. 12. The International Risk Governance Council (IRGC) is an independent non-profit foundation that aims to improve the understanding and management of risks and opportunities by providing insights into systemic risks that have impacts on human health and safety, the environment, the economy and society at large.
13 Ibid. p. 12.

and uncertainty lead in turn to ambiguity about how to respond to such risks, if at all.[14] In combination, these features of systemic risks pose significant regulatory challenges.

This book examines how systemic risks could be regulated to help avoid, or at least reduce, the catastrophic consequences that can result from such risks. It does this by traversing a broad terrain, including analysis of regulatory, risk management and sociological literature, in order to provide insights regarding the complexity of risk and its regulation, particularly in the face of modern, global risks like those caused by climate change. The various facets of risk are discussed, the architecture of regulatory regimes is considered and systems thinking is explored to determine how it can be employed through regulation to effectively manage systemic risks.

While the book is relevant for the regulation of all systemic risks, its primary focus is on systemic risks caused by climate change. An analysis of the regulation of climate change risks to ecosystems in the European Union (EU), the health system in the United Kingdom (UK), the financial system in the United States (US), and systems of infrastructure in Australia is used to help provide context and to determine whether there is room for improvement in the way systems are currently being regulated in response to climate change risks. To set the scene for this book, a brief overview of recent systemic risks, including those caused by climate change, follows.

I Systemic risks

Systemic risks appear to be on the rise. In recent times, we have witnessed the materialisation of systemic risks of a nature, scale and consequence not previously seen.

In late 2019, the emergence of a deadly coronavirus (SARS-CoV-2, or "COVID-19") was first observed in the city of Wuhan, China[15] and has since spread through populations across the globe, producing one of the most significant pandemics in history.[16] The pandemic has already claimed millions of lives[17] and caused significant increases in morbidity.[18] As a result, health systems around the world have been put under extraordinary pressure by the unprecedented surge in the number of people requiring hospitalisation and

14 Ibid. p. 38.

15 *WHO-convened Global Study of Origins of SARS-CoV-2: China Part* (WHO, 2021), p. 9.

16 E.H.P. Frankema & H. Tworek, "Pandemics that changed the world: Historical reflections on Covid-19" (2020) 15(3) *Journal of Global History*, pp. 333–5.

17 The WHO Coronavirus (COVID-19) Dashboard summarises the cumulative COVID-19 cases and deaths and is accessible at: https://covid19.who.int/.

18 L. Piroth et al., "Comparison of the characteristics, morbidity, and mortality of COVID-19 and seasonal influenza: A nationwide, population-based retrospective cohort study" (2021) 9(3) *Lancet Respiratory Medicine*, pp. 251–9.

health care services.[19] Travel bans and border controls used by many governments to help stem the spread of the virus brought transport systems to a grinding halt and caused a massive shock to global supply chains.[20] Quarantine measures, which resulted in the isolation of large swathes of the global population, prevented many people from working and accessing critical services. Socio-economic systems to support the unemployed and the vulnerable have consequently been overwhelmed by those in need.[21] By the time the crisis has finally passed, the COVID-19 pandemic is likely to have affected every nation on earth, including each nation's governments, their businesses and communities, the vast majority of the citizens that make up the global population, as well as a raft of critical systems upon which society depends.

A little more than a decade earlier, in 2008, the global financial system was brought to its knees. A downturn in the US housing market – which followed a credit boom and rapid appreciation of housing asset prices due, in part, to loose banking and prudential regulatory controls – was the catalyst for the so-called "Global Financial Crisis".[22] When complex, risky mortgages were dramatically downgraded by credit rating agencies, a number of key institutions holding large volumes of these assets failed. This triggered the sale of distressed assets, bank withdrawals, and credit freezes. Interlinkages between participants in the global financial system meant that the crisis quickly spread beyond the US to the rest of the world.[23] As a consequence, many countries were plunged into recession, banks across the world suffered massive losses with some bankrupted, and millions of individuals lost their jobs. The Global Financial Crisis has been labelled as the most serious economic crisis since the Great Depression and its effects continue to be felt today.[24]

In what has been characterised as the most important political event in Europe since the fall of the Berlin Wall,[25] "Brexit" has had various

19 V. Haldane et al., "Health systems resilience in managing the COVID-19 pandemic: Lessons from 28 countries" (2021) 27(6) *Nature Medicine*, pp. 964–80.

20 P. Chowdhury et al., "COVID-19 pandemic related supply chain studies: A systematic review" *Transportation Research. Part E, Logistics and Transportation Review* (2021) 148: 102271.

21 *A UN Framework for the Immediate Socio-economic Response to COVID-19* (United Nations, 2020). The United Nations Development Program (UNDP) publishes assessments of the socio-economic impacts of COVID-19 in countries around the world on its website accessible at: https://www.undp.org/coronavirus/socio-economic-impact-covid-19.

22 S. Claessens & L.E. Kodres, *The Regulatory Responses to the Global Financial Crisis: Some Uncomfortable Questions* (International Monetary Fund, 2014), p. 6.

23 IMF, *Global Financial Stability Report – A decade after the global financial crisis: Are we safer?* (IMF, 2018), pp. 56–9; W. Chen, M. Mrkaic, & M.S. Nabar, *The Global Economic Recovery 10 Years After the 2008 Financial Crisis* (2019), p. 5.

24 IMF, n. 23 above, p. 5; C.-Y. Tsang, "Balancing the governance of the modern financial ecosystem: A new governance perspective and implications for market discipline" (2018) 40(2) *Houston Journal of International Law*, pp. 531–612, at p. 1.

25 F. Fabbrini, "Introduction", in *The Law and Politics of Brexit* (Oxford University Press, 2017), p. 1.

repercussions, including for the global trading system. In 2016, the people of the United Kingdom voted by referendum to leave the EU, after being a member since 1973. While provision is made in the Treaty on European Union for withdrawal, at the time the treaty was drafted, there was no serious expectation that such an event would ever occur.[26] However, the Global Financial Crisis coupled with an influx of immigrants from poorer EU States intensified Euroscepticism within the UK and laid the foundation for its withdrawal from the Union.[27] When it finally came to fruition, Brexit sent shock waves throughout the international community. Among other things, Brexit has led to the disruption of trade to, from and through Britain, which is important given that the UK has historically been a major international trading power and was one of the biggest Member States of the EU.[28] Whereas movement of goods between Britain and other countries in the EU was previously free of restriction, this is now subject to new customs procedures, which slow the movement of goods and increase trading costs. Empirical evidence indicates that Brexit played a role in a sharp decline in UK trade in 2021 and that a significant number of firms have experienced delays, increased costs and other administrative barriers in trading with the EU.[29] Noja and colleagues (2021) suggest that the loss of the UK as a significant member of the EU will ultimately lead to a shift in the balance of power regionally and globally[30] and analysis by Graziano and colleagues (2020) indicates that "regime shifts" like Brexit can substantially lower trade and may ultimately lead to "disintegration".[31]

On the surface, these examples – Covid-19, the Global Financial Crisis and Brexit – appear disparate and unrelated. Indeed, they each involve starkly different contexts and discrete causes and consequences. Nonetheless, there is a common feature – namely, they all illustrate the phenomenon of systemic risk. In each case, an adverse event precipitated unexpected and dramatic consequences that propagated throughout an entire system, compromised its operation, and posed a threat to the system's very existence. While the systems affected and the risks to those

26 Ibid. p. 1 referring to C. Hooton & J. Stone, "Brexit: Article 50 was never actually meant to be used, says its author" *Independent* (26 July 2016).
27 P. Schnapper, "Brexit and the risk of European disintegration", in N. da Costa Cabral, J. Renato Gonçalves, & N. Cunha Rodrigues (eds.), *After Brexit* (Springer, 2017), pp. 83–99.
28 P. Della Posta & S.S. Rehman, "Brexit: Origins and future perspectives", in N. da Costa Cabral, J. Renato Gonçalves, & N. Cunha Rodrigues (eds.), *After Brexit* (Springer, 2017), pp. 11–34.
29 J. De Lyon & S. Dhingra, "The impacts of Covid-19 and Brexit on the UK economy: Early evidence in 2021" (2021) *Centre for Economic Performance*, p. 19, at 1, 12.
30 G. Noja, M. Cristea, & A. Yüksel, "Brexit spillovers through international trade and foreign investment: Empirical evidence from EU-27 and the UK" (2021) 68 (5) *Panoeconomicus*, pp. 653–80, at 669.
31 A.G. Graziano, K. Handley, & N. Limão, "Brexit uncertainty and trade disintegration" (2021) 131(635) *The Economic Journal*, pp. 1150–85, at 1183.

systems differ, each example raises questions about the appropriate mechanisms to avoid the consequences that eventually ensued, including regulatory mechanisms.

II Climate change

Systemic risks caused by climate change also appear to be more prevalent. In the recent past and across the globe, a string of disasters likely linked to climate change have left a range of ravaged systems in their wake.

During the so-called "Black Summer" of 2019–20, a period of unusually intense bushfires was experienced in many parts of Australia. The scale, intensity and impacts of these bushfires were unprecedented.[32] There is now a considerable body of research indicating that climate change will exacerbate Australia's exposure to bushfire risks in certain parts of the country by altering key variables – namely, higher temperatures, lower rainfall and increasing drought frequency[33] – all of which are considered to worsen the likelihood, severity, extent and duration of bushfires in Australia.[34] Apart from the lives that were lost during the Black Summer bushfires, millions of hectares of land were burned, destroying systems of infrastructure comprising residential homes and other types of infrastructure,[35] as well as irreversibly damaging vulnerable ecosystems.[36]

A year later, during the northern hemisphere's winter of 2021, the most severe weather seen in 30 years caused intensely cold conditions across the southern parts of the US, including the State of Texas.[37] A recent study indicates a link between global warming caused by climate change and that weather event.[38] The event caused cascading failures of interconnected critical infrastructure –

32 N.J. Abram et al., "Connections of climate change and variability to large and extreme forest fires in southeast Australia" (2021) 2(1) *Communications Earth & Environment*, p. 1.

33 See chapter on "Climate change information for regional impact and for risk assessment" in IPCC, n. 1 above. See, especially, section 12.4.3 on "Regional information on changing climate – Australasia".

34 See, for example, K.J. Hennessy, CSIRO (Australia), & Division of Marine and Atmospheric Research, *Climate Change Impacts on Fire-weather in South-east Australia* (CSIRO Marine and Atmospheric Research, 2005); C. Lucas, K. Hennessy, G. Mills, & J. Bathols, *Bushfire Weather in Southeast Australia: Trends and Projected Climate Change Impacts* (Bushfire Cooperative Research Centre, 2007).

35 A.I. Filkov, T. Ngo, S. Matthews, S. Telfer, & T.D. Penman, "Impact of Australia's catastrophic 2019/20 bushfire season on communities and environment. Retrospective analysis and current trends" (2020) 1(1) *Journal of Safety Science and Resilience*, pp. 44–56; N.J. Abram et al., n. 32 above, p. 2.

36 C.R. Dickman, "Ecological consequences of Australia's 'Black Summer' bushfires: Managing for recovery" (2021) 17(6) *Integrated Environmental Assessment and Management*, pp. 1162–7; N.J. Abram et al., n. 32 above, p. 2.

37 J. Doss-Gollin, D.J. Farnham, U. Lall, & V. Modi, "How unprecedented was the February 2021 Texas cold snap?" (2021) 16(6) *Environmental Research Letters*, 064056, p. 6.

38 J. Cohen, L. Agel, M. Barlow, C.I. Garfinkel, & I. White, "Linking Arctic variability and change with extreme winter weather in the United States" (2021) 373 (6559) *Science*, pp. 1116–21.

namely, gas, electricity, transportation and water infrastructure.[39] Millions of Texans were without light and heat for several days[40] and hundreds died – mostly from hypothermia.[41]

Not long after, during the summer of 2021, parts of western Europe experienced severe flooding caused by heavy rainfall, which was found to be more likely because of climate change and was labelled as a once-in-400-year event.[42] These floods were deadly, resulting in hundreds of lost lives across Germany and Belgium,[43] and thousands more were adversely affected.[44] In addition, many systems of infrastructure were damaged or destroyed, including roads, railway lines, bridges and hospitals. Critical infrastructure, including for the supply of water, gas, electricity, telecommunications and sewerage, was also disrupted or destroyed.[45]

These examples illustrate the consequences of systemic risks caused by climate change, which have been experienced across wide geographic areas, had significant social and economic implications, and affected large numbers of people on an indiscriminate basis. Ominously, it is likely that we will see more of these risks materialising over time.[46] They demand reflection and attention by governments, policy- and law-makers about whether pre-emptive action could have avoided the devastation that eventuated in each case and, if so, what should be done to manage systemic risks that could arise in the future.

Facing up to these risks in an effort to better manage them may be challenging, at least in those countries where climate change and its consequences have not been taken as seriously as they should.[47] Indeed, around the world, climate change has long been controversial.[48] Debate has raged about whether the changing

39 J. Doss-Gollin et al., n. 37 above, p. 1.

40 Ibid. p. 1.

41 Texas Department of State Health Services (2021) *February 2021 Winter Storm-related Deaths – Texas*, accessible at: https://www.dshs.texas.gov/news/updates/SMOC_FebWinterStorm_MortalitySurvReport_12-30-21.pdf.

42 F. Kreienkamp et al., "Rapid attribution of heavy rainfall events leading to the severe flooding in Western Europe during July 2021" (2021) *World Weather Attribution Report*, p. 1, accessible at: https://www.worldweatherattribution.org/wp-content/uploads/Scientific-report-Western-Europe-floods-2021-attribution.pdf.

43 Ibid. p. 2.

44 A. Fekete & S. Sandholz, "Here comes the flood, but not failure? Lessons to learn after the heavy rain and pluvial floods in Germany 2021" (2021) 13(21) *Water*, 3016, p. 1.

45 F. Kreienkamp, et al., n. 42 above, pp. 2 and 38; A. Fekete and S. Sandholz, n. 44 above, pp. 4 and 11.

46 N.J. Abram et al., n. 32 above p. 12; J. Doss-Gollin et al., n. 37 above; A. Fekete & S. Sandholz, n. 44 above, p. 17.

47 At the most recent Conference of the Parties to the United Nations Framework Convention on Climate Change (UNFCCC) in Glasgow in November 2021 (COP26), there were a number of countries that resisted efforts for a rapid move away from fossil fuels, the combustion of which causes the release of greenhouse gases and leads to global warming.

48 See, for example, M. Hulme, *Why We Disagree about Climate Change: Understanding Controversy* (Cambridge University Press, 2013).

climate is linked to human activity and, if so, what action – if any – should be taken to halt further change. This book accepts that climate change is a real phenomenon and, rather than debating its existence, cause and extent, it focuses upon what can be done to deal with the systemic risks caused by climate change.

III Systems thinking

Recognising a risk as systemic is significant because it means that the lens for consideration shifts away from a predominant focus on the specific *causes* and *consequences* of the risk towards the particular *system* at risk. This, in turn, affects the manner in which risk is analysed and managed.

Haimes (2009) suggests that risk to a system is best understood through a systems-based approach, which accounts for the specific characteristics of the system, including its possible states, vulnerability and resilience.[49] Renn (2016) further argues that "[r]isk analysis for systemic risks must focus on interdependencies, ripple and spillover effects, and other nonlinear dynamics that cascade between otherwise unrelated risk domains".[50] These suggestions are logical and make intuitive sense. They imply a decomposition of the system at risk so that the system's complexities, interconnections and interdependencies can be fully understood and properly accounted for when responding to risk.

Systems thinking is an interdisciplinary approach that enhances understanding of the behaviour of a system and, therefore, also assists in managing risks to the system. Arnold and Wade (2015) define systems thinking as "a set of synergistic analytic skills used to improve the capability of identifying and understanding systems, predicting their behaviors, and devising modifications to them in order to produce desired effects."[51] Richmond (1994) describes this way of thinking as "the art and science of making reliable inferences about behaviour by developing an increasingly deep understanding of underlying structure".[52] Among other things, systems thinking provides a framework to describe and investigate a system, particularly the nexus and relationship of the system's components, the boundaries between the system and the external environment, as well as the functioning of the system within the environment of which it is a part.[53]

49 Y.Y. Haimes, "On the complex definition of risk: A systems-based approach" (2009) 29(12) *Risk Analysis*, pp. 1647–54, at 1647.

50 O. Renn, "Systemic risks: The new kid on the block" (2016) 58(2) *Environment: Science and Policy for Sustainable Development*, pp. 26–36, at 31 citing T. Horlick-Jones & J. Sime, "Living on the border: Knowledge, risk and transdisciplinarity" (2004) 36 *Futures*, 441–456.

51 R.D. Arnold & J.P. Wade, "A definition of systems thinking: A systems approach" (2015) 44 *Procedia Computer Science*, pp. 669–78, at 675.

52 Ibid. p. 671 referring to B. Richmond, "Systems dynamics/systems thinking: Let's just get on with it" (1994) *International Systems Dynamics Conference* (UK).

53 S. Strijbos, "Systems thinking" in R. Frodeman, J.T. Klein, & R.C.S. Pacheco (eds.), *The Oxford Handbook of Interdisciplinarity* (Oxford University Press, 2nd edition, 2017), p. 295.

In more practical and specific terms, systems thinking involves two distinct aspects – one involves breaking down a system to identify and understand its component parts and their inter-relationship, and the other involves rising above the detail of the system's elements to better understand how the system functions as a whole. As Grobstein (2007) puts it, "simple things interacting in simple ways can yield surprisingly complex outcomes" and, therefore, "it is necessary to work both downward and upward" when dealing with complex systems.[54]

Regarding the first aspect – which involves a consideration of a system's constituent elements – Gardner and colleagues (2019) suggest that "by exploring the connections between elements, and giving the connections equal status to elements, systems thinking focuses on understanding the inter-relationships, interactions and system boundaries that give rise to, and at the same time, constrain or enable, possibilities for action and change".[55] In relation to the second aspect – which involves seeing the system as a whole – the OECD (2017) suggests that exploring the system, including its components and relationships between them, can lead to a better understanding of how a system is structured and operates in its entirety.[56] The profound understanding of a system that such an approach can yield is likely to affect the manner in which the causes and consequences of risks to the system are identified and examined, which, in turn, helps to ascertain how system outcomes could be improved when risks actually materialise.

Adopting a systems lens in this way means that a narrow and compartmentalised view of a problem is avoided and, rather, account is taken of the entire system in which the problem has emerged before a solution is developed. More specifically, "deconstruction" of the system at risk can help to identify the weak links within the system and provide insights about how such weakness could be addressed, the impact that regulatory intervention could have to resolve the weakness, as well as the implications of such intervention for the system as a whole.

54 P. Grobstein, "From complexity to emergence and beyond: Towards empirical non-foundationalism as a guide for inquiry" (2007) 90(1/2) *Soundings: An Interdisciplinary Journal*, pp. 9–31, at 11–12.

55 K. Gardner, S. Olney, L. Craven, & D. Blackman, "How can systems thinking enhance stewardship of public services?" (2019) *Public Service Research Group, Issues Paper No. 4*, p. 6 referring to R. Abercrombie, E. Harries and R. Wharton, *Systems Change: A Guide to What It Is and How to Do It* (Lankley Chase, New Philosophy Capital (NPC) 2015) and L. Johnston, C. Matteson & D. Finegood, "Systems science and obesity policy: A novel framework for analyzing and rethinking population-level planning" (2014) 104(7) *American Journal of Public Health*, 1270–1278.

56 OECD, *Systems Approaches to Public Sector Challenges: Working with Change* (OECD, 2017), p. 17.

IV A systems approach to regulation

Systems thinking is increasingly used in a variety of fields,[57] such as engineering,[58] epidemiology,[59] and management.[60] A systems approach in these areas has helped to overcome the challenges of complexity by avoiding a limited focus on specific elements of a system or particular events affecting a system. Rather, a broader perspective is adopted that enables an understanding of the system overall.

Although reliance on systems thinking in the public sector is comparatively less well-established, there is, nevertheless, evidence of a growing recognition of the importance of incorporating the precepts of this way of thinking into policy- and law-making, as well as regulatory theory and practice. In 2011, the OECD noted that:[61] "Understanding key characteristics of complex systems is important for anticipating events that may require policy interventions and identifying where those interventions should or could occur for maximum efficiency."

More recently, in 2017, the OECD published a report which considered how a systems approach can help the public sector solve complex or "wicked" problems.[62] The role that systems thinking might play in the context of a government's use of regulation to manage systemics risks has also been specifically recognised. Corbett (2017) states that regulating complex systems involves identifying and exploiting "leverage points", which put "systems on a path toward achieving public policy goals".[63]

Systems thinking offers important benefits for policy- and law-making and for regulation. A detailed understanding of a system's elements, as well as the forces and pressures exerted on the system, can help unpack the role each element plays in the system, how each element could cause risk to the system or be vulnerable to risk, and the consequences for the system if risk were to emerge in the context of one or more elements of the system. In turn, this understanding can assist in determining which of the system's elements is an appropriate candidate for government intervention and also the form that such intervention should take, such as through regulation.

57 D. Cabrera, L. Colosi, & C. Lobdell, "Systems thinking" (2008) 31(3) *Evaluation and Program Planning*, pp. 299–310, at 299.
58 See, for example, J. Monat & T. Gannon, "Applying systems thinking to engineering and design" (2018) 6(3) *Systems*, p. 34.
59 See, for example, S. Galea, M. Riddle, & G.A. Kaplan, "Causal thinking and complex system approaches in epidemiology" (2010) 39(1) *International Journal of Epidemiology*, pp. 97–106.
60 See, for example, M.C. Jackson, "Fifty years of systems thinking for management" (2009) 60(5) *The Journal of the Operational Research Society*, p. S24.
61 OECD, *Future Global Shocks: Improving Risk Governance* (OECD, 2011), p. 58.
62 OECD, n. 56 above.
63 A. Corbett, n. 11 above, pp. 266–7.

Designing regulation that is capable of facilitating adaptation to systemic risks caused by climate change so as to minimise harm is critically important, especially given the concerning, if not alarming, trajectory of global warming[64] and the range of impacts that this is likely to entail.[65] Nonetheless, the dramatic consequences that could befall a system when exposed to risk, as well as the uncertainty and unpredictability of risk outcomes, highlight the challenge for those tasked with regulating systemic risks caused by climate change.

To be effective, the regulatory paradigm must be capable of addressing the key characteristics of systemic risks – namely, complexity, uncertainty and ambiguity. This is no easy task. Applying a systems approach will necessarily involve a mapping of the system at risk, an analysis of the stressors to which the system may be subject, and an examination of the likely interaction between the system and those stressors. This is a big ask of policy- and law-makers and regulators, who may lack the resources and experience to undertake such work. The challenges are likely to be heightened where large-scale systems are at play. Moreover, as noted by Corbett (2017), systems pose distinctive regulatory challenges because, among other things, there will seldom exist any single, clear point of regulatory intervention that can determine or influence system outcomes and regulatory intervention may generate a non-linear, unexpected result given the system's complexity.[66]

Indeed, developing regulatory regimes with the capacity to manage risks to systems that display inherent complexity, profound uncertainty and, at times, overwhelming ambiguity may be beyond the existing portfolio of regulatory options that have been employed by governments to manage risks to date. However, if the kinds of catastrophic consequences associated with systemic risks that have been witnessed in recent times are to be averted, there is no option but to take stock of such risks, recognise their distinctive features, and start thinking more clearly and creatively about regulating them, particularly systemic risks caused by climate change given their existential implications.

V Purpose of this book

Research for this book laid bare the chasm between risk and regulatory thinking and practice. Traditionally, these areas do not cross paths, at least not to any significant extent. Those involved in designing and applying regulation are seldom risk management experts. In a working paper concerning how the US public sector could strengthen risk management, McKinsey (2011) noted that:[67]

64 IPCC, "Future global climate: Scenario-based projections and near-term information" in IPCC, n. 1 above.
65 Various climate change impacts on systems are referred to in IPCC, n. 4 above.
66 A. Corbett, n. 11 above, p. 256.
67 McKinsey, "Strengthening risk management in the US public sector" (2011) *McKinsey Working Papers on Risk*, No. 28, p. 1.

> As the US government seeks to fulfill its purpose of safeguarding the freedom and well-being of its citizens, it operates in a world fraught with risk. Government departments and agencies face enormous risks every day, and the role of most government institutions is implicitly or explicitly rooted in managing risks ... Yet, for a long time, risk management was an often-overlooked player on the grand stage of both business and government. Only recently ... has it come to the forefront of public debate, owing to risk-management failures that had shocking and widespread ripple effects ...

As explained later in this book, regulation is essentially about the management of risk, which emphasises the need to bridge the divide between these two domains. This book seeks to provide such a bridge.

Historically, the concept of systemic risks has been predominantly used to refer to risks arising in the context of financial systems, but is not well-understood in other domains. The book explains what systemic risks are, how they are distinct from other types of risks, and why the concept has general application to a vast array of systems, as well as the variety of risks to which those systems could be exposed. The book includes a discussion of why systemic risks are becoming more prevalent in modern society.

Systems thinking, and its application in a regulatory context, is also explored. Although systems thinking is gradually gaining traction in government, it has mostly been applied to the manner in which a government operates as an enterprise, rather than the way in which it regulates in the public interest. The book considers the role that systems thinking can play in designing regulation and its particular importance in the context of the regulation of systemic risks, especially those caused by climate change.

The book identifies and discusses different approaches to the design of regimes to regulate risk, which have not been the subject of focused attention to date. The book considers how regulation is traditionally used to manage risk, including an examination of the deficiencies of such an approach for the regulation of systemic risks. The book also explores the concept of risk governance, covering its origins and the contexts in which it has been applied, as well as the contribution it can make to enhancing the regulation of systemic risks.

The book puts forward a new paradigm for the design of regimes to regulate systemic risks that is based on the concept of risk governance. This paradigm, which has been developed with systemic risks caused by climate change specifically in mind, has been used in the book as the basis for assessing the way in which systemic risks caused by climate change are currently being regulated in different jurisdictions around the world. The paradigm can also serve as a general guide for the design of regulation for the spectrum of systemic risks that are becoming increasingly commonplace in modern society.

VI Structure of this book

This Introduction provides a brief overview of the key topics and themes covered in this book, which are developed and elaborated in later chapters.

Chapter 2 examines systemic risks. The rise of systemic risks in modern society is discussed, the foundational concept of risk is described, and the distinctive features of systemic risks are explored.

Chapter 3 describes climate change and explains the range of systemic risks that it can cause. The importance of adapting to climate change, including the role of adaptation as a means to manage systemic risks caused by climate change, is discussed. The critical place of regulation among the suite of options to adapt to climate change is considered.

Chapter 4 discusses regulation in more depth to discover the role that regulation can play in managing systemic risks, notwithstanding the challenges. The architecture of a regulatory regime is examined to determine how the components of a regulatory regime could be mobilised to most effectively manage systemic risks.

Chapter 5 defines and illustrates the use of a "traditional approach" to regulate risk by referring to various examples from around the world. The chapter highlights the inadequacies of a traditional approach to the regulation of systemic risks, particularly those caused by climate change.

Chapter 6 considers "risk governance" as an alternative to a traditional approach to the regulation of risk. The key points of distinction between these regulatory approaches are identified. The chapter includes a discussion of why a risk governance approach is likely to be more effective for the regulation of systemic risks, rather than a traditional approach.

Chapter 7 establishes the core components of a new paradigm for the design of a regime to regulate systemic risks, which is based on the concept of risk governance. Reference is made to how those components apply to the regulation of systemic risks caused by climate change. The paradigm is based on a fusion between regulatory theory and practice and risk management principles and approaches.

Chapters 8–11 examine the design of regulatory regimes to address systemic risks caused by climate change from across the world. Regimes in four different jurisdictions are considered – namely, the EU (climate change risks to EU ecosystems), the UK (climate change risks to the UK health system), the US (climate change risks to the US financial system) and Australia (climate change risks to infrastructure systems in Australia). Each regime is considered to determine how systemic risks are regulated and the extent to which the regime aligns with the paradigm for the ideal regulation of systemic risks put forward in Chapter 7.

Using the analysis from the case studies, Chapter 12 identifies lessons that could inform the regulation of the spectrum of systemic risks at international, national and local levels.

Bibliography

Abram, N.J., Henley, B.J., Sen Gupta, A., Lippmann, T.J.R., Clarke, H., Dowdy, A.J., Sharples, J.J., Nolan, R.H., Zhang, T., Wooster, M.J., Wurtzel, J.B., Meissner, K.J., Pitman, A.J., Ukkola, A.M., Murphy, B.P., Tapper, N.J., & Boer, M.M. (2021). "Connections of climate change and variability to large and extreme forest fires in southeast Australia." *Communications Earth & Environment*, 2(1). https://doi.org/10.1038/s43247-020-00065-8.

Arnold, R.D., & Wade, J.P. (2015). "A definition of systems thinking: A systems approach." *Procedia Computer Science*, 44, 669–678. https://doi.org/10.1016/j.procs.2015.03.050.

Cabrera, D., Colosi, L., & Lobdell, C. (2008). "Systems thinking." *Evaluation and Program Planning*, 31(3), 299–310. https://doi.org/10.1016/j.evalprogplan.2007.12.001.

Carboni, J.L., & Milward, H.B. (2012). "Governance, privatization, and systemic risk in the disarticulated state." *Public Administration Review*, 72(s1), S36–S44. https://doi.org/10.1111/j.1540-6210.2012.02670.x.

Centeno, M.A., Nag, M., Patterson, T.S., Shaver, A., & Windawi, A.J. (2015). "The emergence of global systemic risk." *Annual Review of Sociology*, 41(1), 65–85. https://doi.org/10.1146/annurev-soc-073014-112317.

Chen, W., Mrkaic, M., & Nabar, M.S. (2019). "The Global Economic Recovery 10 Years After the 2008 Financial Crisis." *IMF Working Paper, 19/83*. https://www.imf.org/en/Publications/WP/Issues/2019/04/26/The-Global-Economic-Recovery-10-Years-After-the-2008-Financial-Crisis-46711.

Chowdhury, P., Paul, S.K., Kaisar, S., & Moktadir, Md. A. (2021). "COVID-19 pandemic related supply chain studies: A systematic review" *Transportation Research. Part E, Logistics and Transportation Review*, 148: 102271. https://doi.org/10.1016/j.tre.2021.102271.

Claessens, S., & Kodres, L.E. (2014). "The Regulatory Responses to the Global Financial Crisis: Some Uncomfortable Questions." International Monetary Fund. https://www.imf.org/en/Publications/WP/Issues/2016/12/31/The-Regulatory-Responses-to-the-Global-Financial-Crisis-Some-Uncomfortable-Questions-41422.

Cohen, J., Agel, L., Barlow, M., Garfinkel, C. I., & White, I. (2021). "Linking Arctic variability and change with extreme winter weather in the United States." *Science*, 373(6559), 1116–1121. https://doi.org/10.1126/science.abi9167.

Corbett, A. (2017). "A systems approach to regulatory excellence." In C. Coglianese (ed.), *Achieving Regulatory Excellence*. Brookings Institution Press. https://doi.org/10.7864/j.ctt1hfr226.19.

De Lyon, J., & Dhingra, S. (2021). "The impacts of Covid-19 and Brexit on the UK economy: Early evidence in 2021." *Centre for Economic Performance*, 21 (Covid-19 Analysis Series), 19.

Della Posta, P., & Rehman, S.S. (2017). "Brexit: Origins and future perspectives." In N. da Costa Cabral, J. Renato Gonçalves, & N. Cunha Rodrigues (eds.), *After Brexit*, 11–34. Springer International Publishing. https://doi.org/10.1007/978-3-319-66670-9_2.

Dickman, C.R. (2021). "Ecological consequences of Australia's "Black Summer" bushfires: Managing for recovery." *Integrated Environmental Assessment and Management*, 17(6), 1162–1167. https://doi.org/10.1002/ieam.4496.

Doss-Gollin, J., Farnham, D.J., Lall, U., & Modi, V. (2021). "How unprecedented was the February 2021 Texas cold snap?" *Environmental Research Letters*, 16(6), 064056. https://doi.org/10.1088/1748-9326/ac0278.

Fabbrini, F. (2017). "Introduction." In *The Law & Politics of Brexit* (Vol. 1). Oxford University Press. https://doi.org/10.1093/oso/9780198811763.003.0001.

Fekete, A., & Sandholz, S. (2021). "Here comes the flood, but not failure? Lessons to learn after the heavy rain and pluvial floods in Germany 2021." *Water*, 13(21), 3016. https://doi.org/10.3390/w13213016.

Filkov, A.I., Ngo, T., Matthews, S., Telfer, S., & Penman, T.D. (2020). "Impact of Australia's catastrophic 2019/20 bushfire season on communities and environment. Retrospective analysis and current trends." *Journal of Safety Science and Resilience*, 1(1), 44–56. https://doi.org/10.1016/j.jnlssr.2020.06.009.

Financial Stability Board. (2020). "The Implications of Climate Change for Financial Stability." https://www.fsb.org/wp-content/uploads/P231120.pdf.

Frankema, E.H.P., & Tworek, H. (2020). "Pandemics that changed the world: Historical reflections on Covid-19." *Journal of Global History*, 15(3), 333–335. https://doi.org/10.1017/S1740022820000339.

Frodeman, R., Klein, J.T., & Pacheco, R.C.S. (eds.). (2017). *The Oxford Handbook of Interdisciplinarity*. Oxford University Press (Second edition).

Galea, S., Riddle, M., & Kaplan, G.A. (2010). "Causal thinking and complex system approaches in epidemiology." *International Journal of Epidemiology*, 39(1), 97–106. https://doi.org/10.1093/ije/dyp296.

Gardner, K., Olney, S., Craven, L., & Blackman, D. (2019). "How can systems thinking enhance stewardship of public services?" *Public Service Research Group, Issues Paper No. 4*.

Graziano, A. G., Handley, K., & Limão, N. (2021). "Brexit uncertainty and trade disintegration." *The Economic Journal*, 131(635), 1150–1185. https://doi.org/10.1093/ej/ueaa113.

Grobstein, P. (2007). "From complexity to emergence and beyond: Towards empirical non-foundationalism as a guide for inquiry." *Soundings: An Interdisciplinary Journal*, 90 (1/2), 9–31.

Haimes, Y.Y. (2009). "On the complex definition of risk: A systems-based approach." *Risk Analysis*, 29(12), 1647–1654. https://doi.org/10.1111/j.1539-6924.2009.01310.x.

Haldane, V., De Foo, C., Abdalla, S.M., Jung, A.-S., Tan, M., Wu, S., Chua, A., Verma, M., Shrestha, P., Singh, S., Perez, T., Tan, S.M., Bartos, M., Mabuchi, S., Bonk, M., McNab, C., Werner, G.K., Panjabi, R., Nordström, A., & Legido-Quigley, H. (2021). "Health systems resilience in managing the COVID-19 pandemic: Lessons from 28 countries." *Nature Medicine*, 27(6), 964–980. https://doi.org/10.1038/s41591-021-01381-y.

Hennessy, K. J., CSIRO (Australia) & Division of Marine and Atmospheric Research. (2005). *Climate Change Impacts on Fire-weather in South-east Australia*. CSIRO Marine and Atmospheric Research.

Hulme, M. (2013). *Why We Disagree about Climate Change: Understanding Controversy*. Cambridge University Press.

IMF. (2018). *Global Financial Stability Report – A decade after the global financial crisis: Are we safer? International Monetary Fund.*

IPCC. (2014). *Climate Change 2014: Impacts, Adaptation, and Vulnerability. Part A: Global and Sectoral Aspects. Contribution of Working Group II to the Fifth Assessment Report of the Intergovernmental Panel on Climate Change*. Cambridge University Press.

IPCC. (2021). *Climate Change 2021: The Physical Science Basis. Contribution of Working Group I to the Sixth Assessment Report of the Intergovernmental Panel on Climate Change*. Cambridge University Press. https://www.ipcc.ch/report/ar6/wg1/.

IRGC. (2018). *IRGC Guidelines for the Governance of Systemic Risks*. International Risk Governance Center.

Jackson, M.C. (2009). "Fifty years of systems thinking for management." *The Journal of the Operational Research Society*, 60(5), S24.

Kreienkamp, F., Philip, S.Y., Tradowsky, J.S., Kew, S.F., Lorenz, P., Arrighi, J., … Wanders, N. (2021). "Rapid attribution of heavy rainfall events leading to the severe flooding in Western Europe during July 2021." *World Weather Attribution Report*, https://www.worldweatherattribution.org/wp-content/uploads/Scientific-report-Western-Europe-floods-2021-attribution.pdf.

Lucas, C., Hennessy, K., Mills, G. & Bathols, J. (2007). *Bushfire Weather in Southeast Australia: Trends and Projected Climate Change Impacts* (Consultancy Report prepared for The Climate Institute of Australia). Bushfire Cooperative Research Centre.

McKinsey. (2011). "Strengthening risk management in the US public sector." *McKinsey Working Papers on Risk*, No. 28 https://www.mckinsey.com/~/media/mckinsey/dotcom/client_service/risk/working%20papers/28_wp_risk_management_in_the_us_public_sector.ashx.

Monat, J., & Gannon, T. (2018). "Applying systems thinking to engineering and design." *Systems*, 6(3), 34. https://doi.org/10.3390/systems6030034.

Noja, G., Cristea, M., & Yüksel, A. (2021). "Brexit spillovers through international trade and foreign investment: Empirical evidence from EU-27 and the UK." *Panoeconomicus*, 68(5), 653–680. https://doi.org/10.2298/PAN171229008N.

OECD. (2011). *Future Global Shocks: Improving Risk Governance*. Organisation for Economic Co-operation and Development. https://doi.org/10.1787/9789264114586-en.

OECD. (2017). *Systems Approaches to Public Sector Challenges: Working with Change*. Organisation for Economic Co-operation and Development. https://doi.org/10.1787/9789264279865-en.

Piroth, L., Cottenet, J., Mariet, A.-S., Bonniaud, P., Blot, M., Tubert-Bitter, P., & Quantin, C. (2021). "Comparison of the characteristics, morbidity, and mortality of COVID-19 and seasonal influenza: A nationwide, population-based retrospective cohort study." *Lancet Respiratory Medicine*, 9(3), 251–259. https://doi.org/10.1016/S2213-2600(20)30527–0

Renn, O. (2016). "Systemic risks: The new kid on the block." *Environment: Science and Policy for Sustainable Development*, 58(2), 26–36. https://doi.org/10.1080/00139157.2016.1134019.

Richmond, B. (1994). "Systems dynamics/systems thinking: Let's just get on with it." *International Systems Dynamics Conference (UK)*.

Schnapper, P. (2017). "Brexit and the risk of European disintegration." In *After Brexit: Consequences for the European Union* (pp. 83–99). Cham: Springer International Publishing. https://doi.org/10.1007/978-3-319-66670-9_5.

Strijbos, S. (2017). "Systems thinking." In R. Frodeman, J.T. Klein, & R.C.S. Pacheco (eds.), *The Oxford Handbook of Interdisciplinarity*. Oxford University Press (Second edition).

Tsang, C.-Y. (2018). "Balancing the governance of the modern financial ecosystem: A new governance perspective and implications for market discipline." *Houston Journal of International Law*, 40(2), 531–612.

United Nations. (2020). *A UN Framework for the Immediate Socio-economic Response to COVID-19* (UN Executive Office of the Secretary-General (EOSG) Policy Briefs and Papers No. 6; UN Executive Office of the Secretary-General (EOSG) Policy Briefs and Papers, Vol. 6). United Nations. https://doi.org/10.18356/420812ce-en.

WHO. (2021a). *WHO-convened Global Study of Origins of SARS-CoV-2: China Part*. Joint WHO-China Study. World Health Organization

WHO. (2021b). *COP26 Special Report on Climate Change and Health: The Health Argument for Climate Action*. World Health Organization. https://apps.who.int/iris/handle/10665/346168.

Chapter 2

The Rise of Systemic Risks

That the world is different today compared to the past is a truism. Nonetheless, a key point of difference between the past and the present that is worth specifically noting is the degree of interconnectedness and interdependence between countries, economies, communities, institutions and people. Demographic, technological and social changes are among the forces that have created an inter-woven society that is increasingly characterised by a vast array of complex, interconnected systems upon which society has come to heavily depend. These systems, which span numerous sectors and are integrated into many aspects of our everyday lives, also pose systemic risks. As explained in this chapter, these risks are a real cause for concern given our reliance on such systems.

I Modern society and its systems

There are many different types of systems. They include natural systems (such as ecosystems that have existed for millennia), transportation systems (including the network of railway lines established in Britain during the Industrial Revolution) and engineered systems (such as the network of aqueducts comprising pipes, channels and bridges that delivered water to ancient Rome). As is evident from these examples, systems are not new, nor are risks to systems. However, modern systems are different from systems of the past; so too are the risks that confront those systems. Homer-Dixon and colleagues (2015) explain:[1]

> The global economy has expanded nearly 20-fold since the 1950s, as measured by total global GDP, and inputs of resources from natural systems and outputs of waste back into those systems have increased about 7-fold. Many of these natural systems, including forests, fisheries, hydrological cycles, and atmospheric and terrestrial sinks are consequently under enormous strain, and some, such as Earth's climate, are exhibiting a higher frequency of extreme behavior.

1 T. Homer-Dixon et al., "Synchronous failure: the emerging causal architecture of global crisis" (2015) 20(3) *Ecology and Society*, p. 6.

DOI: 10.4324/9781003253464-2

> During the same period, the revolution in information technologies, the quintupling of global trade, and the homogenization of human institutions, culture, and technologies have produced a sharp increase in the connectivity and the speed of operation of human social and economic systems.
>
> The combined result of these changes has been the emergence for the first time in human history of a single, tightly coupled human social-ecological system of planetary scope.

Boin (2010) adds that "[t]he growing complexity of social, corporate, industrial, financial, infrastructural and administrative systems – and the tight coupling between these systems – have produced 'highways for failure'".[2]

Many modern systems extend across the entire globe – like financial, transportation and international trading systems – giving rise to the possibility of risks in one part of the system propagating throughout the world,[3] as occurred during the Global Financial Crisis. Population growth and urbanisation have also led to increasing points of interface between natural and human systems, causing system transformations that would not have otherwise occurred;[4] the zoonotic origins of COVID-19 are a case in point. Typically, many modern systems are themselves interconnected, such as systems for the delivery of energy, water and communication services, leading to the risk of cascading failures across multiple systems from one adverse event,[5] which is what took place following the Texas blackout in 2021.

Modern systems are usually dynamic, because the constituent parts are constantly changing.[6] They are often characterised by tightly coupled and interdependent components, which make them deeply "entangled"[7] and structurally "fragile".[8] Moreover, their behaviour cannot be predicted or explained purely on the basis of an analysis of the system's individual components

2 A. Boin, "Preparing for future crises: Lessons from research", in B.M. Hutter (ed.), *Anticipating Risks and Organising Risk Regulation*. (Cambridge University Press, 2010), p. 233.

3 O. Renn, K. Lucas, A. Haas, & C. Jaeger, "Things are different today: The challenge of global systemic risks" (2019) 22(4) *Journal of Risk Research*, pp. 401–15; D. Helbing, "Globally networked risks and how to respond" (2013) 497(7447) *Nature*, pp. 51–9.

4 E.C. Ellis, "Ecology in an anthropogenic biosphere" (2015) 85(3) *Ecological Monographs*, pp. 287–331.

5 G.H. Kjølle, I.B. Utne, & O. Gjerde, "Risk analysis of critical infrastructures emphasizing electricity supply and interdependencies" (2012) 105 *Reliability Engineering & System Safety*, pp. 80–9.

6 A. Corbett, "A systems approach to regulatory excellence", in C. Coglianese (ed.) *Achieving Regulatory Excellence* (Brookings Institution Press, 2017), p. 259.

7 See, for example, A. Zawadowski, "Entangled financial systems" (2013) 26(5) *Review of Financial Studies*, pp. 1291–323.

8 M.A. Centeno, M. Nag, T.S. Patterson, A. Shaver, & A.J. Windawi, "The emergence of global systemic risk" (2015) 41(1) *Annual Review of Sociology*, pp. 65–85, at 77, 79.

because the degree of interconnection and interdependence of these linked components can lead to non-linear responses to change.[9] Adverse change can result in the system flipping from one state to another state that is normally less favourable.[10]

II The concept of risk

In order to have a clear understanding of what systemic risks are and how they are distinct from other types of risk, it is necessary to start with the concept of risk itself. Surprisingly, there is no commonly accepted definition of risk,[11] neither in the context of academic nor public discourse. Indeed, despite its ubiquity, it is a difficult concept to define.[12] Nevertheless, all conceptions of risk have one core element in common – namely, uncertainty about the future. In particular, risk is typically associated with the possibility that adverse effects may occur in the future as a result of natural events, human activities or a combination of the two.[13] This broad description implies that risk is both a descriptive and a normative concept.[14]

In terms of its *descriptive connotations*, there are two main dimensions of risk – namely, cause and consequence. More specifically, the core of the concept of risk involves a cause-effect relationship between, on the one hand, an event or activity and, on the other hand, adverse effects.[15] In some cases, the *cause* of the risk will be associated with "natural perils" over which people have limited direct control, such as extreme weather events. Natural perils of this kind, which are commonly associated with climate change, are to be contrasted with "human perils" caused directly by humans, such as terrorism, theft or suicide.[16]

In relation to the possible *consequences or effects* if risk materialises, a distinction can be drawn between "pure risks" where there is some chance of loss and no

9 D. Helbing, n. 3 above, pp. 52–3.

10 M.A. Centeno et al., n. 8 above, p. 68.

11 M. Ekberg, "The parameters of the risk society: A review and exploration" (2007) 55(3) *Current Sociology*, pp. 343–66, at 353. A detailed exploration of the definition of risk is contained in Chapter 7 of this book.

12 As discussed further in Chapter 7, various authors have attempted to catalogue the burgeoning list of definitions of risk, but these definitions have not clearly converged over time.

13 L.B. Gratt, "The definition of risk and associated terminology for risk analysis" in J.J. Bonin & D.E. Stevenson (eds.), *Risk Assessment in Setting National Priorities* (Springer US, 1989), p. 676; A. Klinke & O. Renn, "Precautionary principle and discursive strategies: Classifying and managing risks" (2001) 4(2) *Journal of Risk Research*, pp. 159–73, at 159.

14 A. Klinke & O. Renn, n. 13 above, p. 160.

15 Ibid. p. 160.

16 E. Baranoff, P.L. Brockett, and Y. Kahane, *Risk Management for Enterprises and Individuals* (Saylor Academy, 2012), Chapter 1 – The nature of risk: Losses and opportunities.

chance of gain (for example, in the case of a natural disaster) and "speculative risk" where there is a chance to either gain or lose (such as risk associated with a company's investment in a proposed new competitive strategy).[17] In the case of pure risks, which are characteristic of climate change,[18] the nature and severity of harm will depend upon the causal relationship between the stressor giving rise to risk and the consequences for the individual or thing that is exposed to risk.[19] More specifically, the overall impact of a hazard will be the product of the interplay between the exposure of people or things to the hazard and the particular vulnerability of those people or things. The main dimensions of risk – cause and consequence – are illustrated in Figure 2.1 below.

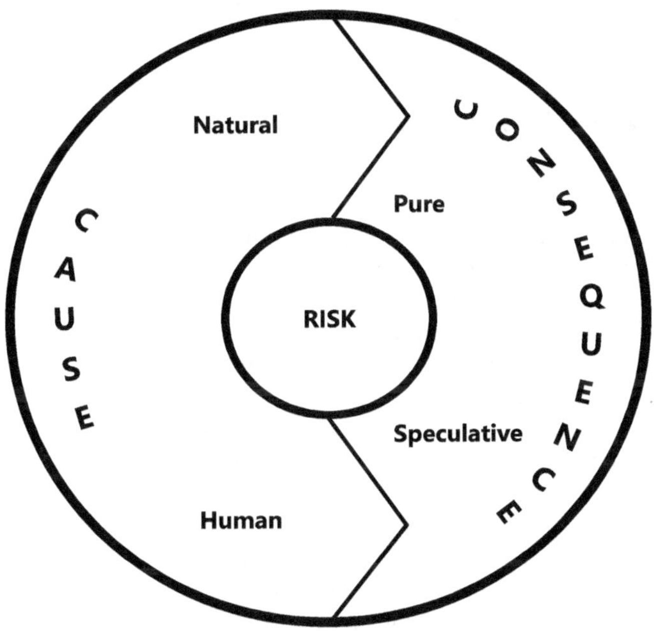

Figure 2.1 The main dimensions of risk[20]

17 Y. Hayashi and K. Kamei, "Risk management" in S. Abe, Y. Kawata, & M. Ozawa (eds.), *Science of Societal Safety: Living at Times of Risks and Disasters* (Springer, 2019), pp. 123–4.

18 It should be noted that not all risks associated with climate change are pure risks. For example, rising temperatures might jeopardise the survival of certain regions where drought and water shortages result but could concurrently benefit other regions by creating a more temperate climate.

19 A. Klinke & O. Renn, "A new approach to risk evaluation and management: Risk-based, precaution-based, and discourse-based strategies" (2002) 22(6) *Risk Analysis*, pp. 1071–94, at 1071.

20 This figure is the author's conceptualisation of risk, based on literature regarding the definition of risk that has been cited so far in this chapter.

Baldwin, Cave and Lodge (2012) suggest that the distinction between cause and consequence has relevance in a regulatory context. In particular, they note that "how risk is perceived and what regulatory solutions are proposed is fundamentally shaped by underlying worldviews and understandings of cause-effect relationships".[21] The specific way in which risk is perceived in a particular societal context may differ depending upon whether, for example, the risk is a natural, pure risk or a human, speculative risk. The perception of risk will, in turn, influence the decision about whether action is needed to avoid or mitigate the adverse effects associated with risk. This involves a *normative* judgement as to whether a response to risk is required[22] and, if so, the specific response necessary to effectively address the risk. This judgement will depend partly upon the person, thing or, for that matter, the system that may be exposed and vulnerable to risk, as well as the cause for and consequences of the risk.

III Distinctive features of systemic risks

Risks faced by systems are "systemic risks". Kaufman and Scott (2003) have put forward a definition of "systemic risk" in the context of financial systems, which has been widely cited in a range of different domains that feature complex systems:[23] "Systemic risk refers to the risk or probability of breakdowns in an entire system, as opposed to breakdowns in individual parts or components, and is evidenced by co-movements (correlation) among most or all parts." Fouque and Langsam (2013) clarify that not all risks that affect systems qualify as systemic risks. Rather, systemic risk refers to "the risk that the ability of the system to function as intended is seriously degraded. It does not refer to the risk that a component or several components fail unless such failure jeopardizes the integrity of the system."[24]

The label "systemic risk" suggests that the predominant focus in identifying such risks is on the system that is threatened and could be compromised. The term does not itself convey the *cause* of the risk – that is, the specific action, event or phenomenon that could disrupt and compromise the system. Nor does it specify the specific *effects* or *consequences* of risk other than the fact that the system is at risk. Instead, the term implies that systemic risk exists because the inherent characteristics of a system make it particularly vulnerable to the range of possible stressors that could bear upon the system.

In the financial sector, systemic risk is generally used in reference to an event that can trigger a collapse across the sector or, even, across the entire

21 R. Baldwin, M. Cave, & M. Lodge, *Understanding Regulation: Theory, Strategy, and Practice* (Oxford University Press, 2012), p. 102.

22 A. Klinke & O. Renn, n. 13 above, p. 160.

23 G.G. Kaufman & K.E. Scott, "What is systemic risk, and do bank regulators retard or contribute to it?" (2003) 7(3) *The Independent Review*, pp. 371–91, at 371.

24 J.-P. Fouque & J.A. Langsam (eds.), *Handbook on Systemic Risk* (Cambridge University Press, 2013), p. xxi.

economy.[25] The trigger event could be at the firm level but severe enough to cause instability in the broader financial system, such as the collapse of the investment banking firm Lehman Brothers in 2008 during the Global Financial Crisis.[26] In contrast, idiosyncratic risk is confined to certain types of investments, assets or sectors – for example, a gas shortage that affects supply of gas by pipeline companies to their customers. This type of risk can be mitigated through diversification of investments.

Apart from the financial sector, the concept of systemic risk is also well-known in the fields of engineering, public health and ecology.[27] An example of systemic risk in an engineering context is a "blackout" caused by a fault in an electricity network, resulting in power outages spanning across extensive geographical areas. In the public health area, systemic risks arise from disease outbreaks that have the capacity to spread quickly and extensively. In relation to ecology, systemic risks may cause ecosystems to undergo a regime shift from one equilibrium to another. In each of these domains, systemic risk has been described as arising in the context of complex systems, which are characterised by features[28] including non-linearities,[29] multiple stable states, hysteresis,[30] contagion,[31] and synchrony.[32] These features of complex systems lead to unexpected and unlikely interactions, which can threaten survival of a system.[33] Moreover, ripple effects are likely to be felt beyond the system.[34] It is the totality of the threat to the system – that is, the possibility that the entire system could collapse when stressed – that distinguishes systemic risks from other types of risk.[35]

25 Another type of risk that exists in the context of financial markets is "systematic risk", which is distinct from "systemic risk". Systematic risk, which is also referred to as "market risk", is the risk inherent in the aggregate market and can be caused by recessions, rises or stagnation of interest rates, and fluctuations of currencies.

26 Lehman Brothers' size and integration in the economy caused its collapse to result in a domino effect that posed a major risk to the US financial system and, eventually, had repercussions for the broader global economy.

27 J. Kambhu, S. Weidman, & N. Krishnan (eds.), *New Directions for Understanding Systemic Risk: A Report on a Conference Cosponsored by the Federal Reserve Bank of New York and the National Academy of Sciences* (National Academies Press, 2007), p. 29.

28 Ibid. pp. 30–3.

29 Non-linear feedback can drive a complex system away from a state of equilibrium and cause non-linear behaviour, including permanent regime shifts.

30 Hysteresis occurs in cases where the system takes a different path to return to its original state.

31 Contagion refers to the cascading effect of systemic risks.

32 Systemic risks may result in synchronised outcomes among component parts, such as cases where similar choices are made by individuals as a result of the systemic risk.

33 M.A. Centeno, M. Nag, T.S. Patterson, A. Shaver, & A.J. Windawi, n. 8 above, p. 68.

34 O. Renn, "Systemic risks: The new kid on the block" (2016) 58(2) *Environment: Science and Policy for Sustainable Development*, pp. 26–36, at 31.

35 Ibid. p. 29.

Aven and Renn (2006, 2012) suggest that systemic risks will generally have trans-boundary reach,[36] are "collectively relevant",[37] and entail a range of social, financial and economic consequences.[38] They are usually characterised by complexity, scientific uncertainty and socio-political ambiguity[39] and often require a co-ordinated effort to manage them by a variety of players extending, at times, beyond conventional country, sectoral, organisational and discipline-based boundaries.[40] According to Schweizer (2019), the concept of "systemic risk" bridges the gap between risk research and complexity science.[41] This is a notable observation, because it signals that the concept is more than a mere descriptor of risks to systems. It also conveys the notion that a systems approach is appropriate to break down complexity and, thereby, assist in identifying aspects of a system that render it vulnerable to external stressors giving rise to systemic risk. This, in turn, can help better target action to manage that risk.

In its 2003 report on global emerging risks in the 21st century, the Organisation for Economic Co-operation and Development (OECD) described the driving forces behind the rise of systemic risks in modern society as including, but not limited to, demographic changes, globalisation, technological innovations and societal change. Demographic changes – such as increasing populations, population density, and urbanisation – place additional stress on the natural environment

36 Trans-boundary risks are those that originate in one country but are capable of resulting in adverse effects in one or more other countries. An example of such a risk in the climate change context is the reduction in the volume of Himalayan glaciers resulting from rising temperatures caused by climate change. See J. Xu et al., "The melting Himalayas: Cascading effects of climate change on water, biodiversity, and livelihoods" (2009) 23(3) *Conservation Biology*, pp. 520–30. This study of the "melting Himalayas" indicates that the cascading effects of rising temperatures and loss of ice and snow in the region are affecting water availability, biodiversity, ecosystem boundaries, and global feedbacks, including monsoonal shifts across Asia, particularly China and India.

37 A risk has "collective relevance" where the possible impact of the risk extends beyond an individual or discrete group of individuals and results in increased vulnerability of groups of the world's population.

38 A. Klinke & O. Renn, "Systemic risks as challenge for policy making in risk governance" (2006) 7(1) *Forum, Qualitative Social Research*, section 2; A. Klinke & O. Renn, "Adaptive and integrative governance on risk and uncertainty" (2012) 15(3) *Journal of Risk Research*, pp. 273–92, at 275.

39 A. Klinke & O. Renn, n. 38 above, section 2; O. Renn, A. Klinke, & M. van Asselt, "Coping with complexity, uncertainty and ambiguity in risk governance: A synthesis" (2011) 40(2) *AMBIO*, pp. 231–46, at 234; A. Klinke & O. Renn, n. 38 above, pp. 275–6; P.-J. Schweizer, "Systemic risks – concepts and challenges for risk governance" (2021) 24(1) *Journal of Risk Research*, pp. 78–93.

40 O. Renn, n. 34 above, p. 32.

41 P.-J. Schweizer, n. 39 above.

and human settlements.[42] Globalisation made possible by, amongst other things, international trade and transport, world-wide markets and globalised communication networks, has created interdependencies that facilitate and expedite the proliferation of adverse impacts across the globe.[43] Technological innovations are now embedded in many facets of life, including in relation to the operation and management of critical infrastructure, failure of which could result in catastrophic disasters.[44] In addition, changes in socio-economic structures, such as market concentration at a global scale, can increase society's vulnerability to shocks and may also alter societal perception of risk.[45]

Arguably, the level of interconnection and interdependence, the speed and scale of interactions, and the tight coupling between constituent system elements have created a new breed of systemic risk,[46] which demands attention because of the extreme consequences that could be inflicted on communities that are dependent on the affected systems. In introducing its 2003 report on global emerging risks, the then OECD Secretary-General explained that:[47]

> Large-scale disasters of the past few years – such as the terrorist attack of September 11, 2001, the appearance of previously unknown infectious diseases, unusually extensive flooding in large parts of Europe, devastating bushfires in Australia and violent ice storms in Canada – have brought home to OECD governments the realisation that something new is happening. Such "mega-risks" have the potential for inflicting considerable damage on the vital systems and infrastructures upon which our societies and economies depend, and create serious difficulties for traditional risk management …

This notion that systemic risk is particularly pronounced in a modern context builds on the work of Ulrich Beck. In his 1986 book – *Risk Society: Towards a New Modernity* – Beck proffered that modern society is a "risk society", where modernisation is producing risks that are felt more widely, and where society is also more concerned about these risks.[48]

42 OECD, *Emerging Systemic Risks in the 21st Century: An Agenda for Action* (OECD, 2003), pp. 10–1; T. Aven & O. Renn, *Risk Management and Governance Concepts, Guidelines and Applications* (Springer, 2010), p. 17. The authors of both publications specifically mention the emission of greenhouse gases as a possible source of additional stress for ecosystems and human settlements.
43 OECD, n. 42 above, p. 12; T. Aven & O. Renn, n. 42 above, p. 17.
44 OECD, n. 42 above, p. 12; T. Aven & O. Renn, n. 42 above, pp. 17–18.
45 OECD, n. 42 above, pp. 12–3. T. Aven & O. Renn, n. 42 above, p. 17. See also OECD, *Future Global Shocks: Improving Risk Governance* (OECD, 2011), pp. 16–22 where the drivers of future "global shocks" that can produce systemic risks are described.
46 M.A. Centeno, M. Nag, T.S. Patterson, A. Shaver, & A.J. Windawi, n. 8 above, p. 66; OECD, n. 42 above, pp. 13–14.
47 OECD, n. 42 above, p. 3.
48 U. Beck, *Risk Society: Towards a New Modernity* (Sage Publications, 1992), pp. 21–4.

> [T]he ecological and high-tech risks that have upset the public for some years now, which will be the focus of what follows, have a new quality. In the afflictions they produce they are no longer tied to their place of origin – the industrial plant. By their nature they endanger *all* forms of life on this planet.

Climate change appears to fall within the spectrum of modern risks contemplated by Beck.[49]

To condense what is known about systemic risks, Table 2.1 summarises their key features and compares them to idiosyncratic risks. More specifically, the table identifies the main context, characteristics and consequences of systemic risks and idiosyncratic risks respectively. In summary, systemic risks may arise in a variety of sectors, they may cause significant damage, harm or loss for the system at risk, and there is likely to be considerable complexity, uncertainty and ambiguity regarding the causes, consequences and responses to manage them.

Homer-Dixon (2006) summarises the challenges associated with managing systemic risks for modern society, by comparing them to those faced by the Romans in ancient times:[50]

> We have an advantage over the Romans that gives us a head start: we understand much better how complex systems around us behave. Rome failed in the end partly because it didn't – and couldn't – understand that it had doomed itself to fail. It couldn't see clearly the multiple stresses converging on it; that it was bounded by the exigencies of its natural world; and that, as complexities and entrenched power accumulated, it was inexorably becoming a static, brittle system. We don't have the same excuse. We know that static, brittle systems don't survive. We also understand that in any complex adaptive system, breakdown, if limited, can be a key part of that system's long-term resilience and renewal.

As explained in the next chapter, climate change can cause systemic risks. At the core of this book is a quest to determine whether, from a regulatory perspective, we are well-equipped to address systemic risks, particularly those caused by climate change, so as to ensure the long-term viability of the systems upon which society depends. Or, rather, is regulation likely to lead to "static, brittle systems" of the kind referred to by Homer-Dixon that are doomed to demise when faced by systemic risks?

49 Ibid. p. 22.

50 T.F. Homer-Dixon, *The Upside of Down: Catastrophe, Creativity and the Renewal of Civilisation* (Text Publishing, 2007), p. 308.

Table 2.1 Key features of systemic risks and idiosyncratic risks

	Systemic risks	*Idiosyncratic risks*
Context	• Interconnectedness • Population concentration • Globalisation • Technological innovations • Increased vulnerability resulting from socio-economic structures • Interface between nature and society	• Confined to limited operations, events or activities
Characteristics	• Complex, uncertain, ambiguous • Collectively relevant • Capacity to spread	• Self-contained • Isolated
Consequences	• Threat to integrity or survival of system • Affect systems upon which societies depend	• Limited consequences • Impact on only certain aspects of a system or discrete segments of society

Bibliography

Abe, S., Kawata, Y., & Ozawa, M. (eds.). (2019). *Science of Societal Safety: Living at Times of Risks and Disasters*. Springer. https://doi.org/10.1007/978-981-13-2775-9.

Aven, T., & Renn, O. (2010). *Risk Management and Governance Concepts, Guidelines and Applications*. Springer.

Baldwin, R., Cave, M., & Lodge, M. (2012). *Understanding Regulation: Theory, Strategy, and Practice*. Oxford University Press. https://doi.org/10.1093/acprof:osobl/9780199576081.001.0001.

Baranoff, E., Brockett, P.L., & Kahane, Y. (2012). *Risk Management for Enterprises and Individuals*. Saylor Academy.

Beck, U. (1992). *Risk Society: Towards a New Modernity*. Sage Publications.

Boin, A. (2010). "Preparing for future crises: lessons from research." In B.M. Hutter (ed.), *Anticipating Risks and Organising Risk Regulation*. Cambridge University Press.

Bonin, J.J., & Stevenson, D.E. (eds.). (1989). *Risk Assessment in Setting National Priorities*. Springer US. https://doi.org/10.1007/978-1-4684-5682-0.

Centeno, M.A., Nag, M., Patterson, T.S., Shaver, A., & Windawi, A.J. (2015). "The emergence of global systemic risk." *Annual Review of Sociology*, 41(1), 65–85. https://doi.org/10.1146/annurev-soc-073014-112317.

Corbett, A. (2017). "A systems approach to regulatory excellence." In C. Coglianese, (ed.), *Achieving Regulatory Excellence*. Brookings Institution Press. https://doi.org/10.7864/j.ctt1hfr226.19.

Ekberg, M. (2007). "The parameters of the risk society: A review and exploration." *Current Sociology*, 55(3), 343–366. https://doi.org/10.1177/0011392107076080.

Ellis, E.C. (2015). "Ecology in an anthropogenic biosphere." *Ecological Monographs*, 85 (3), 287–331. https://doi.org/10.1890/14-2274.1.

Fouque, J.-P., & Langsam, J.A. (eds.). (2013). *Handbook on Systemic Risk*. Cambridge University Press.

Helbing, D. (2013). "Globally networked risks and how to respond." *Nature*, 497(7447), 51–59. https://doi.org/10.1038/nature12047.

Homer-Dixon, T.F. (2007). *The Upside of Down: Catastrophe, Creativity and the Renewal of Civilisation*. Text Publishing.

Homer-Dixon, T., Walker, B., Biggs, R., Crépin, A.-S., Folke, C., Lambin, E.F., Peterson, G.D., Rockström, J., Scheffer, M., Steffen, W., & Troell, M. (2015). "Synchronous failure: The emerging causal architecture of global crisis." *Ecology and Society*, 20(3), 6. https://doi.org/10.5751/ES-07681-200306.

Kambhu, J., Weidman, S. & Krishnan, N. (eds.). (2007). *New Directions for Understanding Systemic Risk: A Report on a Conference Cosponsored by the Federal Reserve Bank of New York and the National Academy of Sciences*. National Academies Press.

Kaufman, G.G., & Scott, K.E. (2003). "What is systemic risk, and do bank regulators retard or contribute to it?" *The Independent Review*, 7(3), 371–391.

Kjølle, G.H., Utne, I.B., & Gjerde, O. (2012). "Risk analysis of critical infrastructures emphasizing electricity supply and interdependencies." *Reliability Engineering & System Safety*, 105, 80–89. https://doi.org/10.1016/j.ress.2012.02.006.

Klinke, A., & Renn, O. (2001). "Precautionary principle and discursive strategies: Classifying and managing risks." *Journal of Risk Research*, 4(2), 159–173. https://doi.org/10.1080/136698701750128105.

Klinke, A., & Renn, O. (2002). "A new approach to risk evaluation and management: risk-based, precaution-based, and discourse-based strategies." *Risk Analysis*, 22(6), 1071–1094. https://doi.org/10.1111/1539-6924.00274.

Klinke, A., & Renn, O. (2006). "Systemic risks as challenge for policy making in risk governance." *Forum, Qualitative Social Research*, 7(1).

Klinke, A., & Renn, O. (2012). "Adaptive and integrative governance on risk and uncertainty." *Journal of Risk Research*, 15(3), 273–292. https://doi.org/10.1080/13669877.2011.636838.

OECD. (2003). *Emerging Systemic Risks in the 21st Century: An Agenda for Action*. Organisation for Economic Co-operation and Development. https://doi.org/10.1787/9789264101234-fr.

OECD. (2011). *Future Global Shocks: Improving Risk Governance*. Organisation for Economic Co-operation and Development. https://doi.org/10.1787/9789264114586-en.

Renn, O. (2016). "Systemic risks: The new kid on the block." *Environment: Science and Policy for Sustainable Development*, 58(2), 26–36. https://doi.org/10.1080/00139157.2016.1134019.

Renn, O., Klinke, A., & van Asselt, M. (2011). "Coping with complexity, uncertainty and ambiguity in risk governance: A synthesis." *AMBIO*, 40(2), 231–246. https://doi.org/10.1007/s13280-010-0134-0.

Renn, O., Lucas, K., Haas, A., & Jaeger, C. (2019). "Things are different today: The challenge of global systemic risks." *Journal of Risk Research*, 22(4), 401–415. https://doi.org/10.1080/13669877.2017.1409252.

Schweizer, P.-J. (2021). "Systemic risks – concepts and challenges for risk governance." *Journal of Risk Research*, 24(1), 78–93. https://doi.org/10.1080/13669877.2019.1687574.

Xu, J., Grumbine, R.E., Shrestha, A., Eriksson, M., Yang, X., Wang, Y., & Wilkes, A. (2009). "The melting Himalayas: Cascading effects of climate change on water, biodiversity, and livelihoods." *Conservation Biology*, 23(3), 520–530. https://doi.org/10.1111/j.1523-1739.2009.01237.x.

Zawadowski, A. (2013). "Entangled financial systems." *Review of Financial Studies*, 26(5), 1291–1323. https://doi.org/10.1093/rfs/hht008.

Chapter 3

Systemic Risks and Climate Change

It is difficult to dispute that climate change is the greatest challenge of our time.[1] Indeed, the Intergovernmental Panel on Climate Change (IPCC, 2021) has found that it is unequivocal that human activity has resulted in global warming[2] and this warming is unprecedented in at least the last 2000 years.[3] The so-called "anthropogenic" causes of climate change have resulted in a range of effects, including increasing temperatures, rising sea levels, and extreme weather events such as fires and floods,[4] and these impacts are being felt in every region across the globe.[5] The future looks more ominous. Temperatures are expected to continue to rise, which will lead to a variety of hazards, including increases in the intensity and frequency of heatwaves, heavy rainfall, droughts and severe storms.[6]

Compelling scientific evidence indicates that, in the past, climate change has directly disturbed the "global system" and will continue to do so into the future.[7] More specifically, the physical hazards caused by climate change are likely to have consequences for a broad array of human and natural systems.[8] Li

1 B. McKibben, *Falter: Has the Human Game Begun to Play Itself Out?* (Schwartz Publishing, 2019).
2 IPCC, "Summary for Policymakers", in *Climate Change 2021: The Physical Science Basis. Contribution of Working Groups I to the Sixth Assessment Report of the Intergovernmental Panel on Climate Change* (Cambridge University Press, 2021), p. 4.
3 Ibid. p. 6.
4 Ibid. pp. 4–9.
5 Ibid. p. 8.
6 Ibid. p. 15.
7 IPCC, "Climate Change 2014: Synthesis Report", in *Contribution of Working Groups I, II and III to the Fifth Assessment Report of the Intergovernmental Panel on Climate Change* (IPCC, 2015); M.A. Centeno, M. Nag, T.S. Patterson, A. Shaver, & A.J. Windawi, "The emergence of global systemic risk" (2015) 41(1) *Annual Review of Sociology*, pp. 65–85, at 77.
8 IPCC, "Technical Summary", in *Climate Change 2014: Impacts, Adaptation, and Vulnerability. Part A: Global and Sectoral Aspects. Contribution of Working Group II to the Fifth Assessment Report of the Intergovernmental Panel on Climate Change*, (Cambridge University Press, 2014), p. 40.

DOI: 10.4324/9781003253464-3

and colleagues (2021) summarise the source, characteristics and consequences of systemic risks caused by climate change:[9]

> [S]ystemic risk induced by climate change results from rapid growth of greenhouse emissions, increasingly complex connections among different socioeconomic systems, and continuous changes in exposure and vulnerability. The fundamental, defining feature of systemic risk is cascading effects. The extent of risk propagation and its duration depend on the characteristics of the various discrete risks that are connected to make up the systemic risk.

As a society, we depend heavily on natural and socio-economic systems that are at risk from climate change. Determining how the existential risk posed by climate change can be effectively managed turns on our practical understanding of its impact on systems. This issue is examined in this chapter.

I The impacts of climate change on systems

Climate change has been described by Corfee-Morlot and colleagues (2011) as encompassing "a complex array of interactions between a large number of human and natural systems across vast spatial and temporal scales which, in turn, challenges scientific assessments and policy efforts that aim to identify and manage these changes".[10] This definition highlights the vastness as well as the sheer complexity of climate change and its impact on systems.

The past and prospective impacts of climate change on the spectrum of the world's systems have been well-documented. In the case of natural systems, many ecosystems have been affected by climate change, leading to loss of biodiversity and consequential impacts on important services provided to humans by these systems.[11] Climate change is projected to compromise freshwater systems as a result of drought and by changing streamflow and water quality, thereby compromising access to drinking water.[12] The impact

9 H.-M. Li, X.-C. Wang, X.-F. Zhao, & Y. Qi, "Understanding systemic risk induced by climate change" (2021) 12(3) *Advances in Climate Change Research*, pp. 384–94, at 393.

10 J. Corfee-Morlot, I. Cochran, S. Hallegatte, & P.-J. Teasdale, "Multilevel risk governance and urban adaptation policy" (2011) 104(1) *Climatic Change*, pp. 169–97, at 172.

11 IPCC, "Summary for Policymakers", in *Climate Change 2014: Impacts, Adaptation, and Vulnerability. Part A: Global and Sectoral Aspects. Contribution of Working Group II to the Fifth Assessment Report of the Intergovernmental Panel on Climate Change* (Cambridge University Press, 2014), pp. 4, 13.

12 IPCC, "Freshwater Resources" in *Climate Change 2014: Impacts, Adaptation, and Vulnerability. Part A: Global and Sectoral Aspects. Contribution of Working Group II to the Fifth Assessment Report of the Intergovernmental Panel on Climate Change* (Cambridge University Press, 2014), p. 232.

of climate change on food systems is expected to be widespread, which will likely adversely affect all aspects of food security.[13]

In terms of human systems, climate change is predicted to intensify existing threats to health and exacerbate current health problems,[14] which will affect the functioning of public health care systems.[15] Climate change will have profound impacts on a wide spectrum of infrastructure systems, including infrastructure for water and energy supply, sanitation and drainage, transport and telecommunication, as well as for the broader built environment in urban settings.[16] The impact of physical hazards caused by climate change on individuals, businesses, communities and governments across the globe and the consequences of changes in law and policy as countries progressively transition to a low-carbon approach will have downstream implications for participants in financial systems, which could result in a range of traditional financial risks such as credit risk, liquidity risk, market risk, and operational risk.[17]

Interactions and interdependencies may also result in a cascade of systemic risk across multiple natural and human systems.[18] This can trigger "tipping points", where a system's structure and characteristics fundamentally change and, at worst, the system collapses completely.[19] Lawrence and colleagues (2020) warn of the "far-reaching effects of climate change impacts that are amplified and generated through feedbacks and virtuous and vicious cycles".[20]

Yet, there is "deep uncertainty" about the trajectories of many climate variables and also about future vulnerability and exposure of human and natural

13 IPCC, "Food Security and Food Production Systems" in *Climate Change 2014: Impacts, Adaptation, and Vulnerability. Part A: Global and Sectoral Aspects. Contribution of Working Group II to the Fifth Assessment Report of the Intergovernmental Panel on Climate Change* (Cambridge University Press, 2014), pp. 488–90.

14 IPCC, "Human Health: Impacts, Adaptation, and Co-Benefits", in *Climate Change 2014 Impacts, Adaptation, and Vulnerability. Working Group II Contribution to the Fifth Assessment Report. Intergovernmental Panel on Climate Change*, (Cambridge University Press, 2014), p. 713.

15 L.F. Wiley, "Climate change adaptation and public health law", in *Research Handbook on Climate Change Adaptation Law* (Edward Elgar Publishing, 2013), pp. 171–213, at 174.

16 IPCC, "Urban Areas", in *Climate Change 2014: Impacts, Adaptation, and Vulnerability. Part A: Global and Sectoral Aspects. Contribution of Working Group II to the Fifth Assessment Report of the Intergovernmental Panel on Climate Change* (Cambridge University Press, 2014), p. 538.

17 Financial Stability Oversight Council, *FSOC Report on Climate-Related Financial Risk* (2021), p. 24.

18 J. Lawrence, P. Blackett, & N.A. Cradock-Henry, "Cascading climate change impacts and implications" (2020) 29 *Climate Risk Management*, 100234, p. 2.

19 P.-J. Schweizer, "Systemic risks – concepts and challenges for risk governance" (2021) 24(1) *Journal of Risk Research*, pp. 78–93, at 82.

20 J. Lawrence, P. Blackett, & N.A. Cradock-Henry, n. 18 above, p. 12.

systems to the impacts of climate change.[21] Small, incremental changes in the climate system may provide no hint about when critical and, potentially, catastrophic tipping points have been reached.[22] This deep uncertainty means that decision-makers involved in responding to the risks posed by climate change may not be able to rely upon the past to guide decisions in the future.[23]

The profound uncertainty that characterises climate change is compounded by complexity. In particular, a system will face risks arising from endogenous factors that are associated with the complexity of the system itself, but these risks will be compounded by those arising from a range of sources of risk that are exogenous to the system, such as climate change. The combination of uncertainty and complexity, coupled with the prospect of inter-system cascading risk, leads to ambiguity about whether and how to respond to the risks associated with climate change, particularly given the broad range of stakeholders affected by these risks and the likely diverse range of values and perceptions of risk among them.[24]

II Climate change adaptation to manage systemic risks

There is a range of available policy responses to address climate change. "Climate change mitigation" focuses on moderating human interference that *causes* anthropogenic climate change. Mitigation measures include restrictions on activity that exacerbate the greenhouse effect, such as the imposition of a tax on emissions-intensive operations (for example, the combustion of coal to produce electricity). In contrast, "climate change adaptation" involves responding to the *consequences* of climate change that are not avoided through climate change mitigation. Specifically, it concerns how the impacts of climate change can be addressed, including measures to deal with the vulnerabilities of individuals, settlements and systems that may be exposed to those impacts. Climate change adaptation is now an important policy priority at a global level, particularly in countries that are especially vulnerable to climate change impacts. However, this has not always been the case.[25]

The United Nations Framework Convention on Climate Change (UNFCCC) is an international convention with near universal membership. Its

21 IPCC, n. 8 above, p. 56; N. Ranger, "Adaptation as a decision making under deep uncertainty", in I. Linkov & T.S. Bridges (eds.), *Climate: Global Change and Local Adaptation* (Springer Netherlands, 2011), pp. 89–122, at 93–5.

22 O. Renn, "Systemic risks: The new kid on the block" (2016) 58(2) *Environment: Science and Policy for Sustainable Development*, pp. 26–36, at 30.

23 N. Ranger, n. 21 above, p. 118.

24 O. Renn, M. Laubichler, K. Lucas, W. Kröger, J. Schanze, R.W. Scholz, & P. Schweizer, "Systemic risks from different perspectives" (2020) *Risk Analysis*, 13657.

25 E.L.F. Schipper, "Conceptual history of adaptation in the UNFCCC process" (2006) 15(1) *Review of European Community and International Environmental Law*, pp. 82–92.

primary aim is to prevent dangerous human interference with the climate system.[26] Even though the UNFCCC's objective, set out in Article 2 of the treaty, references both the need to mitigate climate change ("stabilization of greenhouse gas concentrations in the atmosphere") and to adapt to climate change ("allow ecosystems to adapt naturally to climate change, to ensure that food production is not threatened and to enable economic development to proceed in a sustainable manner"), Schipper (2006) argues that, in the past, climate change mitigation has dominated discourse on climate change policy.[27] Dilling (2015) proffers an explanation – namely, that adaptation was seen as a "morally inferior" choice, pursuit of which amounted to an admission of defeat on the mitigation front.[28]

Nevertheless, recognition of the importance of climate change adaptation has grown over time as the consequences of climate change have become more apparent. Indeed, in 2010, a dedicated committee was established under the auspices of the UNFCCC to deal with adaptation to climate change.[29] Furthermore, in 2015, Parties to the UNFCCC agreed to the landmark "Paris Agreement", which builds upon the UNFCCC and aims to strengthen the global climate change response by, among other things, enhancing countries' adaptive capacity and resilience and reducing vulnerability, with a view to contributing to sustainable development.[30] More recently, at the 2021 UNFCCC Conference of Parties (COP26), a work programme to give effect

26 Article 2 of the UNFCCC provides that: "The ultimate objective of this Convention and any related legal instruments that the Conference of the Parties may adopt is to achieve, in accordance with the relevant provisions of the Convention, stabilization of greenhouse gas concentrations in the atmosphere at a level that would prevent dangerous anthropogenic interference with the climate system. Such a level should be achieved within a time frame sufficient to allow ecosystems to adapt naturally to climate change, to ensure that food production is not threatened and to enable economic development to proceed in a sustainable manner".

27 E.L.F. Schipper, n. 25 above, pp. 84–6.

28 L. Dilling, "Adaptation", in *Research Handbook on Climate Governance* (Edward Elgar Publishing, 2015), pp. 470–9, at 470.

29 In 2010, the UNFCCC Adaptation Committee was established to help promote enhanced action both within and outside the context of the Convention. Since its inaugural meeting in September 2012, the Adaptation Committee has provided technical support and guidance to the UNFCCC Parties on adaptation action and on the means of implementation as well as raising awareness, sharing information and reaching out to stakeholders outside the traditional adaptation community: UNFCCC, *The State of Adaptation under the United Nations Framework Convention on Climate Change*, Thematic Report (UNFCCC Adaptation Committee, 2013), p. 7.

30 Under Article 7(1) of the Paris Agreement, signatories agreed to "establish the global goal on adaptation of enhancing adaptive capacity, strengthening resilience and reducing vulnerability to climate change, with a view to contributing to sustainable development and ensuring an adequate adaptation response in the context of the temperature goal referred to in Article 2". The "temperature goal" is set out in Article 2(1)(a) of the Paris Agreement as follows: "Holding the increase in the global average temperature to well below 2°C above pre-industrial levels and pursuing efforts to limit the temperature increase to 1.5°C above pre-industrial

to the Paris Agreement's global goal on adaptation was launched.[31] These developments are notable as the need for climate change adaptation will only become greater the longer that comprehensive and effective climate change mitigation is delayed. Indeed, it is becoming increasingly clear that pursuit of climate change adaptation is not optional but, rather, is an absolute imperative – not just in the short term, but also in the longer term. As suggested by Pielke and colleagues (2007), policies to adapt to climate change can't come soon enough.[32]

So what is adaptation and what does it involve in practical terms? The UNFCCC and the IPCCC have defined it in similar ways to mean:[33] "Adjustment in natural or human systems in response to actual or expected climatic stimuli or their effects, which moderates harm or exploits beneficial opportunities." This definition is supplemented by numerous others.[34] Collectively, these definitions indicate that adaptation is, at its essence, a form of risk management aimed at moderating or offsetting the negative effects of climate change by reducing exposure and vulnerability.[35] This is an important observation, because it reframes the question about how society can adapt to climate change to one that is more practically centred on how the risks from climate change can be managed.

III Regulating to adapt to climate change

There is a spectrum of adaptive actions that can be taken to manage the risks to human and natural systems caused by climate change.[36] "Anticipatory" responses involve deliberate action that is taken before the impacts of climate change materialise, whereas "reactive" actions are implemented after the

levels, recognizing that this would significantly reduce the risks and impacts of climate change".

31 This work programme is formally referred to as the "Glasgow-Sharm el-Sheik work programme on the global goal of adaptation".

32 R. Pielke, G. Prins, S. Rayner, & D. Sarewitz, "Lifting the taboo on adaptation" (2007) 445(7128) *Nature*, pp. 597–8.

33 "Adaptation" is defined in the UNFCCC Glossary of Key Terms accessible at: https://www4.unfccc.int/sites/NAPC/Pages/glossary.aspx. IPCC, "Annex I: Glossary" in *Global Warming of 1.5°C. An IPCC Special Report on the impacts of global warming of 1.5°C above pre-industrial levels and related global greenhouse gas emission pathways, in the context of strengthening the global response to the threat of climate change, sustainable development, and efforts to eradicate poverty* (2018), p. 542.

34 Schipper, E. L. F., *Climate Change Adaptation and Development: Exploring the Linkages*, Working Paper 107 (Tyndall Centre for Climate Change Research, 2007), pp. 4–5.

35 B. Smit & J. Wandel, "Adaptation, adaptive capacity and vulnerability" (2006) 16 (3) *Global Environmental Change*, pp. 282–92.

36 The UNFCCC definition of "adaptation" clarifies that various types of adaptation can be distinguished, including anticipatory and reactive adaptation, private and public adaptation, and autonomous and planned adaptation.

impacts of climate change have occurred.[37] Some actions may be "public" measures initiated by government, including but not limited to regulation, whereas "private" measures are typically undertaken autonomously by businesses, communities and individuals, rather than in response to changes in policy or law. There may also be "hybrid" public and private measures that are designed and implemented by both government and non-governmental entities.

Examples of the types of adaptive actions that may be available across both human and natural systems are illustrated in Table 3.1. While the various types of adaptive actions are not mutually exclusive, Table 3.1 highlights some important choices for governments in managing risks to human and natural systems caused by climate change. One choice concerns whether or not public measures should be used to adapt to climate change risks. Viscusi (1998) suggests that "[w]e need to identify which risks are important and which are not. Moreover, we have to assess how costly it is to alter the risk so that we can focus our risk reduction efforts on the risks that we can eliminate most

Table 3.1 Types of adaptation to climate change across human and natural systems[38]

	Public measures		*Private measures*	
	Anticipatory	*Reactive*	*Anticipatory*	*Reactive*
Human systems	Building standards to ensure resilience to climate change hazards	Emergency management regulation to facilitate response to extreme weather events	Insurance to protect against personal and property damage arising from extreme weather events	Installation of air conditioning to address extreme heat affecting people and physical infrastructure
Natural systems	Restrictions on land use and management to protect ecosystems and biodiversity that could be affected by climate hazards	Conservation measures to migrate species affected by climate hazards in their previous habitats	Changes in consumption behaviour, such as reduced water consumption, to mitigate the impact of human activity on natural systems	Reforestation on private land to mitigate climate change and restore ecosystems

37 Land use planning to prohibit or restrict land use and development in areas that may be vulnerable to climate change impacts is an example of anticipatory adaptation. Emergency response mechanisms to deal with the aftermath of a major climate change event is an example of reactive adaptation.

38 Table 3.1 is the author's interpretation and adaptation of Figure TS-9 in IPCC, 'Technical Summary' in *Climate Change 2001: Impacts, Adaptation, and Vulnerability. Contribution of Working Group II to the Third Assessment Report of the Intergovernmental Panel on Climate Change* (Cambridge University Press, 2001), p. 62.

cheaply."[39] Leaving management of systemic risks to autonomous adaptation by the private sector could lead to the emergence of a patchwork of incoherent, ineffective or counter-productive measures. More importantly, the far-reaching and potentially catastrophic implications of systemic risks for society if they were to materialise highlight that they are inherently important enough to warrant government intervention. However, the likely complexity as well as the significant costs and effort implied by intervention to address such risks may deter some governments from taking on the challenge, particularly when short-term political cycles are factored into the equation.

In the event that a government does elect to intervene to address systemic risks caused by climate change, another choice to be made concerns the form that public measures should take. In particular, should the intervention take the form of policy initiatives coupled with funding support to incentivise adaptive action or, rather, would mandatory regulatory measures be preferable to advance adaptation? Karassin (2010) suggests that the latter is appropriate when "public goods" are at risk.[40] He defines these as goods that meet the collective wants of society, in contrast to private goods that satisfy individual wants.[41] Systemic risks affect systems upon which society depends, indicating that these systems fall within the general realm of a public good and that enforceable regulation may be necessary to address risks to these systems.

Karassin adds that regulation can help to create structure and predictability notwithstanding the uncertainty associated with climate change and its impacts,[42] it can facilitate prioritisation of issues that might otherwise be contested or for which understanding is limited,[43] and it can be used to change behaviour and mobilise action to reduce loss and harm that could occur in the absence of regulation.[44] This raises a linked choice as to whether government intervention should be anticipatory or reactive. A focus on anticipatory measures can help to avoid climate change risks materialising. In contrast, reactive regulatory measures are those that respond to such risks after they have eventuated. As Ranger (2011) has noted, reactive measures are comparatively more costly and less effective than anticipatory ones.[45]

To summarise, adaptation to climate change is essentially concerned with the management of risks arising from climate change. Climate change adaptation can take various forms, including regulation that is designed and applied by

39 W.K. Viscusi, *Rational Risk Policy: The 1996 Arne Ryde Memorial Lectures* (Clarendon Press; Oxford University Press, 1998), p. 85.

40 O. Karassin, "Mind the gap: Knowledge and need in regulating adaptation to climate change" (2010) 22(3) *Georgetown International Environmental Law Review*, p. 383, at 392.

41 Ibid. p. 392.

42 Ibid. p. 391.

43 Ibid. p. 391.

44 Ibid. pp. 391–2.

45 N. Ranger, n. 21 above, p. 93.

government, and can be utilised in relation to a range of natural and human systems. At least from a theoretical standpoint, there is a compelling case to be made for the use by governments of regulation to anticipate and thereby help to manage systemic risks caused by climate change.

Bibliography

Centeno, M.A., Nag, M., Patterson, T.S., Shaver, A., & Windawi, A.J. (2015). "The emergence of global systemic risk." *Annual Review of Sociology*, 41(1), 65–85. https://doi.org/10.1146/annurev-soc-073014-112317.

Corfee-Morlot, J., Cochran, I., Hallegatte, S., & Teasdale, P.-J. (2011). "Multilevel risk governance and urban adaptation policy." *Climatic Change*, 104(1), 169–197. https://doi.org/10.1007/s10584-010-9980-9.

Dilling, L. (2015). Adaptation. In K. Bäckstrand & E. Lövbrand (eds.), *Research Handbook on Climate Governance* (pp. 470–479). Edward Elgar Publishing. https://doi.org/10.4337/9781783470600.00054.

Financial Stability Oversight Council. (2021). *FSOC Report on Climate-Related Financial Risk*. https://home.treasury.gov/system/files/261/FSOC-Climate-Report.pdf.

IPCC. (2001). Technical Summary. In *Climate Change 2001: Impacts, Adaptation, and Vulnerability. Contribution of Working Group II to the Third Assessment Report of the Intergovernmental Panel on Climate Change*. Cambridge University Press.

IPCC. (2014a). *Climate Change 2014: Impacts, Adaptation, and Vulnerability. Part A: Global and Sectoral Aspects. Contribution of Working Group II to the Fifth Assessment Report of the Intergovernmental Panel on Climate Change*. Cambridge University Press.

IPCC. (2014b). Summary for Policymakers. In *Climate Change 2014: Impacts, Adaptation, and Vulnerability. Part A: Global and Sectoral Aspects. Contribution of Working Group II to the Fifth Assessment Report of the Intergovernmental Panel on Climate Change*. Cambridge University Press.

IPCC. (2014c). Technical Summary. In *Climate Change 2014: Impacts, Adaptation, and Vulnerability. Part A: Global and Sectoral Aspects. Contribution of Working Group II to the Fifth Assessment Report of the Intergovernmental Panel on Climate Change*. Cambridge University Press.

IPCC. (2015). Climate Change 2014: Synthesis Report. In *Contribution of Working Groups I, II and III to the Fifth Assessment Report of the Intergovernmental Panel on Climate Change*. IPCC.

IPCC. (2018). Annex I: Glossary. In *Global Warming of 1.5°C. An IPCC Special Report on the impacts of global warming of 1.5°C above pre-industrial levels and related global greenhouse gas emission pathways, in the context of strengthening the global response to the threat of climate change, sustainable development, and efforts to eradicate poverty*. IPCC.

IPCC. (2021). Summary for Policymakers. In *Climate Change 2021: The Physical Science Basis. Contribution of Working Group I to the Sixth Assessment Report of the Intergovernmental Panel on Climate Change*. Cambridge University Press.

Karassin, O. (2010). "Mind the gap: Knowledge and need in regulating adaptation to climate change." *Georgetown International Environmental Law Review*, 22(3), 383.

Lawrence, J., Blackett, P., & Cradock-Henry, N. A. (2020). "Cascading climate change impacts and implications." *Climate Risk Management*, 29, 100234. https://doi.org/10.1016/j.crm.2020.100234.

Li, H.-M., Wang, X.-C., Zhao, X.-F., & Qi, Y. (2021). "Understanding systemic risk induced by climate change." *Advances in Climate Change Research*, 12(3), 384–394. https://doi.org/10.1016/j.accre.2021.05.006.

McKibben, B. (2019). *Falter: Has the Human Game Begun to Play Itself Out?* Black Inc.

Pielke, R., Prins, G., Rayner, S., & Sarewitz, D. (2007). "Lifting the taboo on adaptation." *Nature*, 445(7128), 597–598. https://doi.org/10.1038/445597a.

Ranger, N. (2011). "Adaptation as a decision making under deep uncertainty." In I. Linkov & T.S. Bridges (eds.), *Climate: Global Change and Local Adaptation*. Springer Netherlands. https://doi.org/10.1007/978-94-007-1770-1_6.

Renn, O. (2016). "Systemic risks: The new kid on the block." *Environment: Science and Policy for Sustainable Development*, 58(2), 26–36. https://doi.org/10.1080/00139157.2016.1134019.

Renn, O., Laubichler, M., Lucas, K., Kröger, W., Schanze, J., Scholz, R.W., & Schweizer, P. (2020). "Systemic risks from different perspectives." *Risk Analysis*, 13657. https://doi.org/10.1111/risa.13657.

Schipper, E.L.F. (2006). "Conceptual history of adaptation in the UNFCCC process." *Review of European Community and International Environmental Law*, 15(1), 82–92. https://doi.org/10.1111/j.1467-9388.2006.00501.x.

Schweizer, P.-J. (2021). "Systemic risks – concepts and challenges for risk governance." *Journal of Risk Research*, 24(1), 78–93. https://doi.org/10.1080/13669877.2019.1687574.

Smit, B., & Wandel, J. (2006). "Adaptation, adaptive capacity and vulnerability." *Global Environmental Change*, 16(3), 282–292. https://doi.org/10.1016/j.gloenvcha.2006.03.008.

Viscusi, W.K. (1998). *Rational Risk Policy: The 1996 Arne Ryde Memorial Lectures*. Clarendon Press; Oxford University Press.

Wiley, L.F. (2013). "Climate change adaptation and public health law." In J. Verschuuren (ed.), *Research Handbook on Climate Change Adaptation Law*. Edward Elgar Publishing. https://doi.org/10.4337/9781781000083.00012.

Chapter 4

The Challenges of Regulating Systemic Risks

Over time, regulation has been used by governments to deal with a broad spectrum of risks. With the rise of cities in ancient civilisations, laws emerged to address health risks from air pollution by curbing the combustion of wood and charcoal for heating within urban centres.[1] Industrialisation corresponded with a dramatic increase in workplace accidents, which eventually led to the development of laws to address occupational risks.[2] In the past century, countries have increasingly had to grapple with how to regulate emerging new technologies, such as nanotechnology, in order to balance benefits and risks to society.[3]

No doubt, every attempt by government to regulate a "new" type of risk has its challenges. There may be popular resistance to the change in behaviour implied by regulation, particularly if long-standing activity that had previously been permitted is limited or prohibited under the proposed law. The costs for regulated entities to comply with new regulation may also result in significant opposition. Furthermore, the use of law to address risks in previously unregulated domains may require policy- and law-makers to anticipate possible yet uncertain developments within the domain as a pre-condition to regulating and to identify potential unintended consequences that could arise from regulation. The latest era of systemic risks – including and perhaps, most notably, those caused by climate change – pose unique regulatory challenges, which are considered in this chapter.

I The regulatory challenge

The Global Financial Crisis is a salient example of the consequences of failing to adequately regulate systemic risks and, as such, it provides useful context for

1 S. Mosley, "Environmental History of Air Pollution and Protection", in M. Agnoletti & S. Neri Serneri (eds.), *The Basic Environmental History* (Springer International Publishing, 2014), pp. 143–69, at 144–8.
2 J. Moses, *The First Modern Risk: Workplace Accidents and the Origins of European Social States* (Cambridge University Press, 2018).
3 G. Hodge, D. Bowman, & K. Ludlow, *New Global Frontiers in Regulation: The Age of Nanotechnology* (Edward Elgar Publishing, 2008). See, in particular, Part 4 – Regulating Nanotechnology: National and Supranational Perspectives.

DOI: 10.4324/9781003253464-4

consideration of the challenges facing policy- and law-makers in doing so. In that case, complex, intricate interconnections between financial institutions and firms around the world coupled with the failure of one key player – Lehman Brothers – led to cascading failures throughout the global financial system. Bell and Hindmoor (2015) suggest that weak, permissive regulation failed to account for the complexity of the entire system.[4] Freixas, Laeven and Peydró (2015) similarly assert that financial regulation had been largely "microfocused" on risk associated with individual financial institutions, rather than the financial system as a whole.[5] Meanwhile, models used within those financial institutions to manage financial risk were in retrospect regarded as "both very precise and very wrong",[6] leading former head of the US Federal Reserve, Alan Greenspan, to admit that "the whole intellectual edifice [of risk management] had collapsed".[7] Claessens and Kodres (2014) explain:[8]

> The crisis has made clear that in spite of what appeared to be individually sound and well supervised financial institutions, risks that were thought to be well diversified, and institutional infrastructures that appeared to be robust, systemic risks nonetheless emerged, went undetected for some time, and then created great havoc.

In the aftermath of the crisis, the regulatory arrangements for financial markets, banking systems and associated prudential frameworks across the globe were overhauled. The reform agenda sought to address the key causes identified as having contributed to the crisis through changes in policy and law, with the aim of avoiding a recurrence of the crisis and its consequences.[9] It was widely recognised that regulatory reforms needed to account for the whole financial system, rather than merely its component parts and individual risk agents. Reforms around firms that were "too big to fail", which were aimed at reducing the probability that such firms would fail, were complemented by mechanisms to limit global contagion, including through system-wide oversight.[10]

4 S. Bell & A. Hindmoor, *Masters of the Universe, Slaves of the Market* (Harvard University Press, 2015), pp. 3–4.
5 X. Freixas, L. Laeven, & J.-L. Peydró, *Systemic Risk, Crises, and Macroprudential Regulation* (MIT Press, 2015), p. 1.
6 S. Bell & A. Hindmoor, n. 4 above, p. 167 referring to M. Blyth, *Austerity: The History of a Dangerous Idea* (Oxford University Press, 2013), p. 37.
7 S. Bell & A. Hindmoor, n. 4 above, p. 167 referring to E. Andrews, 'Greenspan concedes error on regulation', *New York Times* (23 October 2008).
8 S. Claessens & L. E. Kodres, *The Regulatory Responses to the Global Financial Crisis: Some Uncomfortable Questions* (International Monetary Fund, 2014), p. 10.
9 IMF, *Global Financial Stability Report – A Decade after the Global Financial Crisis: Are We Safer?* (IMF, 2018), pp. 59–72.
10 S. Claessens & L. E. Kodres, n. 8 above, p. 8. Among other reforms, an inter-jurisdictional "Financial Stability Board" was established to coordinate the work of national financial authorities and standard setting bodies at an international level.

In the midst of the crisis, the OECD Group on Regulatory Policy met to consider how risk management approaches could be strengthened in order to address "important policy challenges that have been underscored by recent crises", including the problems posed by "risks which transcend sectors and borders".[11] Among the messages that emerged from the meeting was the conclusion that "regulators in the financial sector but also in other sectors will have to pay more attention to the regulation of systemic risks in the future".[12]

To be effective, the regulatory paradigm to address systemic risks must be capable of addressing the key characteristics of this type of risk – namely, complexity, uncertainty and ambiguity. In particular, risks to systems are typically *complex*. Modern systems are themselves complex and may be susceptible to multiple stressors that concurrently affect the system, thereby resulting in an aggregation of risk.[13] The stressors to which such systems may be subject could also be intrinsically complex, like climate change.[14] Moreover, such risks may be difficult to disentangle from risks caused by other stressors, such as demographic change.[15] This compounding complexity indicates that a holistic approach is needed, which accounts for the various drivers of change as well as interdependencies and spill-over effects that could precipitate system failure, and embeds a balancing mechanism to address policy trade-offs. However, such an approach implies a comprehensive, sophisticated and potentially intricate regulatory response that is likely to be extremely challenging to design, let alone implement.

In addition, the *uncertainty* associated with systemic risks, including in relation to the frequency, nature and extent of harm and damage, results in "surprises", which are difficult to anticipate and respond to proactively.[16] A "predict then act" approach is appropriate where comprehensive and accurate historical information is available to inform decisions and actions to manage future risks. However, in cases of profound uncertainty that characterises systemic risks, particularly those caused by climate change, predictions cannot be relied upon with any degree of confidence because it is impossible to correctly anticipate the future state of a system based on conditions prevailing in the past.[17] At one

11 OECD, *Summary of Discussions on Risk and Regulation at the Meeting of the Group on Regulatory Policy* (1–2 December 2008), para. 1.
12 OECD, n. 11 above, para. 6.
13 OECD, *Emerging Systemic Risks in the 21st Century: An Agenda for Action* (OECD, 2003), p. 14.
14 J. Corfee-Morlot, I. Cochran, S. Hallegatte, & P.-J. Teasdale, "Multilevel risk governance and urban adaptation policy" (2011) 104(1) *Climatic Change*, pp. 169–97, at 172.
15 For example, population growth in an area that is vulnerable to certain hazards associated with climate change could intensify climate change risks.
16 OECD, n. 13 above, p. 14.
17 D.R. Johnson, "Integrated risk assessment and management methods are necessary for effective implementation of natural hazards policy" (2021) 41(7) *Risk Analysis*, pp. 1240–7, at 1241.

extreme, accounting for possible yet uncertain surprises may result in over-regulation, particularly if the likelihood that the surprises actually eventuate is very low. However, at the other extreme, the risk of failing to regulate for the unlikely means that catastrophic events could ultimately occur, compromising the affected system in ways that could be permanent and irremediable.

The inherent complexity and uncertainty associated with systemic risks is also likely to produce *ambiguity*, particularly regarding the future that should be planned for among the various scenarios that could possibly ensue. Specifically, differing perceptions of risk and expectations regarding the way in which risks should be responded to, if at all, are likely to arise.[18] In the case of systemic risks of climate change, various studies and surveys have found divergence around the world in climate change awareness and risk perception.[19] Limited action, if any, would be consistent with a low perception of risk, whereas a more extreme response may be demanded by those who perceive risk to be relatively high. Nevertheless, the potentially catastrophic consequences of systemic risks mean that they cannot be ignored – even if public perception is to the contrary – but they may be difficult to mitigate as a matter of practice if relevant stakeholders, including members of the public, are opposed to action. In this way, public perception could act as a barrier to regulatory intervention.

The multi-faceted quandary faced by policy- and law-makers in the context of the regulation of systemic risks has parallels with the so-called "Collingridge dilemma", which was articulated in relation to the regulation of new technology. In his 1980 book – *The Social Control of Technology* – David Collingridge explained the dilemma of decision-making in the face of the kind of deep uncertainty that characterises systemic risks.[20] In the context of the development of technology, the dilemma involves a conflict between an information problem (that is, impacts cannot be easily predicted until technology is extensively developed and widely used) and a power problem (control or change is difficult when technology has become entrenched).

In a similar way, it is difficult to know whether and how to regulate systemic risks because the risks are typically so complex, deeply uncertain and consequently ambiguous. However, failing to regulate them may result in materialisation of risk, which could compromise the operation, and, potentially, the very existence of the

18 OECD, n. 13 above, p. 15.

19 T.M. Lee, E.M. Markowitz, P.D. Howe, C.-Y. Ko, & A.A. Leiserowitz, "Predictors of public climate change awareness and risk perception around the world" (2015) 5(11) *Nature Climate Change*, pp. 1014–20, at 1015.

20 D. Collingridge, *The Social Control of Technology* (Pinter, 1982), p. 11. Collingridge explained the dilemma in the context of developments in technology as follows: "The social consequences of a technology cannot be predicted early in the life of the technology. By the time undesirable consequences are discovered, however, the technology is often so much part of the whole economic and social fabric that its control is extremely difficult. This is the dilemma of control. When change is easy, the need for it cannot be foreseen; when the need for change is apparent, change has become expensive, difficult and time consuming".

systems at risk. Broad options to respond to risk include, at one extreme, a very detailed, prescriptive regulatory regime that explicitly seeks to anticipate and address the range of stressors that could trigger or exacerbate systemic risks and, at the other extreme, a high-level, principles-based regime that is capable of applying and responding to a broad range of scenarios that could affect the system. A prescriptive solution runs the risk of becoming redundant as the system, and the stressors to which it may be subject, evolve and mutate over time, whereas a high-level, principles-oriented solution may lack the specificity to provide adequate guidance to enable effective application of the regime in practice.

In summary, the regulation of systemic risks is challenging for a number of reasons. Such risks relate to *complex* systems that are likely to be vulnerable to a broad range of stressors that may themselves have a complex and aggregate effect on a system. There is *uncertainty* regarding the nature and extent of harm or damage to the system because, among other things, it is difficult to predict which stressor or combination of stressors is likely to affect the system at any given time and, if so, the consequences of those stressors for the system. These features may affect perception of systemic risk among relevant stakeholders, leading to *ambiguity* about how systemic risks should be regulated, if at all. It is evident that the path to be followed in designing a regulatory response to manage such risks is littered with challenges. A critical first step in exploring the contribution regulation could make to tackling systemic risks, including those caused by climate change, is to gain an understanding of what regulation is both in theory and in practice.

II Regulation and risk management

Like "risk", the concept of "regulation" is also contested. There is no consensus regarding the definition of regulation, primarily because, as noted by Levi-Faur (2011), the term is employed for a "myriad of discursive, theoretical and analytical purposes".[21] In apparent recognition of this pluralism, Julia Black (2002) distinguished between "essentialist", "functionalist" and "conventionalist" definitions of regulation.[22]

As the label suggests, an *essentialist definition* of regulation seeks to identify the core "elements"[23] or "central facets"[24] of the concept. Koop and Lodge (2017) have assessed various conceptions of regulation across different disciplines in an effort to identify the "Platonic essence" of regulation.[25] On the basis of their analysis, they conclude that certain features characterise the very essence of

21 D. Levi-Faur, "Regulation and regulatory governance" in D. Levi-Faur (ed.), *Handbook on the Politics of Regulation* (Edward Elgar Publishing, 2011), p. 3.
22 J. Black, "Critical reflections on regulation" (2002) 27 *Australian Journal of Legal Philosophy*, pp. 1–25, at 22.
23 D. Levi-Faur, n. 21 above, p. 5.
24 J. Black, n. 22 above, p. 22.
25 C. Koop & M. Lodge, "What is regulation? An interdisciplinary concept analysis" (2017) 11(1) *Regulation & Governance*, pp. 95–108, at 97.

regulation. In particular, in their view, regulation can be defined as "the intentional intervention in the activities of a target population".[26] Given the range of possible interventions, they argue that "[t]he intervention can be direct and/or indirect, the activities can be economic or non-economic, the regulator may be a public-sector or private sector actor, and the regulatee may equally be a public-sector or private-sector actor".[27] This essentialist view of regulation is broad, encompassing a range of interventions, in a spectrum of contexts, by either public or private sector bodies.

Black (2002) suggests that a *functionalist definition* is based on the "function that regulation performs in society".[28] This definition is linked to what regulation is designed to do.[29] At its most basic level, regulation aims to produce some improvement in the conditions of the world. Coglianese (2012) argues that regulation seeks to make such improvement by "changing individual or organisational behaviour in ways that generate positive impacts in terms of solving societal, environmental and economic problems".[30]

In more theoretical terms, regulation is generally viewed as being designed to serve the "public interest".[31] While the notion of public interest is somewhat amorphous, and is likely to vary depending upon the values and broader context of the society in which the assessment of public interest is made, in general terms, it relates to an interest involving the welfare of the public that warrants recognition, promotion, and protection through regulation.[32] In pursuit of the public interest, a common objective of regulatory frameworks is to prevent or reduce harm.[33] This objective can be achieved not only by altering the behaviour of those at risk of being harmed (for example, by prohibiting or conditioning the use or development of land that is vulnerable to a particular physical hazard, such as bushfires), but also the behaviour of those who manage the environments that produce harm (by, for example, controlling activities or operations to limit the likelihood that a physical hazard like a bushfire will materialise, such as requiring vegetation or "fuel" reduction in areas that are prone to bushfire hazard). This common objective of regulation to prevent or reduce harm can be conceptualised in an alternative way – namely to *manage risk*; that is, to manage the likelihood and consequences of a hazard that produces harm to the public or things that the public values.

26 Ibid. p. 104.
27 Ibid. p. 104.
28 J. Black, n. 22 above, p. 22.
29 D. Levi-Faur, n. 21 above, p. 5.
30 C. Coglianese, *Measuring Regulatory Performance: Evaluating the Impact of Regulation and Regulatory Policy*, Expert Paper No. 1 (OECD, 2012), p. 8.
31 R. Baldwin, M. Cave, & M. Lodge, *Understanding Regulation: Theory, Strategy, and Practice* (Oxford University Press, 2012), p. 15.
32 S.M. King, B.S. Chilton, & G.E. Roberts, "Reflections on defining the public interest" (2010) 41(8) *Administration & Society*, pp. 954–78.
33 R. Baldwin, M. Cave, & M. Lodge, n. 31 above, p. 24.

Indeed, Baldwin, Cave and Lodge (2012) suggest that all regulation can be seen as being inherently about the control of risks of one sort or another.[34] Others have noted that risk is becoming a significant organising principle for government and, by extension, for regulation. For example, Black (2010) states that society may have moved from an era of the "regulatory state" to one of the "risk state" or, at least, to a "risk regulatory state".[35] In this era, regulation may be used to address the broad range of risks that affect society, including economic, health and environmental risks. So, the functionalist definition of regulation indicates that regulation can be used to protect or advance the public interest. In practice, this can be achieved by using regulation to manage risks that could cause harm to members of the public, as well as the things and activities that the public values.

Black (2002) contends that a *conventionalist definition* focuses on the way or ways that the concept of regulation is used in practice.[36] There are various practical facets of regulation that can be used to further articulate this definition. One parameter relates to the form of regulatory intervention. Most traditional definitions of regulation focus on the existence of prescriptive state-made rules, which are supported by the possibility of sanctions for non-compliance.[37] This definition is endorsed by many governments, including the Australian Government, which defines regulation broadly as "any rule endorsed by government where there is an expectation of compliance. It includes legislation, regulations, quasi-regulations, and any other aspect of regulator behaviour that can influence or compel specific behaviour by business and the community".[38]

Within these types of rule-based interventions, Freiberg (2017) notes that there is a spectrum of approaches that are available to give effect to regulatory objectives, ranging from traditional command and control instruments to much more flexible instruments where entities that are the subject of regulation have considerable discretion in complying with regulatory requirements.[39] The rationale for regulation is likely to play a role in determining the approach to regulation and the structure that the regulatory framework eventually takes. For example, regulation that is designed to manage specific, contained risks may result in prescriptive rules because the rules can be carefully targeted at the mitigation of those risks. In contrast, regulation directed at addressing certain types of market failure is more likely to take the form of incentive regulation to

34 Ibid. p. 83.

35 J. Black, "The role of risk in regulatory processes", in R. Baldwin, M. Cave, & M. Lodge (eds.), *The Oxford Handbook of Regulation* (Oxford University Press, 2010), pp. 301–48, at 302.

36 J. Black, n. 22 above, p. 22.

37 C. Koop & M. Lodge, n. 25 above, p. 99.

38 Australian Government, *User Guide to the Australian Government Guide to Regulation* (Office of Best Practice Regulation, 2016), p. 3.

39 A. Freiberg, *Regulation in Australia* (Federation Press, 2017). See Freiberg's description of the various "regulatory methods" in Chapter 6.

motivate behaviour that will lead to more efficient outcomes. Freiberg adds that regulation may extend beyond rule-based interventions to include non-rule-based interventions such as subsidies and grants, informational tools such as disclosure of information, structural interventions such as technological controls, and behaviour change interventions in the form of "nudging" regulated entities based on theories of behavioural economics.[40]

Another facet of regulation relates to the practical effect of regulation. In some cases, regulation may have the effect of prohibiting or limiting activity that could cause harm.[41] Indeed, this appears to be the conventional notion of regulation. However, regulation can also be considered to be "enabling" or "facilitative".[42] In other words, regulation can be used to achieve positive outcomes (such as seeking to ensure the resilience of a system at risk), rather than merely forbidding negative ones (for example, limiting or prohibiting activity that could place a system at risk).

When considered collectively, these definitions of regulation indicate a strong association between risk and regulation. The functionalist definition clarifies that the management of risk is an objective of regulation. The essentialist definition indicates that regulation encompasses the broad range of regulatory interventions that could be used to manage risk. The conventionalist definition makes clear that regulation could be used to prohibit or limit activities that produce or exacerbate risk but may also be used to anticipate risk, seek to minimise risk, as well as ensuring resilience when confronted with risk.

Hutter (2006) contends that regulation and risk are "inextricably connected" but that, in the past, the relationship between the two was assumed, rather than explored in any detail.[43] She suggests that, more recently, the relationship has been explicitly probed with the rise in risk studies and with the recognition that risk is "a characteristic of modernization, with all aspects of modern life being interpreted in terms of risk".[44] In her view:[45] "regulation has come to be defined as controlling and also as a way of managing risks, but whilst the language of risk may be relatively new, the task of regulation as a way of managing risks is not."

The conclusion that regulation is strongly associated with the management of risk has important practical implications. It means that the lens used to design and evaluate every facet of a regulatory regime could and, potentially, should be based around the concept of risk. In more specific terms, the identification and assessment of risk can be used to inform policy choices about the design of regulatory regimes, including the particular legal instruments and rules that give effect to those choices.

40 Ibid. p. 9.
41 R. Baldwin, M. Cave, & M. Lodge, n. 31 above, p. 3.
42 Ibid. p. 3.
43 B. Hutter, "Risk, regulation, and management" in P. Taylor-Gooby & J.O. Zinn (eds.), *Risk in Social Science* (Oxford Scholarship Online, 2006), p. 205.
44 Ibid. p. 206.
45 Ibid. p. 205.

Placing the management of risk at the forefront of a regulatory regime has an important advantage, which is especially relevant in relation to systemic risks. In particular, using a risk perspective when considering societal problems helps to convert what might otherwise be a broad, multi-faceted and amorphous problem into a series of risks that give rise to the problem, which can then be targeted using regulation. Moreover, when viewed through the prism of risk, a regulatory regime does not merely represent a set of rules with which compliance must be secured. Rather, it is a vehicle designed to mitigate the risks that the regime has been established to address.

For some, this might be a novel, if not foreign, conception of regulation. Yet, the use of the notion of risk in this way in a regulatory setting is in line with the coming of age of the concept across many disciplines and areas of practice. In 1998, Chicken and Posner asserted that risk pervades all decision-making.[46] Ten years later, Ekberg (2007) referred to the "omnipresence" of risk, to a heightened sensitivity to the social and political consequences of a risk event, and to the emergence of a "collective risk consciousness".[47] More recently, Murphy (2021) has suggested that risk has assumed a major role in law and public policy with risk "shaping law more extensively than it has in the past, both as formal law and legal practice".[48] Townley (2018) also speaks of the "propagation" of risk across different fields of international law[49] and of the potential of "risk-based law … to help frame solutions to today's most difficult problems".[50]

III Risk and the architecture of a regulatory regime

Assuming then that regulation could have a role in managing systemic risks given the strong association between regulation and risk, the next key question concerns the manner in which regulation can be mobilised to this end.

1 The concept and components of a regulatory regime

In the view of Hood, Rothstein and Baldwin (2001), the concept of a regulatory regime "connotes the overall way risk is regulated in a particular policy domain".[51] Scott (2008) adds that a regulatory regime is "the aggregation of the

46 J.C. Chicken & T. Posner, *The Philosophy of Risk* (Thomas Telford, 1998), pp. 1–2.
47 M. Ekberg, "The parameters of the risk society: A review and exploration" (2007) 55(3) *Current Sociology*, pp. 343–66, at 344, 347.
48 B. Murphy, *Regulating Undercover Law Enforcement: The Australian Experience* (Springer, 2021), pp. 187, 189.
49 S. Townley, "The rise of risk in international law" (2018) 18(2) *Chicago Journal of International Law*, pp. 594–646, at 594.
50 Ibid. p. 646.
51 C. Hood, H. Rothstein, & R. Baldwin, *The Government of Risk: Understanding Risk Regulation Regimes* (Oxford University Press, 2001), p. 8.

activities of those whose actions shape behaviour"[52] and is typically focused on areas in which risk arises, such as environmental hazards, food safety, smoking in public places, or the quality of teaching and research within universities.[53] Windholz (2018) describes a regulatory regime in more specific terms as "the network of actors involved in regulating an issue, the aggregate activities undertaken by them to modify the behaviour of the target audience and the norms, principles, rules and decisionmaking processes according to which, and through which, those activities are coordinated".[54]

These descriptions indicate that, in practice, a regulatory regime relates to a particular policy area or domain. It comprises the various regulatory instruments, mechanisms and rules that are used to control the risk or risks that fall within that domain. The purpose of those regulatory interventions is to modify the behaviour of actors and participants in the domain in order to deliver outcomes that lie at the heart of the regime. For example, building regulatory regimes in jurisdictions around the world regulate building work so that certain minimum standards for the design, construction and operation of buildings are met in order to provide safe and healthy environments for people to live and work.[55] Building regulation was traditionally prescriptive, specifying in some detail how buildings were to be built. However, many jurisdictions have since moved to performance-based regulation, which specifies building outcomes, but building practitioners have considerable flexibility to determine how to achieve those outcomes in practice.[56]

In describing the main dimensions of a regulatory regime, Hood, Rothstein and Baldwin (2001) distinguish between the content and context of the regime.[57] *Regime content* refers to the internal structure of the regime,[58] which may be reflected through the normative objectives, principles, standards, rules and tools that constitute the regime.[59] In contrast, *regime context* refers to the broad setting in which regulation is applied, including the range of external factors that bear upon the regime as a whole.[60] The context includes the different types and levels of risk within the regulatory domain, as well as public

52 C. Scott, "Regulating everything" (2008) (WP/24/2008) *UCD Geary Institute Discussion Paper Series*, p. 7.
53 Ibid. pp. 1, 25.
54 E.L. Windholz, *Governing through Regulation: Public Policy, Regulation and the Law* (Routledge, 2017), p. 74.
55 B.J. Meacham (ed.), "Performance-based building regulatory systems – principles and experiences", *Report of the Inter-Jurisdictional Regulatory Collaboration Committee* (IRCC, 2010), p. 16.
56 Ibid. p. 16.
57 C. Hood, H. Rothstein, & R. Baldwin, n. 51 above, p. 21.
58 D. Levi-Faur, n. 21 above, p. 14.
59 C. Hood, H. Rothstein, & R. Baldwin, n. 51 above, pp. 28–35; P.J. May, "Regulatory regimes and accountability" (2007) 1 *Regulation & Governance*, pp. 8–26, at 9; C. Scott, n. 52 above, p. 7.
60 C. Hood, H. Rothstein, & R. Baldwin, n. 51 above, p. 21.

perception of risk and associated expectations of the regulatory regime.[61] It also includes the way the various actors who produce or are affected by risk are organised and behave.[62] The distinction drawn by Hood and colleagues implies the existence of the *boundary of a regime*, which demarcates the scope of the regime and helps to conceptually and practically distinguish between a regime's content and its context. These elements comprising the "architecture" of a regulatory regime are discussed below.

2 Regulatory boundary

A boundary is a line that marks the edge or limit of something, including something abstract like the subject or sphere of activity. The term implicitly conveys the notion that a boundary is a dividing line, which practically separates one thing from another.

Conceptually, a boundary is useful to define what is governed by a regulatory regime – that is, its ambit – and also serves to distinguish between the context and content for the regime. Used in this way, the idea of a regulatory boundary has the capacity to deliver important political, legal and practical benefits. In particular, a clear boundary can assist in advancing the case for regulation by minimising ambiguity about what is to be regulated. This, in turn, can increase the social acceptability of regulation. Clarity about the boundary for a regulatory regime can also help a regulator apply regulation in practice because the limits of the regulatory regime are well-delineated and, consequently, so too are the regulator's powers.

Using risk as a lens, the regulatory boundary can be used to identify the risks that are sought to be governed by a particular regime and, by implication, those that are excluded from coverage. While this exercise may be relatively straightforward for regimes that regulate discrete, identifiable risks, it is likely to be comparatively much more challenging in the context of the regulation of systemic risks because of the complexity, uncertainty and ambiguity that characterise such risks.

3 Regulatory context

A fundamental axiom that applies to many aspects of life is that "context is everything". This adage applies to the way information is understood and the way in which responses to situations are formulated and implemented. As noted by the social scientist, Alvin Gouldner, in 1955:[63] "Social scientists of the most varying standpoints agree that human action can be rendered meaningful only by relating it to the contexts in which it takes place."

61 Ibid. pp. 21, 28–35; D. Levi-Faur, n. 21 above, p. 14.

62 D. Levi-Faur, n. 21 above, p. 14.

63 A.W. Gouldner, *Wildcat Strike: A Study in Worker-Management Relationships* (Harper & Row, 1965), p. 12.

The axiom also applies in relation to the regulation of risk. Indeed, the regulatory context forms a critical component of a regulatory regime. The capacity of a regulatory regime to successfully manage risk will depend, amongst other things, upon how the context for the regime is assessed, which can then be used as the basis for determining how regulation can be employed to respond to risk. In other words, a regulatory regime is more likely to be effective in mitigating risk if the regulatory response appropriately accounts for the context in which that risk has arisen.

4 *Regulatory content*

In simplistic terms, the content of a regulatory regime is embodied in the rules and requirements that can be relied upon to manage risks governed by the regime. It is a composite of various concrete elements that, together, are designed to manage these risks. These elements include the regulatory approach, objectives, principles and tools. A key issue for the design of regulation to manage systemic risks – as with other types of risks – is to determine how to best combine the available regulatory options in order to maximise the chance of delivering a well-tuned regulatory response to risk when the regime is applied in practice.

Howlett and Mukherjee (2018) refer to the "dilemmas" that can arise when designing regulatory content, particularly when there may be a disconnect between the underlying policy of regulatory design and the practical realities of policy implementation.[64] They suggest that "[e]ven in the best circumstances and with the best of intentions, governments often grapple with complex problems … in complex problem environments that typically evolve and change over time, making it difficult to secure or retain agreement on which policy alternatives are most likely to succeed".[65] The challenges associated with the process of matching regulatory content to policy problems are likely to be particularly marked in the context of regulation of systemic risks, given their complexity, uncertainty and ambiguity.

5 *Depicting the architecture of a regulatory regime*

Differentiating between the content of and context for a regulatory regime is important for a number of reasons. First, the regime's content cannot be developed and applied without understanding the context and, as a corollary, the distinction forces recognition that content of a regulatory regime is actually a product of its context. Second, while a regulatory regime's content and context are distinct, they are inextricably linked. Understanding the interlinkages will

64 M. Howlett & I. Mukherjee, "The importance of policy design: Effective processes, tools and outcomes" in M. Howlett & I. Mukherjee (eds.), *Routledge Handbook of Policy Design* (Routledge, 2018), p. 6.

65 Ibid. p. 4.

help to ensure that regulatory content keeps pace with and responds to the context. Indeed, as risks and perception of them change with time, a shift in the way the relevant regime is structured and applied by the regulator may be needed. Similarly, the content of the regime will have an impact on context, including the behaviour of actors affected by the regime. Recognising and accounting for the interdependent relationship between content and context helps to ensure that the regulatory regime remains relevant and fit-for-purpose over time.

The basic architecture of a regulatory regime is conceptualised in Figure 4.1, using the distinction between regime content and regime context drawn by Hood and colleagues as the basis for this abstraction.[66] The diagram identifies the regime's content and context, separated by the regime's notional boundary, and symbolically illustrates the symbiotic relationship between these two fundamental components of the regime's architecture.

This conceptualisation of the architecture of a regulatory regime provides insights about how a regulatory regime could be designed or evaluated. In particular, it confirms the importance of establishing the regime's boundary, as this will affect both the regulatory context and regulatory content. It highlights the distinction between regulatory context and regulatory content. It also emphasises the need for the regulatory content to be responsive to the regulatory context in terms of how it is designed, as well as applied.

The question remains as to how the components of a regulatory regime can be mobilised to most effectively manage systemic risks, including those caused by climate change. The discussion in the following chapters lays the foundation

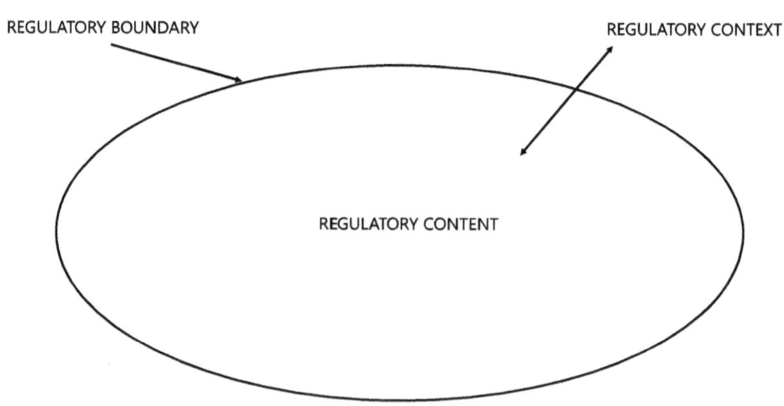

Figure 4.1 Basic architecture of a regulatory regime

66 C. Hood, H. Rothstein, & R. Baldwin, n. 51 above, pp. 21, 28–35.

for establishing a paradigm for the regulation of systemic risks, using the architecture of a regulatory regime as the skeleton for this paradigm. The discussion commences with a consideration of a "traditional approach to the regulation of risk" followed by an examination of "risk governance", which arguably sit at opposite ends of a continuum of options available to policy- and law-makers to regulate risk, including systemic risks.

Bibliography

Baldwin, R., Cave, M., & Lodge, M. (2012). *Understanding Regulation: Theory, Strategy, and Practice*. Oxford University Press. https://doi.org/10.1093/acprof:osobl/9780199576081.001.0001.

Bell, S., & Hindmoor, A. (2015). *Masters of the Universe, Slaves of the Market*. Harvard University Press.

Black, J. (2002). "Critical reflections on regulation." *Australian Journal of Legal Philosophy*, 27(2002), 1–25.

Black, J. (2010). "The role of risk in regulatory processes." In R. Baldwin, M. Cave, & M. Lodge (eds.), *The Oxford Handbook of Regulation*. Oxford University Press. https://doi.org/10.1093/oxfordhb/9780199560219.003.0014.

Chicken, J. C., & Posner, T. (1998). *The Philosophy of Risk*. Thomas Telford.

Claessens, S. & Kodres, L.E. (2014) *The Regulatory Responses to the Global Financial Crisis: Some Uncomfortable Questions*. International Monetary Fund. https://www.imf.org/en/Publications/WP/Issues/2016/12/31/The-Regulatory-Responses-to-the-Global-Financial-Crisis-Some-Uncomfortable-Questions-41422.

Coglianese, C. (2012). *Measuring Regulatory Performance: Evaluating the Impact of Regulation and Regulatory Policy*, Expert Paper No. 1. Organisation for Economic Co-operation and Development.

Collingridge, D. (1982). *The Social Control of Technology*. Pinter.

Corfee-Morlot, J., Cochran, I., Hallegatte, S., & Teasdale, P.-J. (2011). "Multilevel risk governance and urban adaptation policy." *Climatic Change*, 104(1), 169–197. https://doi.org/10.1007/s10584-010-9980-9.

Ekberg, M. (2007). "The parameters of the risk society: A review and exploration." *Current Sociology*, 55(3), 343–366. https://doi.org/10.1177/0011392107076080.

Freiberg, A. (2017). *Regulation in Australia*. Federation Press.

Freixas, X., Laeven, L., & Peydró, J.-L. (2015). *Systemic Risk, Crises, and Macroprudential Regulation*. MIT Press. https://doi.org/10.7551/mitpress/9780262028691.001.0001.

Gouldner, A.W. (1965). *Wildcat Strike: A Study in Worker-Management Relationships*. Harper & Row.

Hodge, G., Bowman, D., & Ludlow, K. (2008). *New Global Frontiers in Regulation: The Age of Nanotechnology*. Edward Elgar Publishing.

Hood, C., Rothstein, H., & Baldwin, R. (2001). *The Government of Risk: Understanding Risk Regulation Regimes*. Oxford University Press. https://doi.org/10.1093/0199243638.001.0001.

Howlett, M., & Mukherjee, I. (eds.). (2018). *Routledge Handbook of Policy Design*. Routledge. https://doi.org/10.4324/9781351252928.

Hutter, B. (2006) "Risk, regulation, and management." In P. Taylor-Gooby & J.O. Zinn (eds.), *Risk in Social Science*. Oxford Scholarship Online.

IMF. (2018). *Global Financial Stability Report – A decade after the global financial crisis: Are we safer?* International Monetary Fund.

Johnson, D.R. (2021). "Integrated risk assessment and management methods are necessary for effective implementation of natural hazards policy." *Risk Analysis*, 41(7), 1240–1247. https://doi.org/10.1111/risa.13268.

King, S.M., Chilton, B.S., & Roberts, G.E. (2010). "Reflections on defining the public interest." *Administration & Society*, 41(8), 954–978. https://doi.org/10.1177/0095399709349910.

Koop, C., & Lodge, M. (2017). "What is regulation? An interdisciplinary concept analysis." *Regulation & Governance*, 11(1), 95–108. https://doi.org/10.1111/rego.12094.

Lee, T.M., Markowitz, E.M., Howe, P.D., Ko, C.-Y., & Leiserowitz, A.A. (2015). "Predictors of public climate change awareness and risk perception around the world." *Nature Climate Change*, 5(11), 1014–1020. https://doi.org/10.1038/nclimate2728.

Levi-Faur, D. (2011). *Handbook on the Politics of Regulation*. Edward Elgar. https://doi.org/10.4337/9780857936110.

May, P.J. (2007). "Regulatory regimes and accountability." *Regulation & Governance*, 1, 8–26. https://doi.org/10.1111/j.1748-5991.2007.00002.x.

Meacham, B.J. (ed.) (2010). "Performance-Based Building Regulatory Systems – Principles and Experiences." In *A Report of the Inter-jurisdictional Regulatory Collaboration Committee*. Inter-jurisdictional Regulatory Collaboration Committee.

Moses, J. (2018). *The First Modern Risk: Workplace Accidents and the Origins of European Social States*. Cambridge University Press. https://doi.org/10.1017/9781108657853.

Mosley, S. (2014). "Environmental history of air pollution and protection." In M. Agnoletti & S. Neri Serneri (eds.), *The Basic Environmental History (Vol.4)*. Springer International Publishing. https://doi.org/10.1007/978-3-319-09180-8_5.

Murphy, B. (2021). *Regulating Undercover Law Enforcement: The Australian Experience*. Springer. https://doi.org/10.1007/978-981-33-6381-6.

OECD. (2003). *Emerging Systemic Risks in the 21st Century: An Agenda for Action*. Organisation for Economic Co-operation and Development. https://doi.org/10.1787/9789264101234-fr.

OECD. (2008). *Summary of Discussions on Risk and Regulation at the Meeting of the Group on Regulatory Policy (1–2 December 2008)*. https://www.oecd.org/gov/regulatory-policy/44806674.pdf.

Scott, C. (2008). "Regulating everything." *UCD Geary Institute Discussion Paper Series, WP/24/2008, 34*.

Townley, S. (2018). "The rise of risk in international law." *Chicago Journal of International Law*, 18(2), 594–646.

Windholz, E.L. (2017). *Governing through Regulation: Public Policy, Regulation and the Law*. Routledge. https://doi.org/10.4324/9781315677286.

Chapter 5

A Traditional Approach to Regulating Risk

Tradition is pervasive in many facets of life. It has been described as "sets of beliefs, customs, teachings, values, practices, or procedures that are transmitted from generation to generation".[1] The attachment to the past implied by tradition is said to derive, at least in part, from "deliberative efficiency".[2] In particular, social conventions, collective habits and models that are based on tradition can be used to facilitate decision-making without the need for excessive deliberation.[3]

Scheffler (2010) explains that tradition is a normative notion, used to legitimate or authorise a decision and may additionally provide guidance regarding the application of tradition in practice.[4] He states that:[5]

> [A] tradition of reasonably long standing may be regarded as a repository of experience and of the kind of wisdom that comes from experience. Traditional practices take shape in light of an accumulation of historical experience, judgment, and perspective that outstrips what any single individual can reasonably aspire to achieve in the course of a lifetime, and someone who adheres to the tradition may gain the advantages of that accumulated experience and judgment.

Given the value that tradition can yield, it is no surprise that it is prominent in a legal context. Indeed, Krygier (1986) characterises law as a "profoundly traditional social practice"[6] that "rests upon mountains of inherited tradition",[7] and asserts that the "legal past is central to the legal present".[8] In fact, legal

1 M. Beckstein, "The concept of a living tradition" (2017) 20(4) *European Journal of Social Theory*, pp. 491–510, at 492.
2 Ibid. p. 494.
3 S. Scheffler, *Equality & Tradition: Questions of Value in Moral and Political Theory* (Oxford University Press, 2010), at p. 292; M. Beckstein, n. 1 above, p. 494.
4 S. Scheffler, n. 3 above, pp. 288, 292.
5 Ibid. p. 292.
6 M. Krygier, "Law as tradition" (1986) 5(2) *Law and Philosophy*, p. 239.
7 Ibid. p. 256.
8 Ibid. p. 241.

DOI: 10.4324/9781003253464-5

tradition is ubiquitous and is a core feature of all legal systems.[9] More specifically, tradition is evident in the way law is interpreted. The history of a legal text, including the circumstances in which it was written and the intention of those responsible for drafting it, may be used to inform the meaning of the text many years after it was drafted.[10] Tradition is also obvious in the way law is applied by judges through the concept of legal precedent, whereby court decisions are considered as authority for subsequent cases dealing with the same or similar legal issues.[11] The role of tradition in regulation is also apparent,[12] including the way risks are regulated.

Various existing types of regulation to manage risks can be traced back to antiquity, such as regulation to address the risks of air pollution,[13] to minimise the risk of sale of unsafe food,[14] and to respond to public hygiene risks.[15] While these ancient forms of regulation may bear limited resemblance to their modern counterparts, there will nonetheless be learnings from each iteration of regulation that helps to inform the manner in which analogous risks are regulated over time.[16] The advantage of this "traditional approach to the regulation of risk" is that it draws upon the past to guide the manner in which law is used to govern the future. But, is this approach appropriate when the future is likely to look very different from the past? This chapter examines whether a traditional approach to the regulation of risk is appropriate for systemic risks – particularly, those caused by climate change – in light of the complexity, uncertainty and ambiguity that characterise these types of risks.

I Defining risk regulation

The notion of a traditional approach to the regulation of risk hinges on the concept of "risk regulation", which is a familiar label used in regulatory discourse. Yet, Hutter (2006) notes that the term is often used without an

9 J.H. Merryman & R. Pérez-Perdomo, *The Civil Law Tradition: An Introduction to the Legal Systems of Europe and Latin America* (Stanford University Press, 2007, 3rd edition), p. 2.

10 M. Greenberg, "Legal interpretation", in E.N. Zalta (ed.), *The Stanford Encyclopedia of Philosophy* (Stanford University, Fall edition, 2021).

11 S. Lewis, "Precedent and the rule of law" (2021) 41(4) *Oxford Journal of Legal Studies*, pp. 873–98.

12 OECD, *Alternatives to Traditional Regulation*, prepared by G. Hepburn accessible at: https://www.oecd.org/gov/regulatory-policy/alternativestoregulation.htm

13 L. Makra, "Anthropogenic air pollution in ancient times", in *Toxicology in Antiquity* (Elsevier, 2019), pp. 267–87.

14 P.B. Hutt, "Government regulation of the integrity of the food supply" (1984) 4 (1) *Annual Review of Nutrition*, pp. 1–21, at 2–3.

15 L. Cilliers, "Public health in Roman legislation" (1993) 36(1) *Acta Classica*, pp. 1–10, at 1–2.

16 J.B. Wiener, "Risk regulation and future learning" (2017) 8(1) *European Journal of Risk Regulation*, pp. 4–9, at 8.

explanation of how it is to be defined.[17] Fisher (2013) further argues that, in fact, risk regulation has various connotations, with "differences of opinion over what exactly is entailed in regulating risk".[18]

A literal interpretation of the term indicates that risk regulation refers to the manner in which risks are regulated. This interpretation is consistent with the conceptualisation of risk regulation by Macrae (2010), who defines it as the spectrum of regulatory activities to control the threat of harmful, adverse or unintended consequences, ranging from formal systems of risk management to more informal mechanisms of social control.[19] Among the range of mechanisms that could be employed by government to manage risk, two discrete levels of risk regulation can be discerned.

At the higher level, risks that could cause harm, damage or loss are regulated by the laws and associated subordinate rules comprising the regulatory regime that governs those risks. Risk regulation in this context refers to how the *design* of the regulatory regime seeks to manage risk. At the lower level, regulators who are established to oversee the regulatory regime *apply* the law and may use formal mechanisms provided for in the regulatory regime to manage risks covered by the regime (such as licensing of regulated entities or issuance of infringement notices for contraventions of the rules), as well as informal mechanisms (in particular, strategies to help prioritise regulatory activities based on relative risk).[20]

While the regulation of risk at both levels pursues the same objective – that is, to manage the risks governed by the regulatory regime – the laws and rules to address risks to society at the higher level set out how the regime could be applied to manage risks covered by the regime, whereas at the lower level the regulator applies those laws and rules in practice to manage specific risks. Implicitly, the risks that are the focus of attention by a regulator at the lower level will be a sub-set of the risks that are identified by policy- and law-makers for coverage at the higher level. This distinction is summarised in Figure 5.1.

17 B. Hutter, "Risk, regulation and management" in P. Taylor-Gooby & J.O. Zinn (Eds.), *Risk in Social Science* (Oxford Scholarship Online, 2006), p. 210.

18 E. Fisher, "Framing risk regulation: A critical reflection" (2013) 4(2) *European Journal of Risk Regulation*, pp. 125–32, at 126.

19 C. MacRae, "Regulating resilience? Regulatory work in high-risk arenas", in B. M. Hutter (ed.), *Anticipating Risks and Organising Risk Regulation* (Cambridge University Press, 2010), pp. 139–60, at 141.

20 This aspect of risk regulation is often referred to as "risk-based regulation", "risk-based approach to regulation" or "risk-based governance". See, for example, J. Black, "The emergence of risk-based regulation and the new public risk management in the United Kingdom" (2005) *Public Law*, p. 512; H. Rothstein, O. Borraz, & M. Huber, "Risk and the limits of governance: Exploring varied patterns of risk-based governance across Europe" (2013) 7(2) *Regulation & Governance*, pp. 215–35.

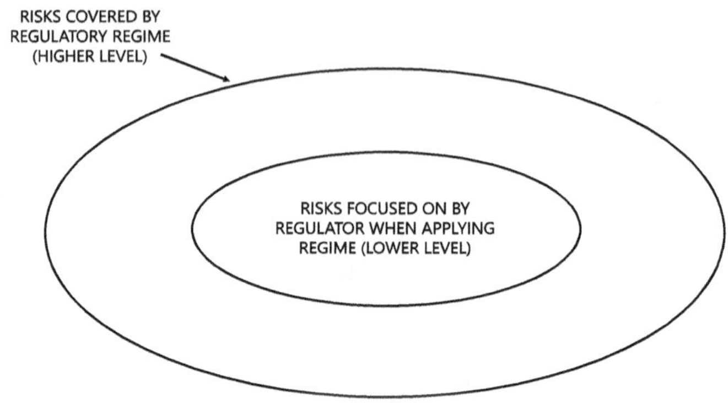

Figure 5.1 Levels of risk regulation

For the purposes of this book, risk regulation is used to refer to the higher level – specifically, the laws and rules comprising a regulatory regime that are designed to avoid or, at least, mitigate risk.

II Common features of a traditional approach to regulating risk

Fisher (2013) considers that there is very little coherence in the way in which risk regulation regimes have developed in practice and attributes "legal, culture, socio-political catalysts, and sheer historical accident" as the varied drivers for the divergent approaches.[21] There is tangible evidence to illustrate these differences.

In particular, there can be differences in the way risks are regulated across regulatory domains in a single jurisdiction. For example, in the UK, the way health and safety risks arising in an occupational context are regulated differs from the regulation of health and safety risks associated with the sale of food.[22] Whereas, the former regime is characterised by a principles-based approach requiring employers to ensure the health and safety of their employees only so far as is reasonably practicable,[23] the food safety regime is prescriptive, resulting in strict liability for selling unsafe food.[24] There are also differences in the regulation of risks across jurisdictions, albeit in the same regulatory domain. For example, while the US regulatory regime is relatively tolerant and permissive in

21 E. Fisher, n. 18 above, p. 126.

22 D. Demeritt, H. Rothstein, A.-L. Beaussier, & M. Howard, "Mobilizing risk: Explaining policy transfer in food and occupational safety regulation in the UK" (2015) 47(2) *Environment and Planning A: Economy and Space*, pp. 373–91, at 377–80.

23 Ibid. p. 377. UK Health & Safety Executive, *A Guide to Health and Safety Regulation in Great Britain*, p. 9 accessible at: https://www.hse.gov.uk/pubns/hse49.pdf.

24 D. Demeritt, H. Rothstein, A.-L. Beaussier, & M. Howard, n. 22 above, p. 378. Food Standards Agency, *The Food Safety Act 1990 – A Guide for Food Business*, (2009 edition) accessible at: https://www.food.gov.uk/about-us/key-regulations

relation to "CRISPR" (an acronym for "clustered regularly interspaced short palindromic repeats" that forms the basis of emerging technology involving the manipulation of DNA of living organisms), the regime in the EU is much more restrictive.[25] Notwithstanding the possible diversity in the approach to the regulation of risk demonstrated by these examples, some common features are discernible. These features, which together help to crystallise and characterise a traditional approach to the regulation of risk, are discussed below.

I Anticipating risks

A key feature of a traditional approach to the regulation of risk is the focus on anticipating risks. Hutter (2017) considers that risk regulation is essentially about the anticipation of risk and preventing its realisation.[26] This implies that risks are somehow known, knowable, expected or predictable. An example of an anticipatory approach to the regulation of risk is the US drug approval process. Drug companies seeking to sell a drug in the US must first test it to prove that the drug is safe and effective for its intended use. The drug can be approved for sale if an independent review establishes that the drug's health benefits outweigh its known risks.[27] The regulatory regime is based on the assumption that the risks associated with a new drug are discoverable through the testing process and approval is granted or refused based on the results of that process.

Hutter notes that the anticipatory role implicit in risk regulation is linked to modern social theorists' conception of risk, which is based on the anticipation of danger and catastrophe.[28] One such theorist – Beck (2006) – suggests that anticipation distinguishes a risk from a catastrophe. He explains that "[r]isk means the anticipation of catastrophe … At the moment at which risks become real … they cease to be risks and become catastrophes".[29] The question arises as to whether regulation has the capacity to anticipate all manner of risks and, thereby, avert the broad spectrum of consequences – including catastrophes – that could arise from these risks. More to the point, to what degree is regulation capable of anticipating systemic risks?

25 A. Asquer & I. Krachkovskaya, "Uncertainty, institutions and regulatory responses to emerging technologies: CRISPR gene editing in the US and the EU (2012–2019)" (2021) 15(4) *Regulation & Governance*, pp. 1111–27, at 1120–3.

26 B.M. Hutter, "A risk regulation perspective on regulatory excellence" in C. Coglianese (ed.), *Achieving Regulatory Excellence* (Brookings Institution Press, 2017), p. 102.

27 A. Dabrowska & S. Thaul, "How FDA approves drugs and regulates their safety and effectiveness" (2018) *US Congressional Research Service*, p. 31.

28 B.M. Hutter, "Anticipating risk and organising risk regulation: current dilemmas", in B.M. Hutter (ed.), *Anticipating Risks and Organising Risk Regulation* (Cambridge University Press, 2010), pp. 3–22, at 4.

29 U. Beck, "Living in the world risk society" (2006) 35(3) *Economy and Society*, pp. 329–45, at 332.

Beck argues that modern society fails to recognise that "rationality, that is the experience of the past, encourages anticipation of the wrong kind of risk, the one we believe we can calculate and control, whereas the disaster arises from what we do not know and cannot calculate".[30] He identifies a "fatal irony" arising from the "futility with which the highly developed institutions of modern society – science, state, business and military – attempt to anticipate what cannot be anticipated".[31]

Beck's concerns about the ability of modern society to "anticipate what cannot be anticipated" have been well illustrated in relation to national security risks. Arguably such risks were not properly anticipated in the US in 2001 in the context of the 9/11 disaster. At the beginning of 2001, US counterterrorism officials were receiving frequent but fragmented reports about threats to national security,[32] including troubling information about a number of individuals who were ultimately responsible for the 9/11 attacks. By spring, the level of reporting of terrorist threats and planned attacks increased dramatically and continued to escalate through the summer months to unprecedented levels.[33] These reports led to repeated advisory notices from US intelligence officials warning of impending Al Qaeda attacks. The system was "blinking red".[34] However, the individual threats posed by those who were responsible for the attacks were not prioritised, allowing the assailants to navigate airport passport control and security checkpoints without obstruction on the day of the 9/11 attacks.[35]

The National Commission on Terrorist Attacks upon the United States found that there had been a lack of institutional imagination, which led to a failure on the part of government leaders to understand the gravity of the threat.[36] This finding is notable. It indicates that, somehow, government's understanding of risk could have been enhanced and, through the application of a degree of "imagination", the tragic events of 9/11 could have been averted. Yet, the nature of systemic risks is such that a system may change drastically, dramatically and, at times, tragically, in the blink of an eye. It is unclear whether regulation is capable of responding to such risks by providing scope for the exercise of imagination. Indeed, can these risks be fully or even partially anticipated in the same way as other risks under a traditional approach to the regulation of risk?

30 Ibid. p. 330.
31 Ibid. p. 329.
32 National Commission on Terrorist Attacks upon the United States, T.H. Kean, & L. Hamilton (eds.), *The 9/11 Commission Report: Final Report of the National Commission on Terrorist Attacks upon the United States* (National Commission on Terrorist Attacks upon the United States, 2004), p. 254.
33 Ibid. pp. 255, 256.
34 Ibid. Chapter 8.
35 Ibid. pp. 1–14.
36 Ibid. pp. 339–40.

2 A linear process

Another characteristic feature of a traditional approach to the regulation of risk is the reliance on a linear process when managing risk. Fisher explains that, for many lawyers and policy-makers, risk regulation is understood to be "a linear process of risk assessment and risk management".[37] Macrae (2010) adds that "[r]isk regulation has been conceptualised as a rational and sequentially structured process, based on stages of information gathering, standard setting and behaviour modification".[38] Fisher (2013) also states that "[t]he linear model of risk regulation has come to dominate much of the policy and legal discourse".[39] This linear process is evident in biosecurity legislation in Australia. The *Biosecurity Act 2015*, which applies at the federal level, is designed to protect Australia from biosecurity risks, including risks associated with goods that are sought to be imported into Australia from other countries. An assessment of biosecurity risk is undertaken based on available information about the goods that are sought to be imported. A permit authorising the importation of goods can be granted, refused or made subject to conditions depending upon the outcome of the assessment of biosecurity risk.[40]

Macrae (2010) argues that a linear approach is based on the assumption that risk regulation is a "rational decision-making process" and that risks can be anticipated and prevented on the basis of forecasting and planning.[41] In essence, this regulatory option is a "precautionary and forward-looking" tool,[42] which involves the anticipation and assessment of future hazards or perils and the use of pre-emptive action to reduce or mitigate the risks which the regulatory framework is designed to address. This "quintessentially managerial" approach[43] is attractive because it suggests that the "messiness" of risk can be contained through a structured and regimented approach.[44] However, the OECD (2003) suggests that "simplistic" and "standardised" approaches may not properly account for complex human factors, including risk perception and associated patterns of behaviour.[45] Moreover, a traditional approach to the regulation of risk that is based on a linear process for the assessment and response to risk may not be appropriate for complex systems that cannot be easily modelled and for which past data cannot be relied upon to regulate for the future.

37 E. Fisher, n. 18 above, p. 126 citing OECD Working Party on Regulatory Management and Reform, *Risk and Regulation: Issues for Discussion* GOV/PGC/REG (2006)1.
38 C. MacRae, n. 19 above, p. 139.
39 E. Fisher, n. 18 above, pp. 125, 127.
40 See Chapter 3 of the *Biosecurity Act 2015* (Cth).
41 C. MacRae, n. 19 above, p. 142.
42 Ibid. p. 139.
43 V. Heyvaert, "Governing climate change: Towards a new paradigm for risk regulation" (2011) 74(6) *The Modern Law Review*, pp. 817–44, at 834.
44 E. Fisher, n. 18 above, p. 129.
45 OECD, *Emerging Systemic Risks in the 21st Century: An Agenda for Action* (OECD, 2003), p. 16.

3 *Lack of feedback between content and context*

Yet another apparent feature of a traditional approach to the regulation of risk is the absence of a feedback mechanism between regulatory content and regulatory context. As observed by Hutter (2010), the "extent to which risks are effectively anticipated depends on context and the domain in which they are situated".[46] Tellingly, major regulatory failures often eventuate where the applicable regulatory regime has become static and not sufficiently responsive to the context in which it applies.[47] The Global Financial Crisis is a case in point.

Implicitly, a regulatory regime that fails to adequately account for context will be unable to effectively respond to risk. The so-called "pacing problem" is concerned with how law can remain relevant and applicable despite changes in context.[48] The prospect of regulation being outpaced by context is particularly pronounced in cases where the content of the regulatory regime is not sufficiently flexible to account for and accommodate a changing context and where the context is prone to dramatic change, as in the case of emerging technology. In that area, concerns have been expressed about the failure of law to control risks associated with new technologies, and about the risk of technological developments making existing regulation obsolete.[49] These concerns have played out in relation to the regulation of "FinTech" (financial technology for the delivery of financial services),[50] blockchain (to facilitate a range of transactions, including those involving international trade),[51] and "AI" (the use of artificial intelligence for decision-making and other tasks performed by humans).[52]

46 B.M. Hutter, n. 28 above, p. 17.

47 There is evidence of regulatory regimes becoming out of touch with the relevant contexts in the findings of various Royal Commissions established in Australia in the recent past dealing respectively with "Institutional responses to child sexual abuse" (2012, accessible at: https://www.childabuseroyalcommission.gov.au/final-report), "Misconduct in the banking, superannuation and financial services industry" (2017, accessible at: https://www.royalcommission.gov.au/banking) and "Aged care quality and safety" (2021, accessible at: https://agedcare.royalcommission.gov.au/publications/final-report).

48 L.B. Moses and M. Zalnieriute, "Law and technology in the dimension of time" in S. Ranchordás & Y. Roznai (eds.), *Time, Law, and Change: An Interdisciplinary Study* (Hart Publishing, 2020), p. 303.

49 G.E. Marchant, "The growing gap between emerging technologies and the law" in G. E. Marchant, B.R. Allenby, & J.R. Herkert (eds.), *The Growing Gap Between Emerging Technologies and Legal-Ethical Oversight* (Springer Netherlands, 2011), pp. 27–8.

50 P. Chirulli, "FinTech, RegTech and SupTech: Institutional challenges to the supervisory architecture of the financial markets" in I.H.-Y. Chiu & G. Deipenbrock (eds.), *Routledge Handbook of Financial Technology and Law* (Routledge/Taylor & Francis Group, 2021).

51 E. Ganne, "Blockchain for trade: When code needs law" (2021) 115 *AJIL Unbound*, pp. 419–24.

52 M.A. Jackson, "Regulating AI", in M. Jackson & M. Shelly (eds.), *Legal Regulations, Implications, and Issues Surrounding Digital Data* (IGI Global, 2020), pp. 159–81.

There is, of course, an inherent tension between the need for law to provide stability and certainty for regulated entities as well as the public at large, while also remaining relevant and responsive to change. Stability and certainty may be more successful in driving regulatory outcomes in the short-to-medium terms. However, responsiveness to change to avoid disconnection between regulation and the contextual landscape is likely to enhance regulatory outcomes in the long run. As to which of these imperatives should prevail, this will depend on the particular issues and risks at play.

4 Limited interdisciplinarity

Risk is, at least partly, a mental construct because the concept is linked to the actual experience of harm that has been produced by risk in the past.[53] The way in which harm is experienced and the associated way in which risk is perceived will be affected by the particular physical, political, cultural and social context within which the risk has arisen. It is, therefore, no surprise that Fisher (2013) considers that the regulation of risk is "inherently interdisciplinary" because the "regulation of a threat requires the integration of science, law, and socio-political discourses".[54] The fundamental interdisciplinarity of regulation is particularly obvious in areas where recourse to disciplines such as science, medicine, psychology and economics is useful to fully understand the issues and risks of concern and to determine how law could be used to respond – such as environmental law, heritage protection law, medical negligence law, and technology law. As noted by Sullivan (2002), "law, though a discipline, is not and never has been an *autonomous* discipline".[55]

Yet, Mišćenić and Raccah (2016) question the inclination and capacity of law-makers to undertake "a thorough examination of economic, social, environmental and other important impacts of their legislative drafts" leading to "regulations either offering inadequate protection or restricting market freedom".[56] Their comments relate to the law-making process but the absence of an interdisciplinary approach in this process will also likely be reflected in the law itself. Similarly, Fisher suggests that creating a comprehensible framework that accounts for interdisciplinarity is challenging, particularly given that policy-makers seldom have expertise in all areas involved in regulating a threat.[57] She considers that these

53 OECD, n. 45 above, p. 67.
54 E. Fisher, n. 18 above, p. 126.
55 K. M. Sullivan, "Interdisciplinarity" (2002) 100(6) *Michigan Law Review*, pp. 1217–26, at 1219.
56 E. Mišćenić & A. Raccah (eds.), *Legal Risks in EU Law: Interdisciplinary Studies on Legal Risk Management and Better Regulation in Europe* (Springer International Publishing, 2016), p. viii–ix.
57 E. Fisher, n. 18 above, p. 126.

difficulties are likely to be magnified in the context of global risks because "there is considerable divergence between legal cultures in how risk is framed and regulated and, thus, how issues of coherence and interdisciplinarity are to be managed".[58] Indeed, Fisher's concerns about the scope for interdisciplinary analysis in the context of risk regulation are apposite for systemic risks, particularly those caused by climate change, which Lorenzoni, Pidgeon and O'Connor (2005) describe as "genuinely interdisciplinary".[59]

There is no evidence to suggest that policy- and law-makers are ideologically resistant to the notion of embedding interdisciplinarity in regulation. Rather, the fact that applying interdisciplinarity in practice can be complex, challenging and the outcomes uncertain may militate against doing so because, among other things, regulation strives to provide clarity, certainty and predictability for those who are subject to it.[60]

III Regulatory architecture for a traditional approach

Arguably, the features discussed above characterise a traditional approach to the regulation of risk and help to paint a picture of how a regulatory regime reflecting such an approach is likely to be constructed – that is, its architecture.

In relation to the *boundary* of such a regime, a traditional approach to the regulation of risk assumes that risk can be anticipated, which implies that the regulatory boundary is clear, with a focus on specific and identifiable risks. Such an approach is based on a linear decision-making process that is aimed at stabilising the status quo by minimising the prospect that risk materialises. The *content* of the regime is likely to involve prescription about how risks are to be assessed and managed. Finally, the absence of a clear feedback mechanism between regulatory content and context and an interdisciplinary perspective, indicates that the scope for the changes in *context* of a regulatory regime to influence regulatory content will be limited. This conceptualisation of the architecture of a regime that embeds a traditional approach to regulation is depicted in Figure 5.2.

58 Ibid. p. 126.

59 I. Lorenzoni, N.F. Pidgeon, & R.E. O'Connor, "Dangerous climate change: The role for risk research" (2005) 25(6) *Risk Analysis*, pp. 1387–98, at 1388. The OECD also recommends an interdisciplinary approach because "[r]isk is multi-dimensional – a variety of factors influence the nature of hazards as well as exposure and vulnerability to them. Such factors are bound to become even more diverse in a world where obstacles to the movement of people, goods, capital and information are reduced while physical, informational and economic linkages are multiplying. Tailoring a given risk policy to the environment in which that risk arises entails integrating, to the extent possible, the complex interactions of these factors."; OECD, n. 45 above, pp. 259–60.

60 See, for example, H. Avila, *Certainty in Law* (Springer Berlin Heidelberg, 2016).

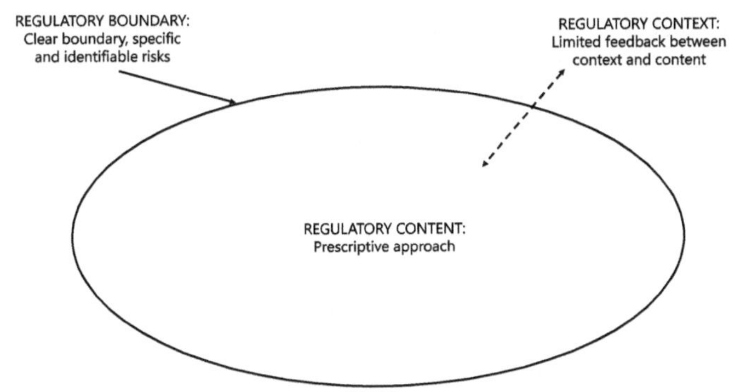

Figure 5.2 Architecture for a traditional approach to the regulation of risk

IV The special case of systemic risks caused by climate change

The features that characterise a traditional approach to the regulation of risk indicate that such an approach is not suitable for systemic risks, particularly systemic risks caused by climate change. In fact, Heyvaert (2011) considers that climate change "calls for regulatory responses that look very different from conventional risk regulation".[61] There are various arguments in support of this rallying call.

The focus of a traditional approach on the *anticipation of risk* is misplaced in the case of systemic risks caused by climate change because the future is likely to look very different from the past. While regulation that is based on predictions about the future may be better than backward-looking risk assessments, there is a real chance that the reality will still deviate from these predictions. This leads to the unhelpful conclusion that regulation of systemic risks caused by climate change should be tethered neither to knowledge about the past nor to assumptions about the future.

Moreover, Heyvaert suggests that regulation that is based on the present is equally problematic. She argues that traditional risk regulation is geared towards stabilising outcomes by optimising the circumstances in which risks might arise so as to minimise the prospect of risks materialising.[62] At a more practical level, this means that risk regulation sets the standards with which risky activity must comply based on current conditions to avoid deviation from the status quo and, thereby, minimises the risk of harm.[63] However, in the case of systemic risks, like those associated with climate change, stabilisation may actually increase a system's vulnerability.[64] Accordingly, Heyvaert suggests that, in the area of climate change, stabilisation of this kind is the

61 V. Heyvaert, n. 43 above, p. 837.
62 Ibid. p. 835. Heyvaert refers to risk regulation's "stabilising propensities".
63 An example where risk regulation has been employed in this way across various jurisdictions is the regulation of the quality of goods through product standards to ensure that they are safe for use by consumers.
64 M. van Rijswick & W. Salet, "Enabling the contextualization of legal rules in responsive strategies to climate change" (2012) 17(2) *Ecology and Society*, p. 18.

opposite of what is needed – namely, systemic change.[65] In a similar vein, Craig (2010) exclaims "stationarity is dead – long live transformation"![66] She suggests that, when addressing climate change risks through regulation, "regulatory goals and the legal mechanisms for accomplishing them will have to be centered on the concept of change itself'.[67] Maintenance of the status quo that could result from a traditional approach to the regulation of risk may, in fact, be counter-productive by embedding vulnerability or increasing the chance of exposure to climate change risks.[68]

The relevance of a *linear process* to manage systemic risks caused by climate change is also questionable. It is hard to imagine how the complexity and uncertainty surrounding climate change hazards, the systems that could be affected by such hazards, and the compounding and dynamic effect of the interaction between hazards and systems can be accommodated within a linear model. Even if this were to be practically possible, the capacity of regulation to successfully incorporate such a model is in doubt. Risks are generally narrowly constructed around the *causes* of risk in traditional risk regulation frameworks.[69] This enables risk regulation to be targeted at those causes so that they can be eliminated.[70] The identification of causes enables their representation as "attributable and manageable".[71] In contrast, climate change adaptation is concerned with identifying and responding to the *consequences* of climate change (such as the consequences associated with bushfires or floods).[72] Furthermore, Heyvaert suggests that risk regulation frameworks are typically characterised by compartmentalisation of risks into "discrete, manageable segments",[73] where individual risks are dealt with on a case-by-case basis with only "limited regard for cumulative or synergistic impacts".[74] In Heyvaert's view, such an approach does not allow for accumulation of risk,[75] which is particularly important in the context of systemic risks, where a single event can result in cascading losses.[76]

In relation to the lack of *interaction between context and content* under a traditional approach to the regulation of risk, there are serious questions about its likely efficacy in addressing systemic risks caused by climate change, which may well evolve and

65 V. Heyvaert, n. 43 above, p. 818.
66 R.K. Craig, "'Stationarity is dead' - Long live transformation: Five principles for climate change adaptation law" (2010) 34(1) *The Harvard Environmental Law Review*, p. 9.
67 Ibid. p. 30.
68 V. Heyvaert, n. 43 above, p. 818.
69 Ibid. p. 820.
70 For example, the regulation of noise pollution typically focuses on the source of noise, such as construction sites.
71 V. Heyvaert, n. 43 above, p. 836.
72 The difference between climate change "adaptation" and climate change "mitigation" is discussed in Chapter 3. Mitigation, rather than adaptation, is focused on addressing the causes of climate change.
73 V. Heyvaert, n. 43 above, p. 820 citing N. Pidgeon & C. Butler, "Risk analysis and climate change" (2009) 18(5) *Environmental Politics*, pp. 670–88, at 673.
74 V. Heyvaert, n. 43 above, p. 835.
75 Ibid. pp. 835–6.
76 Ibid. p. 836.

mutate over time – sometimes dramatically. A lag between regulation and context could have catastrophic consequences. Among other things, the disconnect between regulatory content and context may mean that the available regulatory tools are inappropriate or ineffective. Heyvaert argues that the compartmentalisation of risk in typical risk regulation frameworks results in "instrumental parsimony", where each risk is addressed through a single or narrow set of control techniques.[77] In her view, regulatory strategies to address climate change must facilitate "integration and orchestration, over itemisation and compartmentalisation".[78] Similarly, Craig suggests that "climate change adaptation law should be based on principled flexibility" and that application of such an approach will "necessarily encompass a broad and creative range of adaptation decisions and actions".[79]

The challenges of applying *interdisciplinarity* under a traditional approach to the regulation of risk are particularly problematic in a climate change context. The inherent complexity of many systems that are vulnerable to climate change risks means that determining whether and how to intervene to manage risk is itself a complex and confounding task.[80] Schipper and colleagues (2021) explain that interdisciplinarity enables a diversity of perspectives when considering how a problem should be addressed and that this interdisciplinarity is indispensable in the context of climate change because "the nature of the challenge lies well beyond the scope of any single discipline".[81]

In summary, there are grounds to argue that a traditional approach to the regulation of risk is not appropriate for the regulation of systemic risks, particularly those caused by climate change. Heyvaert goes so far as to contend that we are "confronted with the reality that the established risk regulation paradigm is of limited use to understand and guide regulatory efforts to control the most important environmental risk of our time".[82] The predictive, linear model that characterises a traditional approach to the regulation of risk may have worked well for specific, contained and unambiguous risks. However, the challenges presented by modern systemic risks, like those caused by climate change, are different in nature, scale and scope from other risks. Moreover, the uncertainty associated with such risks suggests that the deference to the past that is implicit in the traditional approach is misplaced for systemic risks like climate change where the past is unlikely to be a good guide for the future. As noted by Heyvaert, there is a real danger in "mapping new problems onto old formats".[83]

77 Ibid. p. 836.
78 Ibid. p. 818.
79 R.K. Craig, n. 66 above, p. 18.
80 A. Corbett, "A systems approach to regulatory excellence", in C. Coglianese (ed.), *Achieving Regulatory Excellence* (Brookings Institution Press, 2017), pp. 255–69, at 255–6.
81 E.L.F. Schipper, N.K. Dubash, & Y. Mulugetta, "Climate change research and the search for solutions: Rethinking interdisciplinarity" (2021) 168(3–4) *Climatic Change*, p. 18, at 5.
82 V. Heyvaert, n. 43 above, p. 818.
83 Ibid. p. 837.

This does not mean that it is time to altogether abandon a traditional approach to the regulation of risk in the field of climate change adaptation nor, for that matter, for systemic risks more broadly. Rather, a new paradigm of risk regulation is needed that addresses the challenges posed by systemic risks. The next chapter explores whether risk governance can be used as the foundation for this new paradigm.

Bibliography

Asquer, A., & Krachkovskaya, I. (2021). "Uncertainty, institutions and regulatory responses to emerging technologies: CRISPR gene editing in the US and the EU (2012–2019)." *Regulation & Governance*, 15(4), 1111–1127. https://doi.org/10.1111/rego.12335.

Avila, H. (2016). *Certainty in Law*. Springer Berlin Heidelberg.

Beck, U. (2006). "Living in the world risk society." *Economy and Society*, 35(3), 329–345.

Beckstein, M. (2017). "The concept of a living tradition." *European Journal of Social Theory*, 20(4), 491–510. https://doi.org/10.1177/1368431016668185.

Black, J. (2005). "The emergence of risk-based regulation and the new public risk management in the United Kingdom." *Public Law*, 2005*(Autumn)*, 512–549.

Chiu, I. H.-Y., & Deipenbrock, G. (eds.). (2021). *Routledge Handbook of Financial Technology and Law*. Routledge/Taylor & Francis Group.

Cilliers, L. (1993). "Public health in Roman legislation." *Acta Classica*, 36(1), 1–10.

Coglianese, C. (ed.). (2017). *Achieving Regulatory Excellence*. Brookings Institution Press.

Corbett, A. (2017). "A systems approach to regulatory excellence." In C. Coglianese, (ed.), *Achieving Regulatory Excellence*. Brookings Institution Press. https://doi.org/10.7864/j.ctt1hfr226.19.

Craig, R.K. (2010). "'Stationarity is dead' – long live transformation: Five principles for climate change adaptation law." *The Harvard Environmental Law Review*, 34(1), 9.

Dabrowska, A. & Thaul, S. (2018). "How FDA approves drugs and regulates their safety and effectiveness." *US Congressional Research Service*, 31.

Demeritt, D., Rothstein, H., Beaussier, A.-L., & Howard, M. (2015). "Mobilizing risk: Explaining policy transfer in food and occupational safety regulation in the UK." *Environment and Planning A: Economy and Space*, 47(2), 373–391. https://doi.org/10.1068/a140085p.

Fisher, E. (2013). "Framing risk regulation: A critical reflection." *European Journal of Risk Regulation*, 4(2), 125–132. https://doi.org/10.1017/S1867299X00003299.

Ganne, E. (2021). "Blockchain for trade: When code needs law." *AJIL Unbound*, 115, 419–424. https://doi.org/10.1017/aju.2021.64.

Greenberg, M. (2021). "Legal interpretation." In E.N. Zalta (ed.), *The Stanford Encyclopedia of Philosophy* (Fall 2021). Stanford University. https://plato.stanford.edu/archives/fall2021/entries/legal-interpretation/.

Heyvaert, V. (2011). "Governing climate change: Towards a new paradigm for risk regulation." *The Modern Law Review*, 74(6), 817–844. https://doi.org/10.1111/j.1468-2230.2011.00874.x.

Hutt, P. B. (1984). "Government regulation of the integrity of the food supply." *Annual Review of Nutrition*, 4(1), 1–21. https://doi.org/10.1146/annurev.nu.04.070184.000245.

Hutter, B. (2006). "Risk, regulation and management." In P. Taylor-Gooby & J.O. Zinn (eds.), *Risk in Social Science*. Oxford Scholarship Online.

Hutter, B.M. (2010). "Anticipating risk and organising risk regulation: Current dilemmas." In B.M. Hutter (ed.), *Anticipating Risks and Organising Risk Regulation*. Cambridge University Press. https://doi.org/10.1017/CBO9780511761553.002.

Hutter, B.M. (2017). "A risk regulation perspective on regulatory excellence." In C. Coglianese (ed.), *Achieving Regulatory Excellence*. Brookings Institution Press.

Jackson, M.A. (2020). "Regulating AI." In M. Jackson & M. Shelly (eds.), *Legal Regulations, Implications, and Issues Surrounding Digital Data*. IGI Global. https://doi.org/10.4018/978-1-7998-3130-3.ch009.

Krygier, M. (1986). "Law as tradition." *Law and Philosophy*, 5(2). https://doi.org/10.1007/BF00190762.

Lewis, S. (2021). "Precedent and the rule of law." *Oxford Journal of Legal Studies*, 41(4), 873–898. https://doi.org/10.1093/ojls/gqab007.

Lorenzoni, I., Pidgeon, N.F., & O'Connor, R.E. (2005). "Dangerous climate change: The role for risk research." *Risk Analysis*, 25(6), 1387–1398. https://doi.org/10.1111/j.1539-6924.2005.00686.x.

MacRae, C. (2010). "Regulating resilience? Regulatory work in high-risk arenas." In B. M. Hutter (ed.), *Anticipating Risks and Organising Risk Regulation*. Cambridge University Press. https://doi.org/10.1017/CBO9780511761553.008.

Makra, L. (2019). "Anthropogenic air pollution in ancient times." In P. Wexler (ed.), *Toxicology in Antiquity*. Elsevier. https://doi.org/10.1016/B978-0-12-815339-0.00018-4.

Marchant, G.E., Allenby, B.R., & Herkert, J.R. (eds.). (2011). *The Growing Gap Between Emerging Technologies and Legal-Ethical Oversight(Vol.7)*. Springer Netherlands. https://doi.org/10.1007/978-94-007-1356-7.

Merryman, J.H., & Pérez-Perdomo, R. (2007). *The Civil Law Tradition: An Introduction to the Legal Systems of Europe and Latin America*. Stanford University Press (3rd edition).

Miščenić, E., & Raccah, A. (eds.). (2016). *Legal Risks in EU Law: Interdisciplinary Studies on Legal Risk Management and Better Regulation in Europe*. Springer International Publishing. https://doi.org/10.1007/978-3-319-28596-2.

Moses, L.B. & Zalnieriute, M. (2020). "Law and technology in the dimension of time." In Ranchordás, S., & Roznai, Y. (eds.). (2020). *Time, Law, and Change: An Interdisciplinary Study*. Hart Publishing. https://doi.org/10.5040/9781509930968.

National Commission on Terrorist Attacks upon the United States, Kean, T.H., & Hamilton, L. (eds.). (2004). *The 9/11 Commission Report: Final Report of the National Commission on Terrorist Attacks upon the United States*. National Commission on Terrorist Attacks upon the United States.

OECD. (2003). *Emerging Systemic Risks in the 21st Century: An Agenda for Action*. Organisation for Economic Co-operation and Development. https://doi.org/10.1787/9789264101234-fr.

OECD. *Alternatives to Traditional Regulation*, prepared by G. Hepburn. https://www.oecd.org/gov/regulatory-policy/42245468.pdf.

Pidgeon, N., & Butler, C. (2009). "Risk analysis and climate change." *Environmental Politics*, 18(5), 670–688. https://doi.org/10.1080/09644010903156976.

Rothstein, H., Borraz, O., & Huber, M. (2013). "Risk and the limits of governance: Exploring varied patterns of risk-based governance across Europe." *Regulation & Governance*, 7(2), 215–235. https://doi.org/10.1111/j.1748-5991.2012.01153.x.

Scheffler, S. (2010). *Equality & Tradition: Questions of Value in Moral and Political Theory*. Oxford University Press. http://site.ebrary.com/id/10392197.

Schipper, E.L.F., Dubash, N.K., & Mulugetta, Y. (2021). "Climate change research and the search for solutions: Rethinking interdisciplinarity." *Climatic Change*, 168(3–4),18. https://doi.org/10.1007/s10584-021-03237-3.

Sullivan, K.M. (2002). "Interdisciplinarity." *Michigan Law Review*, 100(6), 1217–1226.

Taylor-Gooby, P., & Zinn, J.O. (eds.). (2006). *Risk in Social Science*. Oxford Scholarship Online.

van Rijswick, M., & Salet, W. (2012). "Enabling the contextualization of legal rules in responsive strategies to climate change." *Ecology and Society*, 17(2). https://doi.org/10.5751/ES-04895-170218.

Wiener, J. B. (2017). "Risk regulation and future learning." *European Journal of Risk Regulation*, 8(1), 4–9. https://doi.org/10.1017/err.2017.2.

Chapter 6

A Governance Approach to Regulating Risk

The concept of "governance" is familiar to the public, not-for-profit, and private sectors alike. While definitions and manifestations are varied, governance in each of these contexts broadly refers to the mechanisms by which organisations are managed, controlled and held to account. Organisations strive for "good governance" to enhance performance, to ensure responsiveness to change, and to preserve the confidence of stakeholders. Essentially, when done well, governance can yield positive outcomes.

Governance is a relatively less well-established feature of regulation. Indeed, the conventional understanding of governance relates to organisations and it may not be readily apparent how this conception of governance can be applied in a regulatory setting. Nonetheless, once stock is taken of the fact that all regulation is, at its essence, about the management of risk, the potential scope for application of governance principles to regulation becomes clearer. In particular, governance – and, more specifically, "risk governance" – could be the means by which risks are managed or controlled by regulation. This chapter examines the concept of risk governance, how such a concept could be expressed in regulatory form, and how it differs from a traditional approach to the regulation of risk.

I Defining risk governance

Risk governance appeared in academic discourse at the turn of the millennium and involves the "marriage" of governance and risk research by translating the substance and core principles of governance to risk and risk-related decision-making.[1]

The concept of governance is familiar in the corporate sphere. From the time of the world's first publicly listed companies, such as the Dutch East

1 IRGC, *White Paper on Risk Governance: Toward an Integrative Framework* (IRGC, 2005), p. 22; M.B.A. van Asselt & O. Renn, "Risk governance" (2011) 14(4) *Journal of Risk Research*, pp. 431–49, at 432; A. Klinke & O. Renn, "The coming of age of risk governance" (2021) 41(3) *Risk Analysis*, pp. 544–57, at 544.

DOI: 10.4324/9781003253464-6

India Trading Company established in 1602, governance has been understood to mean the *mechanisms by which a company is directed, controlled and evaluated*, principally by the company's board of directors, to help ensure the long-term success of the company. This concept of corporate governance has endured until now. For example, the Corporate Governance Council for the Australian Stock Exchange (ASX) defines it as "the framework of rules, relationships, systems and processes within and by which authority is exercised and controlled within corporations. It encompasses the mechanisms by which companies, and those in control, are held to account".[2] In practice, corporate governance includes measures to ensure effective management and supervision, promote information transparency, and enable democratic consideration of stakeholder interests in order to "avoid rudimentary management deficiencies and promote mutual gains".[3] However, there has been much water under the bridge since the concept of governance in a corporate context was first conceived.

The notion of governance has been applied in other areas, including international relations, environmental management and the social sciences.[4] In these areas, governance is linked to the *structure and processes associated with distributed authority and collective decision-making aimed at realising collective goals*. More specifically, at a national level, governance has been described as collective decision-making involving governmental and non-governmental actors, whereas at a global level, collective decision-making includes State and non-State actors.[5] Cottrell and Trubek (2012) have argued that a growing emphasis on solving problems in the "global space" has resulted in "collective problem solving [which implies] a dynamic process that involves the common identification of a problem, formation of a consensus that it ought to be solved, and the mobilization of appropriate expertise and resources to do so".[6]

2 ASX, *Corporate Governance Principles and Recommendations* (ASX, 4th edition, 2019), p. 1. The OECD defines the concept of corporate governance more broadly as involving "a set of relationships between a company's management, its board, its shareholders and other stakeholders. Corporate governance also provides the structure through which the objectives of the company are set, and the means of attaining those objectives and monitoring performance are determined": OECD, *G20/OECD Principles of Corporate Governance* (OECD Publishing, 2015), p. 9.

3 V. Stein & A. Wiedemann, "Risk governance: Conceptualization, tasks, and research agenda" (2016) 86(8) *Journal of Business Economics*, pp. 813–36, at 816.

4 O. Renn, A. Klinke, & M. van Asselt, "Coping with complexity, uncertainty and ambiguity in risk governance: A synthesis" (2011) 40(2) *AMBIO*, pp. 231–46, at 232; M.B.A. van Asselt & O. Renn, n. 1 above, p. 434; A. Klinke & O. Renn, n. 1 above, p. 544; IRGC, n. 1 above, p. 22.

5 IRGC, n. 1 above, p. 22.

6 M.P. Cottrell & D.M. Trubek, "Law as problem solving: Standards, networks, experimentation, and deliberation in global space" (2012) 21(2) *Transnational Law & Contemporary Problems*, p. 363.

Van Asselt and Renn (2011) suggest that the increasing uptake of governance across a range of domains and disciplines:

> is best understood as a response to new challenges, such as globalization, increased international cooperation … societal changes, including the increased engagement of citizens … an augmenting complexity of policy issues, and the resulting difficulty in taking decisions with confidence and legitimacy.[7]

They further argue that a governance approach helps to "enlarge the perspective on policy and politics by acknowledging that government is not the only and may not even be the most important actor in managing and organizing society".[8] Collectivity is at the heart of this concept of governance, which distinguishes it from the corporate notion of the term.

Turning now to risk governance, there are various explanations of this concept but they can be reduced to the notion that risks are managed in an inclusive, collective and interactive manner. Klinke and Renn (2019) have proffered the following definition: "risk governance marks out institutional structures and sociopolitical processes that guide and restrain collective activities of a group, society, or international community to influence or direct the course of events or people's behaviour when dealing with risk issues."[9] Van Asselt and Renn (2011) clarify that it is more than just a way of describing how collective decisions are made because it "provides a conceptual as well as normative basis for how to deal responsibly with uncertain, complex and/or ambiguous risks".[10] They suggest that the shift from a risk management approach to a risk governance approach is based on the premise that "many risks cannot be calculated on the basis of probability and effects alone and that models that build on that assumption are not just inadequate, but also constitute an obstacle to responsibly dealing with risk".[11] In this regard, Klinke and Renn (2019) assert that decisions about risks are rightly the outcome of a "mosaic of interactions" between a broad range of actors, including government, technical experts, scientists, corporate actors and actors from civil society[12] and, therefore, have a "democratic foundation" which is "inclusive and participatory".[13]

Renn, Klinke and Van Asselt (2011) further argue that risk governance involves "a process of continuous and gradual learning and adjustment that

7 M.B.A. van Asselt & O. Renn, n. 1 above, p. 434.
8 Ibid. p. 434.
9 A. Klinke & O. Renn, n. 1 above, p. 545.
10 M.B.A. van Asselt & O. Renn, n. 1 above, p. 432.
11 Ibid. pp. 431–2.
12 IRGC, n. 1 above, p. 15.
13 A. Klinke & O. Renn, n. 1 above, p. 456; IRGC, n. 1 above, p. 15.

permits a prudent handling of complexity, scientific uncertainty and/or socio-political ambiguity".[14] Such a process aims at generating a comprehensive appreciation of risk through a holistic, inclusive and contextualised approach, which forms the basis of the response to risk.[15] It seeks to integrate a range of perspectives and complementary analyses to help understand the diverse, uncertain, and complex realities of risk problems, but does not exclude the incorporation of, or conclusions based upon, a theoretical perspective.[16] In the words of Klinke and Renn (2019): "[t]he risk governance architecture calls for a new thinking ... in governing risk, a pragmatic way ... that corresponds to the dynamic nature of technical and social change".[17]

Risk governance is to be distinguished from "risk management" because it encompasses the broader decision-making context within which responses to risk are identified and implemented.[18] Risk governance is also distinct from "risk-based governance", which concerns arrangements to ensure that "resources are allocated on the basis of calculations of risk".[19] Additionally, risk governance should not be confused with "adaptive management".[20] Whereas the decision-making process in the context of an adaptive management approach focuses on adapting to a changing context, the focus in risk governance is squarely on the mitigation of risk.

II A risk governance model for systemic risks

Risk governance in the abstract is appealing, particularly the flexible and inclusive decision-making process implied by the concept. However, the question of whether risk governance is likely to be useful to regulate systemic

14 O. Renn, A. Klinke, & M. van Asselt, n. 4 above, p. 236.

15 IRGC, n. 1 above, p. 11; O. Renn, A. Klinke, & M. van Asselt, n. 4 above, p. 236.

16 T.H. Lippert, *NATO, Climate Change, and International Security: A Risk Governance Approach* (Palgrave Macmillan, 2019), p. 83.

17 A. Klinke & O. Renn, n. 1 above, p. 555.

18 O. Renn, *Risk Governance: Coping with Uncertainty in a Complex World* (Routledge, 2017), p. 8.

19 K. Krieger, "The limits and variety of risk-based governance: The case of flood management in Germany and England" (2013) 7(2) *Regulation & Governance*, pp. 236–57, at 236 referring to H. Rothstein, P. Irving, T. Walden, & R. Yearsley, "The risks of risk-based regulation: Insights from the environmental policy domain" (2006) 32(8) *Environment International*, pp. 1056–65. See also H. Rothstein, O. Borraz, & M. Huber, "Risk and the limits of governance: Exploring varied patterns of risk-based governance across Europe" (2013) 7(2) *Regulation & Governance*, pp. 215–35.

20 Adaptive management has been described as implying a "process of structured decision-making, in which both the decision-making process and institutions allow continual learning, re-evaluation, and revision in response to new information and surprises": J. McDonald & M.C. Styles, "Legal strategies for adaptive management under climate change" (2014) 26(1) *Journal of Environmental Law*, pp. 25–53, at 30.

risks, particularly those caused by climate change, depends upon how the concept is articulated in practical terms.

A risk governance model has been specifically designed by the International Risk Governance Council (IRGC) to address systemic risks – namely the "IRGC Framework".[21] In 2003, the Swiss State Secretariat for Education and Research recommended to the Swiss Parliament that the IRGC be established as a neutral, independent, international body to bridge what was perceived to be increasing gaps between science, technological development, decision-makers and the public. More specifically, it was intended that the organisation would serve as a catalyst for improvements in the design and implementation of effective risk governance strategies.[22] The IRGC's primary mission is to develop the concept of risk governance, with a view to helping policy- and decision-makers to better understand, anticipate and respond to risks – particularly, systemic risks.[23]

The IRGC Framework, which was first put forward in 2005,[24] was developed by the IRGC in response to a growing concern about the capacity of governments and regulators to effectively manage systemic risks.[25] A fundamental premise underlying the Framework is that handling systemic risks requires a holistic, interdisciplinary and inclusive approach.[26] It is based on the concept of risk governance by seeking to integrate a range of perspectives and complementary analyses when responding to risk and, in so doing, accounts for the complexity, uncertainty and ambiguity of systemic risks.[27]

I Features

The IRGC Framework draws on traditional risk management frameworks,[28] but includes certain features that are intended to distinguish it from those frameworks and provide a stronger foundation to address systemic risks. These features are discussed below.

21 IRGC, n. 1 above, p. 19; IRGC, *Introduction to the IRGC Risk Governance Framework, Revised Version* (EPFL International Risk Governance Center, 2017), p. 1.
22 See the IRGC's website accessible at: https://irgc.org/about/history/.
23 IRGC website, ibid.; IRGC, n. 1 above, p. 19.
24 The details of the IRGC Framework were first set out in the 2005 White Paper (IRGC, n. 1 above). The Framework has evolved since then (see IRGC, n. 21 above; IRGC, *IRGC Guidelines for the Governance of Systemic Risks* (EPFL International Risk Governance Center, 2018)). However, the key elements of the Framework set out in the 2005 White Paper have been retained in subsequent versions.
25 IRGC website, n. 22 above; IRGC, n. 1 above, p. 19.
26 IRGC, n. 1 above, pp. 19, 49, 61; IRGC, n. 21 above, p. 7.
27 A. Klinke & O. Renn, n. 1 above, p. 545.
28 IRGC, n. 1 above, p. 18.

Broad notion of risk governance

The Framework is based on a broad notion of risk governance, which extends beyond "technical" risk management and includes stakeholder involvement and collaborative decision-making.[29] The premise for this feature is that it will help with understanding complexity, reducing uncertainty and handling ambiguity.[30] The Framework contemplates involvement of all relevant stakeholders in the decision-making process, including civil society.[31] Such a collaborative approach is intended to assist in detecting actual or potential deficits within the risk governance process[32] and thereby enable decision-makers to act on the basis of the best available information, while accounting for societal values and interests associated with the risks at issue.[33]

Structured decision-making process

The Framework focuses on process, rather than prescribing outcomes. It includes a structured yet flexible decision-making process, which is intended to enable a comprehensive approach in responding to risk.[34] The process comprises inter-linked elements that are largely sequential in application – namely, risk pre-assessment, risk appraisal, risk characterisation and evaluation, risk management, and communication and stakeholder engagement.[35] These elements are illustrated in Figure 6.1. Collectively, they are aimed at gaining a

29 O. Renn & K.D. Walker (eds.), *Global Risk Governance: Concept and Practice Using the IRGC Framework* (Springer Netherlands, 2008), p. xxiv–xxv.

30 IRGC, n. 21 above, p. 7.

31 O. Renn & K.D. Walker (eds.), n. 29 above, p. xxv.

32 The IRGC has identified two "clusters" of risk deficits, which respectively concern the assessment and understanding of risks ("Cluster A") and the management of risks ("Cluster B"). In relation to Cluster A, the deficits relate to gathering and interpreting knowledge, dealing with disputed, potentially biased or subjective knowledge, dealing with knowledge related to systems and their complexity, and acknowledging that knowledge and understanding are never complete or adequate. Regarding Cluster B, the deficits concern preparing and deciding on risk management strategies and policies, formulating responses, resolving conflicts and deciding to act, and developing organisational capacities for responding and monitoring. See IRGC, *Risk Governance Deficits: Analysis, Illustration and Recommendations, Policy Brief* (International Risk Governance Council, 2010).

33 IRGC, n. 21 above, p. 7.

34 Ibid. p. 7.

35 The 2017 IRGC Synthesis additionally includes "context" as a cross-cutting issue for the IRGC Framework – specifically, the social, institutional, political and economic context; organisational capacity; and risk culture within a particular societal context. There are likely to be significant variations of "context" (as defined by the IRGC) between countries around the world. As the intent of this book is to make broad observations about the governance of systemic risks regardless of where they are geographically situated, the "context" element of the IRGC Framework is not considered further in this book.

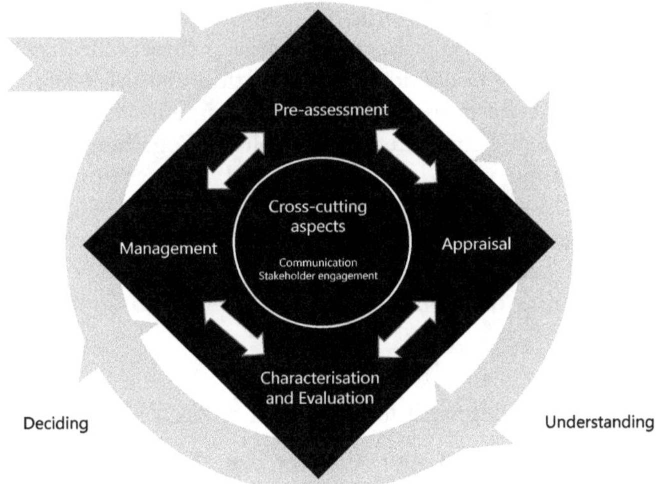

Figure 6.1 Illustration of the IRGC Framework[37]

thorough understanding of risk, ultimately leading to the identification of a response to risk that is based on that understanding.[36]

As is evident from Figure 6.1, the IRGC Framework distinguishes between, on the one hand, an "assessment sphere" (on the right-hand side) where the emphasis is on understanding risk and, on the other hand, a "management sphere" (on the left-hand side), where decision-making and implementation of a response to risk occur.[38] The primary rationale for this distinction is to maximise the objectivity of and accountability for both sets of activities.[39] However, the pre-assessment, characterisation and evaluation phases are in both spheres to ensure that they involve the combined efforts of those respectively responsible for understanding and managing risk. Risk communication and stakeholder engagement is positioned at the centre of the Framework to reflect its crucial role throughout the entire process.[40]

The distinction in the IRGC Framework between understanding and deciding how to respond to risk is a modern interpretation of Herbert Simon's seminal work, *Administrative Behaviour* (1947), in which he put forward a decision-making framework that challenged the pre-existing assumption that finding a solution to

36 IRGC, n. 21 above, p. 10.
37 Ibid. p. 9. This figure is a representation of the IRGC Framework.
38 IRGC, n. 1 above, p. 65; IRGC, n. 21 above, p. 10.
39 IRGC, n. 21 above, p. 10.
40 Ibid. p. 10; O. Renn & K.D. Walker (eds.), n. 29 above, p. xxv.

a problem involves a rational process.[41] Specifically, Simon described the process of developing public policy as comprising three main stages – namely, "intelligence" involving collection of information about a problem, "design" of the criteria to be used to assess the information, and "choice" between the various options available to address the problem. Together, these stages were designed to overcome what has been referred to as "bounded rationality", resulting from a lack of contextual information that leads to sub-optimal decisions.[42] In a similar way, the decision-making process in the IRGC Framework has been designed to deliver an appropriate response to systemic risks, notwithstanding the complexity, uncertainty and ambiguity associated with these risks.[43]

Holistic, interdisciplinary approach to assessment of risk

The IRGC Framework recognises the fact that the management of systemic risk is inherently interdisciplinary. Accordingly, it provides for a "risk assessment", which is a scientific or technical exercise, as well as a "concern assessment", which is a social science activity aimed at providing a comprehensive diagnosis of concerns and expectations that individuals, groups or different cultures may link to risk.[44] Together, the technical risk assessment and the concern assessment seek to enable a holistic assessment of risk.[45]

Communication and public engagement

The IRGC Framework emphasises the crucial role of "communication" and "stakeholder engagement". This includes not only informing people about a risk or a risk management decision, but also establishing a two-way dialogue with relevant stakeholders at all stages of the risk management process, to help

41 For the most recent edition, see H.A. Simon, *Administrative Behavior: A Study of Decision-Making Processes in Administrative Organizations* (Free Press, 4th edition, 1997).

42 Simon describes bounded rationality as "the idea that the choices people make are determined not only by some consistent overall goal and the properties of the external world, but also by the knowledge that decision makers do and don't have of the world, their ability or inability to evoke that knowledge when it is relevant, to work out the consequences of their actions, to conjure up possible courses of action, to cope with uncertainty (including uncertainty deriving from the possible responses of other actors), and to adjudicate among their many competing wants. Rationality is bounded because these abilities are severely limited". See H.A. Simon, "Bounded rationality in social science: Today and tomorrow" (2000) 1(1) *Mind & Society*, pp. 25–39, at 25.

43 IRGC, n. 1 above, p. 19; IRGC, n. 21 above, p. 7.

44 IRGC, n. 21 above, pp. 13–5; O. Renn, "White Paper on Risk Governance: Towards an Integrative Framework" in O. Renn & K.D. Walker (eds.), n. 29 above, p. 25.

45 IRGC, n. 21 above, pp. 13–16.

ensure meaningful engagement with stakeholders.[46] This participative approach to decision-making is intended to ensure that choices about how to respond to risk are informed by factual knowledge, as well as an understanding of interests, concerns and beliefs held by relevant stakeholders.[47]

2 Objectives and outcomes

The primary objective of the IRGC Framework is to support decision-makers to make "better" decisions about the management of systemic risks, which fully account for the characteristics of these risks.[48] Application of the IRGC Framework does not yield discrete outcomes but, rather, leads to risk management strategies which vary depending upon the relative degree of complexity, uncertainty and ambiguity associated with risks being analysed.[49] As the complexity and uncertainty of systemic risks increase, the focus under the IRGC Framework is increasingly on ensuring that the system at risk can withstand and is resilient to hazards. Discourse and deliberation among relevant stakeholders are emphasised as the ambiguity of risks intensifies.[50] These risk management strategies can be viewed as a means to guide systems towards desired states, rather than managing risks to systems in a rigid, linear way.[51]

The emphasis on system resilience in the Framework's risk management strategies is notable because it implicitly recognises the difficulties associated with anticipating systemic risks given their inherent complexity and uncertainty. Even if these types of risks could be predicted, Wildavsky (1988) suggests that anticipatory strategies reduce the capacity to cope with the unexpected because they can lead to extreme risk aversion and wasted resources and argues that it is more appropriate to build resilience.[52] Holley and Sofronova (2017) note that the concept of resilience has received widespread attention across many different fields:[53]

46 Ibid. pp. 27–30.
47 IRGC (2012) *An Introduction to the IRGC Risk Governance Framework* (IRGC, Geneva, Switzerland), p. 6.
48 IRGC, n. 21 above, p. 7.
49 Ibid. p. 24.
50 Ibid. pp. 24–5.
51 A. Corbett, "A systems approach to regulatory excellence", in C. Coglianese (ed.) *Achieving Regulatory Excellence* (Brookings Institution Press, 2017), p. 256.
52 B.M. Hutter, "Anticipating risk and organising risk regulation: current dilemmas", in B.M. Hutter (ed.), *Anticipating Risks and Organising Risk Regulation* (Cambridge University Press, 2010), pp. 3–22, at 11–12 referring to A. Wildavsky, *Searching for Safety* (University of California Press, Berkeley, 1988).
53 C. Holley & E. Sofronova, "New environmental governance: Adaptation, resilience and law", in B. Hutter (ed.), *Risk, Resilience, Inequality and Environmental Law* (Edward Elgar Publishing, 2017), pp. 129–46, at 129.

> In an "age of catastrophe", [resilience] has become the touchstone for responding to global problems and risks that are complex, difficult to predict and often have uneven social effects. Formidable environmental challenges, like water scarcity, biodiversity loss and degraded natural resources, exemplify these complex global problems. Climate change and its cascading risks of potentially unpredictable magnitude will only exacerbate the complexity of these challenges.

The distinction drawn by Wildavsky between anticipation and resilience three decades ago highlights the choice facing policy- and law-makers today. More specifically, a choice must be made between, on the one hand, attempting to predict risks and to prevent them from arising (anticipation) and, on the other hand, building capacity to cope quickly and effectively with dangers once they have arisen (resilience).[54]

The prominence of discourse and deliberation in the IRGC Framework's risk management strategies is also important and is linked to the ambiguity associated with systemic risks, which arises from complexity and uncertainty. Klinke and Renn (2002) explain that, in cases of ambiguous risks, strategies aimed at clarifying facts will not be enough and, rather, need to extend to "building up consciousness, building confidence, strengthening trustworthiness in regulatory bodies, and initiating collective efforts of institutions for taking responsibility".[55] The main objectives of this aspect of the IRGC risk management strategies are to allow an identification of the diverse concerns and values of relevant stakeholders,[56] to understand their tolerance of risk,[57] and to facilitate trade-offs between competing interests.[58]

3 *Practical application*

Since its creation, the IRGC Framework has been applied in several contexts involving a variety of different types of risks, particularly risks that could be considered systemic in nature. In particular, the Framework has been applied to technological risks,[59] as well as risks to public health and safety,[60] the natural

54 C. Briault, "Risk society and financial risk" in B.M. Hutter, *Anticipating Risks and Organising Risk Regulation* (Cambridge University Press, 2010), p. 26.
55 A. Klinke & O. Renn, "A new approach to risk evaluation and management: Risk-based, precaution-based, and discourse-based strategies" (2002) 22(6) *Risk Analysis*, pp. 1071–94, at 1088–9.
56 Ibid. p. 1071.
57 Ibid. p. 1073.
58 Ibid. p. 1086.
59 The IRGC Framework has been applied to nanotechnology. See M. Roco, M. and O. Renn, "Nanotechnology risk governance" in O. Renn & K.D. Walker (eds.), n. 29 above.
60 The IRGC Framework has been applied to genetically modified foods in Europe, acrylamide in foodstuffs in Germany, raw milk soft cheeses in the United States and vector-borne diseases. See J. Tait, "Risk governance of genetically modified

environment[61] and critical infrastructure.[62] Additionally, there are examples where the IRGC Framework has been applied to risks arising from climate change. In particular, Lippert (2016) considered whether application of the IRGC Framework could assist the North Atlantic Treaty Organization (NATO) to more effectively address the security consequences associated with climate change.[63] Lass and colleagues (2011) also used the IRGC Framework to assist in identifying adaptation measures to respond to European heatwaves caused by increasing temperatures linked to climate change.[64]

Within the spectrum of risk domains where the IRGC Framework has been used, the IRGC Framework has largely been applied *retrospectively* to determine whether outcomes might have changed had the Framework been applied in the first instance to problems for which the responses to risk are known. For example, the IRGC Framework was used in a 2008 case study to historically examine how genetically modified (GM) crops had been dealt with in Europe and to determine whether and how the Framework might have made a difference to the governance of those crops if it had been applied. Tait argued that application of the IRGC Framework revealed that the precautionary approach adopted by European regulators towards GM crops prioritised societal concerns about potential risks that could have arisen at some future date over technical evidence of harm because societal concerns expressed by influential minority groups were given undue weight.[65]

Lass and colleagues (2011) also applied the IRGC Framework to *prospectively* develop a risk governance system to assist in improving existing heat health early warning systems in place in Europe to respond to heatwaves. The authors argued that the risk governance system they developed, which was modelled

crops – European and American perspectives", S. Bonneck, "Acrylamide risk governance in Germany", A.J. Knight, M.R. Worosz, E.C.D. Todd, L.D. Bourqin, and C.K. Harris, C. K "Listeria in raw milk soft cheese: A case study of risk governance in the United States using the IRGC Framework" in O. Renn & K. D. Walker (eds.), n. 29 above and K. Schmidt, K.M. Dressel, M. Niedrig, M. Mertens, S.A. Schüle, & M.H. Groschup, "Public health and vector-borne diseases – a new concept for risk governance" (2013) 60(8) *Zoonoses and Public Health*, pp. 528–38.

61 The IRGC Framework has been applied to the construction of the Nagara River Estuary Barrage in Japan and to nature-based tourism. See N. Okada, "Nagara River Estuary Barrage conflict" and K. Kuenzi and J. McNeely, "Nature-based tourism" in O. Renn & K.D. Walker (eds.), n. 29 above.

62 The IRGC Framework has been applied to threats to energy security in the Baltic arising from, among other things, access to critical infrastructure. See, D. Warner North, "Energy security for the Baltic region" in O. Renn & K.D. Walker (eds.), n. 29 above.

63 T.H. Lippert, n. 16 above.

64 W. Lass, A. Haas, J. Hinkel, & C. Jaeger, "Avoiding the avoidable: Towards a European heat waves risk governance" (2011) 2(1) *International Journal of Disaster Risk Science*, pp. 1–14.

65 J. Tait, above n. 60, pp. 133, 143, 147.

on the IRGC Framework, "can be universally applied" and "can give structure to processes of social learning that decrease the vulnerabilities of societies to weather extremes".[66] Roco and Renn (2008) also asserted that the IRGC Framework could be used in a proactive manner to develop a risk governance framework for nanotechnology,[67] while Lippert (2019) applied the Framework to determine whether the application of the IRGC Framework to security challenges associated with climate change could "add policy-relevant insight for NATO if applied in real-time to a phenomenon as broad, uncertain, and complex as climate change-induced security risk".[68]

The IRGC Framework has been employed in various *non-regulatory contexts* to determine how actors and stakeholders affected by the relevant risk or set of risks could change behaviour to mitigate risk in the absence of government intervention to enforce such behaviour change. For example, a 2013 case study analysed how the IRGC Framework could be used to assess the risks associated with vectors responsible for the transmission of pathogen-causing diseases such as dengue fever, yellow fever, Japanese encephalitis, tick-borne encephalitis and West Nile fever with a high impact on the health of humans and animals.[69] Schmidt and colleagues argued that the Framework could assist authorities in understanding the risks to public health, to analyse the potential impact of the vector-borne diseases on public health, and help to manage, control and communicate important risk-related issues to the public.[70] Similarly, in the climate change context, the application of the IRGC Framework was used to analyse past evidence of heat-related mortality and to determine how Europe could better adapt to heatwaves without the need for a rule-based response.[71] The Framework has also been employed in various *regulatory contexts* to help determine how risks could be governed, using regulation as a core mechanism to do so. In particular, in a 2008 case study involving the health risks associated with raw milk soft cheeses in the US, the authors suggested that the IRGC Framework could be used to help develop regulatory standards to manage these risks.[72]

Notwithstanding the value derived from the application of the IRGC Framework evident in these cases, critics have stated that the Framework is highly conceptual and idealised, rather than practical[73] and that the manner in which the IRGC Framework could be employed through regulation is unclear.[74] The

66 W. Lass, A. Haas, J. Hinkel, & C. Jaeger, n. 64 above, p. 12.
67 M. Roco and O. Renn, n. 59 above, p. 300.
68 T.H. Lippert, n. 16 above, p. 2.
69 K. Schmidt et al., n. 60 above, p. 528.
70 Ibid. p. 537.
71 W. Lass, A. Haas, J. Hinkel, & C. Jaeger, n. 64 above, p. 1.
72 A. Knight et al., n. 60 above, p. 215.
73 O. Renn and A. Jäger, "Synopsis of critical comments on the IRGC Risk Governance Framework" in O. Renn & K.D. Walker (eds.), n. 29 above, p. 120; Å. Boholm, H. Corvellec, & M. Karlsson, "The practice of risk governance: Lessons from the field" (2012) 15(1) *Journal of Risk Research*, pp. 1–20, at 15.
74 O. Renn and A. Jäger, ibid. p. 121.

sentiment underlying these criticisms appears to be that the Framework is neither specific nor prescriptive enough to assist in designing regulatory responses to manage systemic risks. This leads to a conundrum. Risk governance offers flexibility and fluidity, which appear indispensable to address systemic risks. Yet, the question arises as to whether regulation is well-suited to such a seemingly unstructured form implicit in a risk governance approach.

III Risk governance in regulatory form

A recent pivot away from traditional approaches towards "new governance" is evident in public administration and, more specifically, in regulation. Lobel (2004) speaks of "The fall of regulation and the rise of governance",[75] which involves a shift from command-and-control regulation to a collaborative, multi-party approach to achieving regulatory objectives. In more practical terms, Lobel (2012) explains that:[76]

> Instead of focusing on substantive prohibitions and adversarial enforcement, new governance approaches attempt to involve private parties in the legal process. … Instead of regulating the details of behaviour, agencies increasingly use broad policy goals such as "risk management" and allow the regulated industries to implement and interpret these mandates.

In line with Lobel's theory about the rise of governance in a regulatory context, there is evidence of increased emphasis on governance in the area of environmental regulation. Ruhl (2011) suggests that a growing number of environmental law scholars have gravitated towards new governance, which turns "away from the familiar model of command-style, fixed-rule regulation by administrative fiat, and toward a new model of collaborative, multiparty, multi-level, adaptive, problem-solving governance".[77] More generally, Levi-Faur (2011) sees the promise that "regulatory governance" offers for the "financial, ecological, legitimation, and moral crises" of our time, but he considers that this regulatory option has not been accorded sufficient attention.[78] He notes that "[m]uch of the academic and public discussion of regulation

75 O. Lobel, "The renew deal: The fall of regulation and the rise of governance in contemporary legal thought" (2004) 89(2) *Minnesota Law Review*, pp. 342–370.

76 O. Lobel, "New governance as regulatory governance", in D. Levi-Faur (ed.), *Oxford Handbook of Governance* (Oxford University Press, 2012), p. 71.

77 J.B. Ruhl, "General design principles for resilience and adaptive capacity in legal systems – with applications to climate change adaptation" (2011) 89(5) *North Carolina Law Review*, pp. 1373–403, at 1397 citing B.C. Karkkainen, "'New governance' in legal thought and in the world: Some splitting as antidote to overzealous lumping" (2004) 89(2) *Minnesota Law Review*, p. 471.

78 D. Levi-Faur, "Regulation & regulatory governance" in D. Levi-Faur, *Handbook on the Politics of Regulation* (Edward Elgar Publishing, 2011), pp. 16–17.

nowadays deals with the governance of regulation itself (or regulating regulation) rather than governance via regulation".[79] In his view:[80]

> [t]he demand for better, fairer, more efficient, and more participatory systems of governance promises that regulatory governance will continue to capture the imagination of scholars and dominate the agenda of policy makers. While regulatory governance is hardly a new feature of the social sciences, the issue still attracts less systematic and theoretical attention than it deserves.

This sentiment is echoed by Cottrell and Trubek (2012) who suggest that a "broadened notion of how law operates in a global space" is needed to "respond to the increased complexity and volatility of many transnational regulatory arenas".[81] Systemic risks caused by climate change fall within the types of problems that could benefit from a governance approach of the kind represented in the IRGC Framework.

So, what have we learnt about the use of the IRGC Framework to date and its potential application to the regulation of systemic risks, particularly those associated with climate change? The case studies where the IRGC Framework has been applied in practice tend to indicate that the Framework has the capacity to overcome the key challenges associated with regulating systemic risks. The Framework could aid in unpacking complexity, particularly during "pre-assessment" where risk is sought to be fully understood. It could assist in addressing deep uncertainty through a comprehensive and contextualised "appraisal" of risk. In addition, the risk governance process could help to resolve ambiguity regarding whether and how risks should be mitigated and, thereby, avoid a failure of collective imagination by accounting for the societal context in which risks arise through the "concern assessment" and also through ongoing "communication" and "stakeholder engagement". The risk governance process could also assist in ensuring the appropriateness of the response to risk by appropriately "characterising" and "evaluating" risk and ensuring a fit-for-purpose strategy to regulate risk through the "risk management" process. In summary, each element of the IRGC Framework could, theoretically, play an important role in addressing the key characteristics of systemic risks – namely, complexity, uncertainty and ambiguity.

There is a clear emphasis on risk in all elements of the IRGC Framework. This single-minded focus helps to centre the governance process on the management of risk. The varied risk domains in which the IRGC Framework has been practically applied lend support to the view that the Framework is versatile, flexible and has the capacity to be adapted to a broad range of risks and a

79 Ibid. p. 14.
80 Ibid. pp. 16–7.
81 M.P. Cottrell & D.M. Trubek, n. 6 above, p. 363.

variety of systems that may be vulnerable to risk. Although systemic risks share some common features, the systems and the various stressors that could put them at risk are likely to vary greatly. This militates against a one-size-fits-all approach to responding to risk. In this regard, one of the most apparent advantages of the IRGC Framework is that it reflects a process to assess systemic risks, which helps to identify an appropriate risk management strategy based on the characteristics of those risks, rather than a specific solution.

Parsons (1995) advocates "stagist" frameworks of the kind reflected in the sequential, inter-linked elements of the decision-making process in the IRGC Framework. He argues that "we need some way in which complexity can be reduced to a more manageable form".[82] Likewise, Althaus, Bridgman and Davis (2018) argue that a stagist model assists the complex business of policy-making "because it is an organising device with a range of strengths".[83] In particular, such a model allows disaggregation of "complex phenomena into manageable steps", enables "synthesis of existing knowledge", and provides "a map of how things could be done and how, perhaps, they might be done better".[84] Windholz (2017) adds that "[e]mploying a systematic approach and stepping through each of the key stages … should produce better outcomes, or put another way, prevent obvious (and not so obvious) mistakes that may cause regulatory failure."[85] These benefits are compelling, particularly where the stakes for "getting it wrong" are significant, as in the case of the systemic risks and the catastrophic consequences they could entail.

Set out in Table 6.1 is a summary of the key features of the two options for the regulation of risk considered in this book – namely, a traditional approach to the regulation of risk and risk governance. The contrast between the two approaches is strikingly apparent. The comparison in Table 6.1 indicates that risk governance is broader, more flexible, inclusive and responsive, compared to the traditional approach to the regulation of risk. These features are significant. Risk governance implies a process where a diverse range of voices will be heard in determining the response to risk, rather than basing the response to risk on pre-existing knowledge. It involves a holistic, interdisciplinary assessment of risk, which is not necessarily confined to expert opinions. It is designed to deliver risk management strategies that are matched to risk characteristics, rather than yielding a response that is fixed and rigid. Described as such, risk governance appears to hold great promise for the regulation of systemic risks, particularly systemic risks caused by climate change, which are characterised by complexity, uncertainty and ambiguity.

82 D.W. Parsons, *Public Policy: An Introduction to the Theory and Practice of Policy Analysis* (Edward Elgar, 1995), p. 80.
83 C. Althaus, P. Bridgman, & G. Davis, *The Australian Policy Handbook: A Practical Guide to the Policy-making Process* (Allen & Unwin, 6th edition, 2018), p. 44.
84 Ibid. pp. 44, 53.
85 E.L. Windholz, *Governing through Regulation: Public Policy, Regulation and the Law* (Routledge, 2017), pp. 83–4.

Table 6.1 Comparison of a traditional approach to risk regulation and risk governance

	Traditional approach to the regulation of risk	*Risk governance*
Definition	Regulatory response to control threat of harmful, adverse or unintended consequences	Decision-making process to identify and evaluate options to prevent harm to humans and what they value
Premise	Risk can be anticipated and assessed	Risk may not always be capable of being anticipated and this will affect the manner of assessment
Objective	Anticipating risk and preventing it from materialising	Obtaining a comprehensive understanding of risks and identifying options to manage risk
Decision-making	Narrow, rational and linear	Broad, collective and inclusive
Assessment of risk	Standardised and contained	Contextualised and interdisciplinary
Outcomes	Stabilisation of outcomes and the status quo	Risk management strategies to guide system to a desired state

IV Regulatory architecture for a risk governance approach

Using these insights about risk governance and how it is distinct from a traditional approach to the regulation of risk, the architecture of a regulatory regime based on risk governance begins to emerge. The *boundary* of such a regime will not necessarily be fixed but, instead, will depend upon the regime's context, which may vary over time. The *context* will be assessed using a holistic and inclusive decision-making process. The *content* for such a regime will not prescribe a specific regulatory response; rather it will enable a strategy to be developed to guide a system to a desired state based on the outcome of the assessment of context. This conceptualisation of the architecture of a regime that embeds a risk governance approach to regulation is depicted in Figure 6.2.

This depiction of a regulatory regime is in stark contrast to the architecture for a traditional approach to the regulation of risks.[86] Yet, it just may be the way forward for the regulation of systemic risks. Corbett (2017) argues that regulators must "abandon the quest for a single, fixed point of control, such as by crafting the one 'best' rule".[87] Instead, the regulatory response has more to do with "strategically, profoundly, madly, letting go and dancing with the system",[88] an

86 A traditional approach to the regulation of risk, including a depiction of the architecture of such an approach, is discussed in Chapter 5.

87 A. Corbett, n. 51 above, p. 256.

88 Ibid. p. 256 citing D. Meadows, *Thinking in Systems: A Primer* (Chelsea Green, 2008), p. 188.

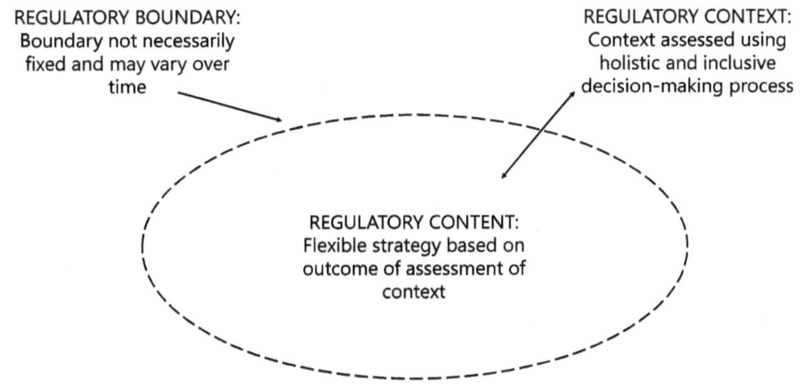

Figure 6.2 Architecture for a governance approach to the regulation of risk

approach that is, at least theoretically, enabled through governance of risk. This rather romantic conceptualisation of how systemic risks should be regulated is captivating and, intuitively, makes sense. The question remains whether regulation is capable of taking such a fluid and, seemingly, amorphous form and, if so, what does such a regulatory regime look like in practice? This question is addressed in the next chapter.

Bibliography

Althaus, C., Bridgman, P., & Davis, G. (2018). *The Australian Policy Handbook: A Practical Guide to the Policy-making Process*. Allen & Unwin (6th edition).

ASX (2019). *Corporate Governance Principles and Recommendations* (4th edition). https://www.asx.com.au/documents/asx-compliance/cgc-principles-and-recommendations-fourth-edn.pdf.

Boholm, Å., Corvellec, H., & Karlsson, M. (2012). "The practice of risk governance: Lessons from the field." *Journal of Risk Research*, 15(1), 1–20. https://doi.org/10.1080/13669877.2011.587886.

Briault, C. (2017). "Risk society and financial risk." In B.M. Hutter (ed.), *Anticipating Risks and Organising Risk Regulation* (pp. 3–22). Cambridge University Press. https://doi.org/10.1017/CBO9780511761553.003.

Corbett, A. (2017). "A systems approach to regulatory excellence." In C. Coglianese (ed.), *Achieving Regulatory Excellence*. Brookings Institution Press. https://doi.org/10.7864/j.ctt1hfr226.19.

Cottrell, M.P., & Trubek, D.M. (2012). "Law as problem solving: Standards, networks, experimentation, and deliberation in global space." *Transnational Law & Contemporary Problems*, 21(2), 359.

Holley, C., & Sofronova, E. (2017). "New environmental governance: Adaptation, resilience and law." In B. Hutter (ed.), *Risk, Resilience, Inequality and Environmental Law*. Edward Elgar Publishing. https://doi.org/10.4337/9781785363801.00016.

Hutter, B.M. (2010a). "Anticipating risk and organising risk regulation: Current dilemmas." In B.M. Hutter (ed.), *Anticipating Risks and Organising Risk Regulation*. Cambridge University Press. https://doi.org/10.1017/CBO9780511761553.002.

Hutter, B.M. (2010b). *Anticipating Risks and Organising Risk Regulation*. Cambridge University Press.

IRGC. (2005). *White Paper on Risk Governance: Toward an Integrative Framework*. International Risk Governance Council.

IRGC. (2010). *Risk Governance Deficits: Analysis, Illustration and Recommendations, Policy Brief*. International Risk Governance Council.

IRGC. (2017). *Introduction to the IRGC Risk Governance Framework, Revised Version*. EPFL International Risk Governance Center. https://doi.org/10.5075/EPFL-IRGC-233739.

IRGC. (2018). *IRGC Guidelines for the Governance of Systemic Risks*. EPFL International Risk Governance Center. https://doi.org/10.5075/EPFL-IRGC-257279.

Karkkainen, B.C. (2004). "'New governance' in legal thought and in the world: Some splitting as antidote to overzealous lumping." *Minnesota Law Review*, 89(2), 471.

Klinke, A., & Renn, O. (2002). "A new approach to risk evaluation and management: Risk-based, precaution-based, and discourse-based strategies." *Risk Analysis*, 22(6), 1071–1094. https://doi.org/10.1111/1539-6924.00274.

Klinke, A., & Renn, O. (2021). "The coming of age of risk governance." *Risk Analysis*, 41(3), 544–557. https://doi.org/10.1111/risa.13383.

Krieger, K. (2013). "The limits and variety of risk-based governance: The case of flood management in Germany and England." *Regulation & Governance*, 7(2), 236–257. https://doi.org/10.1111/rego.12009.

Lass, W., Haas, A., Hinkel, J., & Jaeger, C. (2011). "Avoiding the avoidable: Towards a European heat waves risk governance." *International Journal of Disaster Risk Science*, 2(1), 1–14. https://doi.org/10.1007/s13753-011-0001-z.

Levi-Faur, D. (2011). "Regulation and regulatory governance." In D. Levi-Faur (ed.), *Handbook on the Politics of Regulation*. Edward Elgar Publishing. https://doi.org/10.4337/9780857936110.

Lippert, T. H. (2019). *NATO, Climate Change, and International Security: A Risk Governance Approach*. Palgrave Macmillan. https://doi.org/10.1007/978-3-030-14560-6.

Lobel, O. (2004). "The renew deal: The fall of regulation and the rise of governance in contemporary legal thought." *Minnesota Law Review*, 89(2), 342.

Lobel, O. (2012). "New governance as regulatory governance." In D. Levi-Faur (ed.), *The Oxford Handbook of Governance*. Oxford University Press. https://doi.org/10.1093/oxfordhb/9780199560530.013.0005.

McDonald, J., & Styles, M. C. (2014). "Legal strategies for adaptive management under climate change." *Journal of Environmental Law*, 26(1), 25–53. https://doi.org/10.1093/jel/equ003.

Meadows, D. (2008). *Thinking in Systems: A Primer*. Chelsea Green.

OECD (2015). *G20/OECD Principles of Corporate Governance*. OECD Publishing. https://www.oecd.org/corporate/principles-corporate-governance/.

Parsons, D.W. (1995). *Public Policy: An Introduction to the Theory and Practice of Policy Analysis*. Edward Elgar.

Renn, O. (2017). *Risk Governance: Coping with Uncertainty in a Complex World*. Routledge. https://nls.ldls.org.uk/welcome.html?ark:/81055/vdc_100061934738.0x000001.

Renn, O., Klinke, A., & van Asselt, M. (2011). "Coping with complexity, uncertainty and ambiguity in risk governance: A synthesis." *AMBIO*, 40(2), 231–246. https://doi.org/10.1007/s13280-010-0134-0.

Renn, O., & Walker, K.D. (eds.) (2008). *Global Risk Governance: Concept and Practice Using the IRGC Framework (Vol.*1). Springer Netherlands. https://doi.org/10.1007/978-1-4020-6799-0.

Rothstein, H., Borraz, O., & Huber, M. (2013). "Risk and the limits of governance: Exploring varied patterns of risk-based governance across Europe." *Regulation & Governance*, 7(2), 215–235. https://doi.org/10.1111/j.1748-5991.2012.01153.x.

Rothstein, H., Irving, P., Walden, T., & Yearsley, R. (2006). "The risks of risk-based regulation: Insights from the environmental policy domain." *Environment International*, 32(8), 1056–1065. https://doi.org/10.1016/j.envint.2006.06.008.

Ruhl, J. B. (2011). "General design principles for resilience and adaptive capacity in legal systems – With applications to climate change adaptation." *North Carolina Law Review*, 89(5), 1373.

Schmidt, K., Dressel, K.M., Niedrig, M., Mertens, M., Schüle, S.A., & Groschup, M.H. (2013). "Public health and vector-borne diseases – a new concept for risk governance." *Zoonoses and Public Health*, 60(8), 528–538. https://doi.org/10.1111/zph.12045.

Simon, H.A. (1997). *Administrative Behavior: A Study of Decision-Making Processes in Administrative Organizations*. Free Press (4th edition).

Simon, H.A. (2000). "Bounded rationality in social science: Today and tomorrow." *Mind & Society*, 1(1), 25–39. https://doi.org/10.1007/BF02512227.

Stein, V., & Wiedemann, A. (2016). "Risk governance: Conceptualization, tasks, and research agenda." *Journal of Business Economics*, 86(8), 813–836. https://doi.org/10.1007/s11573-016-0826-4.

van Asselt, M.B.A., & Renn, O. (2011). "Risk governance." *Journal of Risk Research*, 14 (4), 431–449. https://doi.org/10.1080/13669877.2011.553730.

Wildavsky, A. (1988). *Searching for Safety*. Berkeley: University of California Press.

Windholz, E.L. (2017). *Governing through Regulation: Public Policy, Regulation and the Law*. Routledge. https://doi.org/10.4324/9781315677286.

Chapter 7

A New Paradigm for Regulating Systemic Risks

Without question, determining how to regulate risk is difficult. A host of political, economic and social issues may complicate the design of regulatory regimes. The challenges are likely to be particularly pronounced for the regulation of systemic risks. The natural and human systems that society depends on and could be at risk are complex. Each system will be subject to a broad variety of stressors and the possible consequences of those stressors for a system will necessarily be system-specific. The level of complexity, uncertainty and ambiguity associated with modern systemic risks surpasses what most policy- and law-makers have previously had to contend with.

The notion of applying a risk governance approach to the regulation of systemic risks is compelling. It conveys the idea that, despite the complexity, uncertainty and ambiguity associated with such risks, they could, nonetheless, be better analysed, understood and brought under control through a holistic, inclusive, structured yet flexible process that can be tailored to a particular system and the associated context. This is a deviation from a traditional approach to the regulation of risk and would entail a paradigm shift in the way risk is regulated. This chapter provides a much-needed practical pathway by setting out how an ideal regime to regulate systemic risks could be designed based on the concept of risk governance.

I Regulatory ideals and building blocks

1 The concept of a regulatory ideal

An ideal conveys the notion of perfection. In a regulatory context, an ideal can be conceived as a standardised way to regulate a particular type of risk in the most appropriate and effective way.[1] So described, the concept of a regulatory

1 This conception of a regulatory ideal is distinct from a model law. Whereas a model law sets out a series of specific provisions that represent the way in which a particular subject or domain is to be regulated (such as the UNCITRAL Model Law on International Commercial Arbitration, which is designed to assist States to ensure that their laws are consistent with requirements for international commercial arbitration to help facilitate international trade and investment), a regulatory

DOI: 10.4324/9781003253464-7

ideal is not well-developed. This is unsurprising. Pursuing perfection in a regulatory context may seem ill-fated given the complex public policy problems regulation is called on to address, not least of which are the problems arising from systemic risks, like those caused by climate change.

Indeed, in discussing "Challenges for Australia and the Commonwealth in Regulatory Reform" in 2001, Gary Banks, the former Chair of Australia's Productivity Commission dismissed outright the possibility that regulation can achieve anything near perfection through a regulatory ideal, with an allusion to global risks as the context for his statement:[2]

> [R]egulation is essential to the proper functioning of a society. Rules create order and the basis for stability and progress. They shape incentives and influence how people behave and interact. And they help societies deal with otherwise intractable economic, social and environmental problems. All that goes without saying. The more important issues have to do with the appropriate form and content of regulation. While good regulation is essential to achieving desirable social and economic goals, bad regulation can undermine a society's capacity to achieve both. …
>
> There is no such thing, however, as a single regulatory ideal or a permanent general solution to a society's regulatory needs. All regulation has both positive and negative attributes, and most of it tends to obsolescence as a country's economic, social or environmental circumstances evolve. This is particularly evident today with the rapid changes brought on by technological advances and global integration.

No doubt, Banks' scepticism about the existence of regulatory ideals is shared by many policy- and law-makers, as well as regulators, who have witnessed regulatory regimes that had once been fit-for-purpose become outdated, irrelevant, unhelpful or, at worst, causative of regulatory failure. Concerns may be linked, at least in part, to a perceived difficulty for regulation to "abstract from the specific to the general" because legislation typically focuses on the "particular and the temporary" rather than the "permanent and the general".[3] Banks' comments also appear to suggest that regulatory ideals cannot exist, because regulatory regimes are destined to become obsolete as the context for their application changes over time.

ideal is of a higher order and sets out the key elements of the regulatory architecture in more abstract and generic terms.

2 NSW Regulation Review Committee, *Proceedings of the International Conference on Regulation Reform Management and Scrutiny of Legislation, Parliament House, Sydney, 9–13 July 2001* (Parliament of New South Wales, 2001), p. 48.

3 Van Rijswick, M. and Salet, W. (2012) "Enabling the contextualization of legal rules in responsive strategies to climate change", *Ecology and Society*, Vol. 17, No. 2, 18–25, p. 21.

Arguably, these concerns are not insurmountable. Regulation that is based on principles rather than prescription and is capable of evolving as the context changes instead of remaining static and immutable could form the foundation of an ideal to regulate risks, particularly systemic risks. The next section discusses the specific building blocks that could help lay the foundation for the establishment of such an ideal.

2 Regulatory building blocks

The concept of a "building block" is used metaphorically in a broad variety of contexts to describe a basic component or unit from which something is constructed or built up. It has been used in a regulatory context to describe the basis for the economic regulation of energy utilities, which are generally characterised as natural monopolies.[4] The "building block" model in this context is so-called because, under the regulatory framework, the allowed total revenue of a regulated utility is based on underlying building blocks, which are used to determine the utility's total cost of service. This approach is aimed at minimising the risk of the utility's exploitation of market power by charging excessive prices for its services.

The concept of building blocks is also relevant more broadly for regulation. In particular, it is useful to explain that a regulatory regime is characterised by a bottom-up approach, where the building blocks form the foundation of a regulatory regime, reflect the core attributes of the regime, and together describe the benchmark or regulatory ideal that has been established under the regime. In fact, a building blocks approach has been advocated to address climate change mitigation,[5] but a similar approach for climate change adaptation has not previously been put forward.

An appreciation and understanding of the building blocks for a regulatory regime can be helpful and instructive, whether in the context of designing a new regulatory regime or evaluating a pre-existing one. That is not to say that the regulation of risk is a mechanical, formulaic process that can be standardised through the exclusive reliance on pre-established building blocks. On the

4 The concept of a "building block" model was coined by the Australian Competition and Consumer Commission when determining the total revenue requirement for a regulated gas pipeline: J. Dimasi, "Rethinking utility regulation in Australia", *Monash Business Policy Forum* (Monash Business School, 2015) p. 13 accessible at: https://www.monash.edu/__data/assets/pdf_file/0005/906971/rethinking-utility-regulation-in-australia-final.pdf.

5 R. Falkner, H. Stephan, & J. Vogler, "International climate policy after Copenhagen: Towards a 'building blocks' approach" (2010) 1(3) *Global Policy*, pp. 252–62; R.B. Stewart, M. Oppenheimer, & B. Rudyk, "Building blocks for global climate protection" (2013) 32(2) *Stanford Environmental Law Journal*; E. Waters, "Why the world should act like children: Using the building blocks method to combat climate change, beginning with methane" (2017) 51(4) *University of Richmond Law Review*, pp. S27–42.

contrary, determining whether a particular risk should be regulated and, if so, how, is typically complex, and most risks will require a bespoke regulatory response. This is especially the case in the context of systemic risks, the regulation of which is likely to vary depending upon the particular system at risk as well as the risks faced by that system. Nevertheless, the core characteristics of a risk can be helpful in guiding identification of the ideal elements of a regime to regulate that risk – that is, the regime's building blocks.

3 Constructing an ideal for the regulation of systemic risks

A risk governance framework has been specifically designed for the management of systemic risks – namely, the IRGC Framework.[6] Accordingly, this Framework would seem to be a good starting point for identifying the building blocks of an ideal for the regulation of systemic risks. Nonetheless, significant challenges associated with the translation of the elements of the IRGC Framework into regulatory language exist and are linked, firstly, to the way the IRGC Framework has been articulated and, secondly, to the manner in which regulation is conventionally used to regulate risk.

In relation to the first challenge, the authors of the IRGC Framework have themselves noted that "[t]he IRGC framework is … first and foremost a broad conceptual framework incorporating a set of key principles for sound risk governance" and "was never intended to be a comprehensive manual that would provide solutions for all types of problems".[7] Moreover, those that have sought to employ the IRGC Framework in practice have noted that it is highly conceptual and idealised, rather than practical, and that the manner in which it could be implemented through regulation is unclear.[8] Regarding the second challenge, while policy- and law-makers have had a long history of employing a traditional approach to the regulation of risk, explicit reference to and reliance on risk governance as the basis for regulating risks is less common and obvious.

Accordingly, this book takes up where the IRGC Framework leaves off. It seeks to interpret the IRGC Framework in a regulatory context to determine how to design an effective regime to regulate systemic risks, like those caused by climate change, using an approach that is based upon risk governance, rather than a traditional approach to the regulation of risk. The main elements of the IRGC Framework are translated into regulatory "building blocks" which are then assembled into an ideal for the regulation of systemic risks.

6 Chapter 6 describes the main features of the IRGC Framework and discusses how the IRGC Framework has been applied in practice to a variety of systemic risks.

7 O. Renn and K. Walker, "Lessons learned: A re-assessment of the IRGC Framework on Risk Governance" in O. Renn & K.D. Walker (eds.), *Global Risk Governance: Concept and Practice Using the IRGC Framework* (Springer Netherlands, 2008), pp. 337, 354.

8 O. Renn and A. Jäger "Synopsis of critical comments on the IRGC Risk Governance Framework" in O. Renn & K.D. Walker, ibid. pp. 120, 121.

Pre-assessment

The first element of the IRGC Framework is known as "pre-assessment". The emphasis during pre-assessment is on understanding risk, which occurs before risk is assessed and responded to.[9] The various issues that may be associated with a certain problem are filtered in order to identify the risk to be addressed.[10] More practically, boundaries of the risk are established, which enables risk to be defined and framed.[11] The results of pre-assessment establish the basis for how risk is assessed and managed during subsequent aspects of the risk governance process.[12]

Appraisal

The second element of the IRGC Framework – "appraisal" – involves the development and synthesis of knowledge about the risk, which is then used for risk management later in the process.[13] Risk appraisal comprises both a technical "risk assessment" and a "concern assessment", which together are aimed at providing a comprehensive assessment of risk.[14]

In the context of the technical risk assessment, the source of harm is linked with likely consequences.[15] On the source side, the "risk agent" is considered – that is, the hazard that has the potential to cause harm. On the impact side, the "risk absorbing system" is assessed – that is, the system that could be vulnerable and exposed to the risk agent.[16] The concern assessment involves an assessment of different stakeholders' perceptions, opinions and concerns about risk, including an analysis of the benefits and risks that stakeholders may associate with a hazard, its causes and consequences.[17] It also enables an assessment to be made about the likely reactions of stakeholders to a response to risk that could arise from differences in stakeholder objectives and values, or from inequities that could arise from options to respond to risk.[18]

Characterisation and evaluation

The third element of the IRGC Framework focuses on "characterisation and evaluation" of risk, which occurs once it has been appraised, and helps

9 IRGC, *White Paper on Risk Governance: Toward an Integrative Framework* (IRGC, 2005), pp. 13, 23.
10 Ibid. p. 13.
11 IRGC, *Introduction to the IRGC Risk Governance Framework, Revised Version* (EPFL International Risk Governance Center, 2017), pp. 9, 11.
12 Ibid. p. 11.
13 Ibid. p. 13.
14 Ibid. p. 13.
15 IRGC, n. 9 above, p. 14; IRGC, n. 11 above, p. 14.
16 IRGC, n. 11 above, p. 14.
17 Ibid. p. 19.
18 IRGC, *An Introduction to the IRGC Risk Governance Framework* (IRGC, 2012), p. 13.

form judgements about risk and the need to take action to mitigate the risk, if at all.[19]

"Characterisation" uses the knowledge generated from the risk appraisal to determine which category of systemic risk applies – namely, "simple", "complex", "uncertain", "ambiguous", or a combination.[20] Such categorisation depends upon three factors – namely, the degree of difficulty of establishing the cause–effect relationship between a risk agent and the potential consequences for the risk absorbing system, the reliability of this relationship, and the degree of controversy with regard to both what the risk means in practice for the risk absorbing system as well as the values to be applied when judging whether or not something needs to be done about it.[21] These considerations respectively concern the relative degree of complexity, uncertainty and ambiguity associated with risk. In turn, risk characterisation informs "risk evaluation", which involves making a judgement about the acceptability of risk (where risk reduction is considered unnecessary), tolerability of the risk (where risk reduction may be required), or intolerability of risk (where risk avoidance will be needed).[22]

Management

The fourth element of the IRGC Framework – "risk management" – uses the output generated from the previous elements of the risk governance process to determine the response to risk.[23] Risk management includes the generation, assessment, evaluation and selection of appropriate risk reduction options.[24] The IRGC Framework proposes generic risk management strategies applicable to each category of systemic risk put forward by the IRGC – specifically, simple, complex, uncertain and ambiguous risks.[25]

Communication and stakeholder engagement

The IRGC Framework emphasises "communication" and "stakeholder engagement", which concern how stakeholders, including civil society, understand risk and participate in the risk governance process.[26] These elements are

19 IRGC, n. 11 above, p. 9.
20 Ibid. p. 8. These categories of systemic risks, and their associated characteristics, are discussed later in this chapter.
21 IRGC, n. 9 above, p. 12.
22 Ibid. p. 14; IRGC, n. 11 above, p. 20.
23 O. Renn, "White Paper on Risk Governance: Towards an Integrative Framework" in O. Renn & K.D. Walker (eds.), n. 7 above, p. 32; O. Renn, A. Klinke, & M. van Asselt, "Coping with complexity, uncertainty and ambiguity in risk governance: A synthesis" (2011) 40(2) *AMBIO*, pp. 231–46, at 241.
24 IRGC, n. 9 above, p. 14; IRGC, n. 11 above, p. 23.
25 IRGC, n. 9 above, p. 15; IRGC, n. 11 above, pp. 24–5.
26 IRGC, n. 11 above, p. 8; O. Renn & K.D. Walker (eds.), n. 7 above, p. xxv.

"cross-cutting", in the sense that they are intended to apply throughout the risk governance process.[27]

"Communication" involves the exchange of risk-related data, information and knowledge with stakeholders.[28] By developing a common understanding of risk and possible responses, effective communication helps establish trust in the risk management process.[29] "Stakeholder engagement" helps to generate insights for the risk governance process. More specifically, such engagement assists in identifying, assessing and accounting for stakeholders' risk perceptions and concerns in decisions involving the management of risks, which can increase the effectiveness, fairness and acceptability of risk management decisions.[30]

4 Building blocks for the regulation of systemic risks

While most of the elements comprising the IRGC Framework described above are relatively unfamiliar in a regulatory context, the components of the architecture of a regulatory regime – boundary, context and content – can be used to identify how those elements could be used in a regulatory setting for the design of a regime to regulate systemic risks.[31]

Regulatory boundary

The "pre-assessment" element of the IRGC Framework is relevant for the establishment of the boundary of a regulatory regime to address systemic risks, which concerns the regime's ambit or scope. The boundary serves to demarcate the risks covered by the regime and to identify the aspects of those risks that will be subject to regulatory control.

Two building blocks derive from the pre-assessment element of the IRGC Framework. The first involves *defining risks* that are at the heart of the problem that is sought to be addressed by the regulatory regime (Building Block 1: Risk Definition). This exercise is not merely a question of labelling the risk or risks to be regulated under a regime but, rather, involves describing risk in a manner that will facilitate its assessment and management.

Even if risk is clearly defined for the purposes of establishing the regulatory boundary, identifying the aspect of the risk that is the best candidate for the exercise of regulatory control will not necessarily be readily apparent. Therefore, the second building block to establish the boundary of a regime to regulate systemic risk involves *deconstructing risk* (Building Block 2: Risk Deconstruction) – that is, understanding the system at risk as well as the process giving rise to risk. This

27 IRGC, n. 11 above, pp. 27–30.
28 Ibid. p. 27.
29 Ibid. p. 27.
30 Ibid. pp. 8, 31.
31 These components of a regulatory regime were discussed in Chapter 4.

building block is useful to identify the aspect of the risk for which regulatory intervention is considered necessary.

Regulatory context

Various elements of the IRGC Framework have been translated into a series of building blocks to establish the regulatory context for the management of systemic risks, which then forms the foundation for the response to risk.

The "appraisal" element of the IRGC Framework provides for a comprehensive approach to the assessment of risk to arrive at the best possible estimate of the harm that could be produced by risk. Accordingly, the third building block involves the establishment of the *regulatory approach towards risk assessment* (Building Block 3: Risk Assessment). Under the IRGC Framework, risk appraisal encompasses consideration of the hazard giving rise to risk, as well as exposure and vulnerability to risk. Contextualising the assessment of risk by accounting for these factors helps to ensure the relevance and effectiveness of a response to risk, given the specific context within which the risk has arisen. Therefore, the fourth building block, which is closely linked to the third, involves *accounting for the context within which the risk is to be assessed* and is also based on the "appraisal" element of the IRGC Framework (Building Block 4: Risk Contextualisation).

"Characterisation and evaluation" under the IRGC Framework are aimed at enabling judgments to be made about the risk in order to determine whether a response to risk is needed. The fifth proposed building block is based on the "characterisation" element of the IRGC Framework and involves *identifying the risk's characteristics* (Building Block 5: Risk Characterisation). The sixth building block, based on the "risk evaluation" element of the IRGC Framework, involves identification of the *regulatory mechanisms to evaluate risk* to determine whether risk is acceptable, tolerable or intolerable and, ultimately, to determine whether or not a regulatory response is needed to mitigate risk (Building Block 6: Risk Evaluation).

The cross-cutting elements in the IRGC Framework include "communication" and "stakeholder engagement", which address the questions of whether, when and how the public (and other relevant stakeholders) should be involved in the risk management process. These components are referred to collectively in this book as "public participation" to convey the notion that the public are critically important stakeholders in the governance of systemic risks, rather than mere bystanders. The seventh building block involves *embedding public participation in the regulatory regime* to ensure that relevant stakeholders – particularly, the public – are kept informed about the risk management process as well as ensuring appropriate engagement of those stakeholders in the process (Building Block 7: Public Participation).

Regulatory content

The final element of the IRGC Framework, "management", is used to define the regulatory content for systemic risks and is the eighth building block (Building

Block 8: Risk Management). More specifically, it involves using the output from the previous elements of the risk management process to identify the appropriate regulatory approach, objectives, principles and tools that should be included in the design of a regulatory regime to manage systemic risks.

Assembly

In summary, translation of the IRGC Framework into building blocks for an ideal regime to regulate systemic risks involves understanding the objectives of each element of the IRGC Framework and using that understanding to articulate regulatory building blocks that could collectively form the architecture of the ideal regulatory regime. This translation process is summarised in Table 7.1.

Table 7.1 Summary of translation of the IRGC Framework into building blocks of an ideal to regulate systemic risks

<table>
<tr><th colspan="2">Risk Governance</th><th colspan="2">Regulatory Risk Governance</th></tr>
<tr><th>Elements of IRGC Framework</th><th>Objectives of elements of IRGC Framework</th><th>Regulatory building blocks</th><th>Aspect of architecture of regulatory regime</th></tr>
<tr><td>"Pre-assessment"</td><td>To identify and understand risks for which a response might be required</td><td>• Building Block 1: Risk Definition
• Building Block 2: Risk Deconstruction</td><td>Boundary of the Regulatory Regime</td></tr>
<tr><td>"Appraisal"</td><td>To undertake a comprehensive assessment of risk, which accounts for the context in which the risk has arisen</td><td>• Building Block 3: Risk Assessment
• Building Block 4: Risk Contextualisation</td><td rowspan="3">Context for the Regulatory Regime</td></tr>
<tr><td>"Characterisation and Evaluation"</td><td>To help characterise the risk and form judgements about whether a response to risk is required</td><td>• Building Block 5: Risk Characterisation
• Building Block 6: Risk Evaluation</td></tr>
<tr><td>"Communication" and "Stakeholder engagement"</td><td>To inform and engage relevant stakeholders, including the public, in relation to whether and how risk should be managed</td><td>• Building Block 7: Public Participation</td></tr>
<tr><td>"Management"</td><td>To define the options available to manage risks</td><td>• Building Block 8: Risk Management</td><td>Content of the Regulatory Regime</td></tr>
</table>

The various building blocks are also illustrated in Figure 7.1 to demonstrate how these building blocks can be practically assembled in order to establish an ideal to regulate systemic risks. The first set of building blocks (Building Block 1: Risk Definition and Building Block 2: Risk Deconstruction) comprise those that are used to *establish the boundary of a regime to regulate systemic risk*. The second set of building blocks (Building Block 3: Risk Assessment, Building Block 4: Risk Contextualisation, Building Block 5: Risk Characterisation, Building Block 6: Risk Evaluation, and Building Block 7: Public Participation) concern the *regulatory mechanisms to provide context for a regime to regulate systemic risk*. The third set of building blocks, which are embedded in Building Block 8: Risk Management, reflect the various *options available to respond to systemic risks through the regulatory content*.

Figure 7.1 is a high-level representation of risk governance in regulatory form. It articulates an ideal to regulate systemic risks based on the guidance provided by the IRGC Framework and has important benefits. First, it provides a vision of the architecture of a regime to regulate systemic risks, which places risk at the heart of the regime. Second, it implicitly outlines a systematic process to design or evaluate a regime to regulate systemic risks by specifying the key components one would expect to see in such a

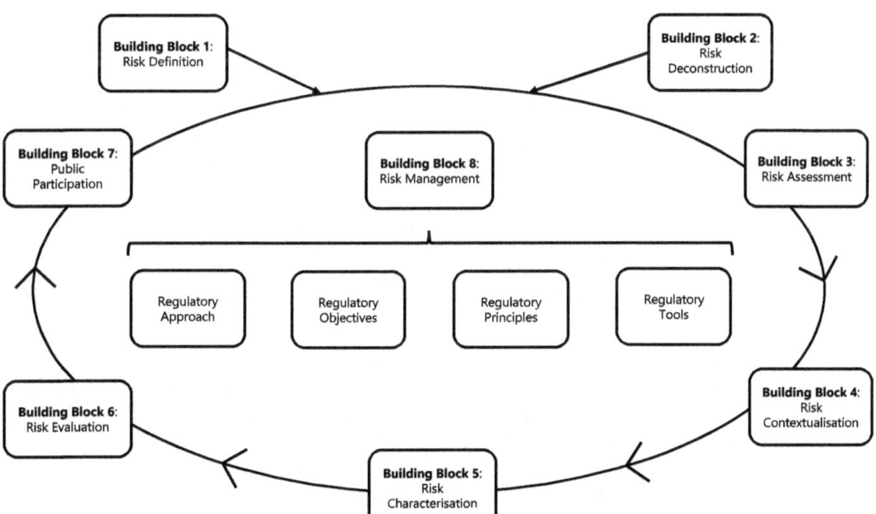

Figure 7.1 Building blocks of an ideal for the regulation of systemic risks[32]

32 This diagram is based on the author's interpretation of how governance could be applied in a regulatory context to manage systemic risks using the IRGC Framework as a guide for the development of the individual building blocks.

regime. Thirdly, and most importantly, it provides a means by which regulation can be used to unpack the complexity associated with systemic risks, respond to uncertainty in an intelligent and strategic way, and help resolve ambiguity regarding whether and how systemic risks should be mitigated. Each set of building blocks is explored in the following three sections of this chapter.

II Boundary of an ideal regime to regulate systemic risks

Using risk as a lens for the design of regulation can help clarify regulatory scope. In particular, the boundary of a regulatory regime can be prescribed to include risks that are sought to be governed by a particular regime. While this exercise may be relatively straightforward for regimes that govern discrete, identifiable risks, it is likely to be comparatively much more challenging in the context of the regulation of systemic risks because of the complexity, uncertainty and ambiguity that characterise these risks. As explained below, the building blocks concerning risk definition and risk deconstruction can be helpful when establishing the boundary for the regulation of systemic risks.

1 Risk definition

In the case of many societal problems that require government intervention, it may not be practically possible nor prudent from a cost-benefit perspective to seek to minimise or eliminate all risk that underpins the problem using regulation.[33] Viewing a regulatory regime through the prism of risk enables a line to be drawn between those risks to be covered by the regime, and those excluded from coverage. However, the capacity of risk as a concept to yield this benefit when designing a regulatory regime hinges on how risk is defined.

The linkage between a clear definition of risk and sound decision-making has been made by a number of academic writers. For example, Kaplan and Garrick (1981), who are proponents of a much-cited definition of risk,[34] assert

33 Productivity Commission, *Rethinking Regulation*, Report of the Taskforce on Reducing Regulatory Burdens on Business (Commonwealth of Australia, 2006), p. 146. The Productivity Commission notes in this report that "it is important to recognise that not all 'problems' will justify (additional) government action. For example, it will generally make more sense to accept a certain level of risk than to implement measures that seek to minimise or eliminate all risk".

34 Kaplan and Garrick suggest that risk can be quantitatively expressed as a set of "triplets", which combine consideration of the scenario giving rise to risk, the probability of that scenario, and the consequence of the scenario: S. Kaplan & B.J. Garrick, "On the quantitative definition of risk" (1981) 1(1) *Risk Analysis*, pp. 11–27, at 13. Aven (2020) suggests that this paper by Kaplan and Garrick is one of "[t]hree of the most influential scientific works in the risk field": T. Aven, "Three influential risk foundation papers from the '80s and '90s: Are they still state-of-the-art?" (2020) 193 *Reliability Engineering & System Safety*, p. 106680.

that "[r]ational decision-making requires ... a clear and quantitative way of expressing risk so that it can be properly weighed, along with all other costs and benefits, in the decision process".[35] Fischhoff, Watson and Hope (1984) also suggest that "an explicit and accepted definition of risk is essential" for the effective management of risk.[36] The clear implication of these comments is that, to be practically useful for the purposes of risk management, defining risk goes beyond merely labelling the risk in qualitative terms – such as, the "risk of a natural disaster occurring" or the "risk of critical infrastructure failing" – and extends to expressing risk in a way that enables its assessment.

Indeed, searching for a response to mitigate risk without a clear and practically useful definition of risk is akin to feeling around aimlessly in the dark in the hope that light will miraculously be shed on the problem that precipitated the search in the first place. As stated by Slovic (1999):[37]

> Whoever controls the definition of risk controls the rational solution to the problem at hand. If you define risk one way, then one option will rise to the top as the most effective, or the safest, or the best. If you define it another way, perhaps incorporating qualitative characteristics and other contextual factors, you likely get a different ordering of your action solutions.

Perhac (1996) similarly claims that "because the manner in which we conceive of risk goes a long way in determining how risk management is conceived and carried out, any definition of risk has important normative elements".[38]

Indeed, the manner in which risk is defined in a regulatory context has important repercussions. Apart from facilitating the identification of risks that are governed by the regulatory regime, the definition of risk implicitly dictates the analytical framework for assessment of risk (which is relevant to the establishment of the *regime's context*) and affects the determination of whether a regulatory response to risk is needed (which concerns the *regime's content*). Failure to clearly define risk can lead to ambiguity about how risk is to be assessed and can, ultimately, lead to excessive responses to risk (that is, over-regulation) or inadequate responses to risk (under-regulation).

The first proposed building block of an ideal to regulate systemic risks is Building Block 1: Risk Definition and is highlighted in Figure 7.2. This building block is particularly important in the context of systemic risks because of the complex, multi-faceted nature of systems and the broad range of stressors and associated risks

35 S. Kaplan & B.J. Garrick, n. 34 above, p. 11.

36 B. Fischhoff, S.R. Watson, & C. Hope, "Defining risk" (1984) 17(2) *Policy Sciences*, pp. 123–39, at 123.

37 P. Slovic, "Trust, emotion, sex, politics, and science: Surveying the risk-assessment battlefield" (1999) 19(4) *Risk Analysis*, pp. 689–701, at 699.

38 R.M. Perhac, "Defining risk: Normative considerations" (1996) 2(2) *Human and Ecological Risk Assessment: An International Journal*, pp. 381–92, at 381.

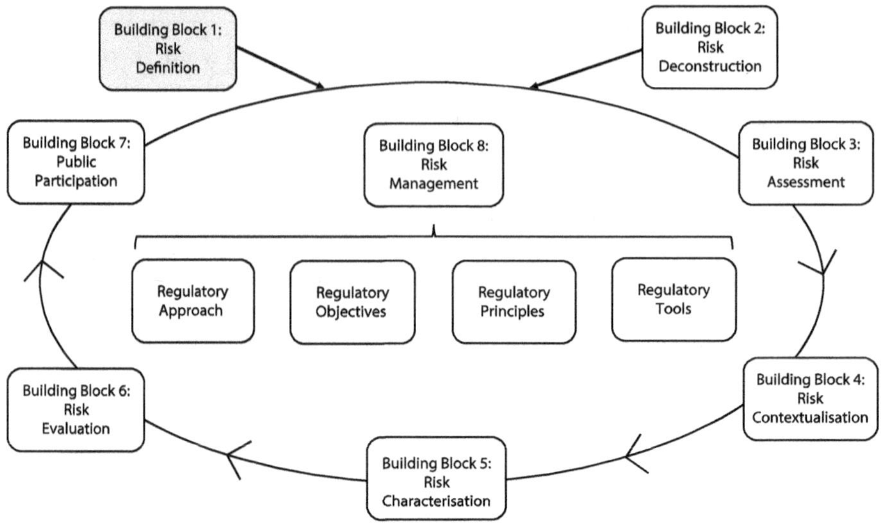

Figure 7.2 Building Block 1: Risk Definition

to which they might be subject. The exercise of defining risk in this context can help to filter the various sources of risk that could be faced by a system and better target regulatory intervention to mitigate systemic risk.

Core definitions of risk

Defining risk is inherently controversial because, as explained by Fischhoff, Watson and Hope (1984), it may "affect the outcome of policy debates, the allocation of resources … and, the distribution of political power in society".[39] A broad range of risk definitions have been suggested by academics across various disciplines to help systematise the approach to defining risk. Moreover, several attempts have been made to catalogue these definitions, including by Brooks (2003),[40] Kelman (2003),[41] Aven and Renn (2009),[42] Aven, Renn and Rosa (2011)[43] and Aven (2013a).[44] These catalogues reflect a multitude of risk definitions that have been advanced over several decades. Of note, there is no

39 B. Fischhoff, S.R. Watson, & C. Hope, n. 36 above, p. 124.

40 N. Brooks, "Vulnerability, risk and adaptation: A conceptual framework", *Working Paper 38* (Tyndall Centre for Climate Change Research, 2003), p. 7.

41 I. Kelman, "Defining Risk", (2003)2 *FloodRiskNet Newsletter*, Winter, p. 7.

42 T. Aven & O. Renn, "On risk defined as an event where the outcome is uncertain" (2009) 12(1) *Journal of Risk Research*, pp. 1–11, at 1–2.

43 T. Aven, O. Renn, & E.A. Rosa, "On the ontological status of the concept of risk" (2011) 49(8–9) *Safety Science*, pp. 1074–9, at 1075.

44 T. Aven, "On the meaning and use of the risk appetite concept" (2013) 33(3) *Risk Analysis*, pp. 462–8, at 462–3.

clear evidence that the various definitions of risk are converging with time. Rather, the spectrum of definitions illustrates how subjective the concept of risk appears to be.[45]

Nevertheless, there are three main observations that can be made about the various definitions of risk that have been put forward so far. The first is that each definition can be viewed as a distinct attempt to break down the particular problem that has given rise to risk. This is an important point because it recognises that there may be different ways to view a problem and, therefore, different approaches to defining risk, each of which may lead to a different response. Second, despite the diversity of definitions, there are certain common concepts that feature prominently in many of them – namely, probability, consequences and uncertainty. Third, three core definitions of risk appear to dominate the debate about how risk should be defined, with each relying upon one or more of the common concepts that have been mentioned. These core definitions are each linked to a different articulation of the problem giving rise to risk and are explored in turn below.[46]

The first core definition of risk focuses on "what can go wrong?". This definition concerns a situation or event where something of human value (including humans themselves) is at stake and where the outcome is uncertain, such as a flood or storm surge.[47] Expressed in this way, risk is equated with the uncertain adverse event or outcome. This definition assumes that the adverse event or outcome can be objectively identified but accepts that there is uncertainty as to the occurrence of the event or outcome.[48] An example is emergency management legislation, the application of which is triggered by a natural disaster or other type of emergency that can, in turn, activate a raft of powers to address the emergency.[49]

45 Aven notes that "[i]t is not realistic to expect everyone to agree on one definition of risk": T. Aven, "The concepts of risk and probability: An editorial" (2013) 15(2) *Health, Risk & Society*, pp. 117–22, at 121.

46 The articulation of each of these risk definitions broadly corresponds to the three questions Kaplan and Garrick (1981) suggest should be used to analyse and, thereby, define risk – namely "(i) What can happen? (i.e. What can go wrong?) (ii) How likely is it that that will happen? (iii) If it does happen, what are the consequences?": S. Kaplan & B.J. Garrick, n. 34 above, p. 13. Another way to view these questions (or variants of them) is that they each break down the problem giving rise to risk to differing degrees.

47 T. Aven, "Risk analysis and management. Basic concepts and principles" (2009) 4 (1 (12)) *Reliability: Theory & Applications*, pp. 57–73, at 60; E.A. Rosa, "Meta-theoretical foundations for post-normal risk" (1998) 1(1) *Journal of Risk Research*, pp. 15–44, at 28; IRGC, n. 9 above, p. 19.

48 E.A. Rosa, n. 47 above, p. 28.

49 For example, the definition of "emergency" in section 3 of *Emergency Management Act 2013 (Victoria)* includes "an earthquake, flood, wind-storm, or other natural event". Among other things, under section 23 of the Act, a state of disaster may be declared in the event of an emergency, which triggers a range of powers that can be exercised to address the emergency.

The second core definition responds to the question "how likely is it that things will go wrong?". This definition equates risk with the probability of an undesirable event or adverse outcome[50] and assumes that the probability associated with that event or outcome can be calculated. For example, an "annual exceedance probability" (AEP) is often attached to the risk of a flood of a given magnitude – such as a 1% chance of a flood occurring in a given period. In some jurisdictions, regulatory requirements to help manage flood risks apply if the applicable AEP is likely to be exceeded.[51]

The third core definition addresses the question "if something is likely to go wrong, what will the consequences be?". This definition implies that risk is the measure of the probability of an adverse event or outcome (such as a major flood), as well as the consequences associated with such events or outcomes (including the number of injuries or deaths and the type and severity of other damage that could be caused).[52] It appears to be widely used in practice including in a regulatory context.[53] Regulatory requirements can be triggered if the overall risk, which accounts for both probability and consequence of harm, is relatively high.[54]

Each of the core risk definitions that have been identified represents a distinct effort to come to terms with uncertainty. However, the capacity of each definition to facilitate assessment and management of systemic risks will ultimately depend upon how well uncertainty can be resolved in the particular context in which systemic risks have arisen. This issue is considered further below in relation to the definition of systemic risks caused by climate change.

Definitions of risk in the climate change context

An obvious place to start in identifying the appropriate definition of the risks caused by climate change is the work of the IPCC. It is notable that the IPCC's definition of risk has evolved over time, which appears to reflect a change in its approach to address uncertainty associated with climate change risks.

50 T. Aven, n. 47 above, p. 61.

51 D.W. Davis, "Contemporary risk analysis as the foundation of a new national flood standard" (2004) *Reducing Flood Losses: Is the 1% Chance (100-Year) Flood Standard Sufficient?* accessible at: https://www.nrcs.usda.gov/Internet/FSE_DOCUMENTS/16/nrcs143_009401.pdf.

52 T. Aven, n. 47 above, p. 58; T. Aven, O. Renn, & E.A. Rosa, n. 43 above, p. 1075.

53 This definition of risk is referred to in an international standard – AS ISO 31000:2018. Risk management – Guidelines, p. 1. It is referenced in a broad range of risk management strategies and related documents produced by private and public sector bodies alike.

54 For example, clause 116(2)(f) of Queensland's *Coal Mining Safety and Health Regulation 2017 (Qld)* requires the establishment of procedures for the transportation of explosives for a surface mine to "reduce the probability and consequence of incidents".

In the context of the IPCC's Fifth Assessment Report, a probabilistic approach to the definition of risk was apparent. In particular, in the *Guidance Note for Lead Authors of the IPCC Fifth Assessment Report on Consistent Treatment of Uncertainties* (2010), the IPCC stated that "risk is a function of probability and consequence".[55] Similarly, the report on *Impacts, Adaptation and Vulnerability for the IPCC's Fifth Assessment Report* (2014) notes that "[r]isk is often represented as probability of occurrence of hazardous events or trends multiplied by the impacts if these events or trends occur".[56] This definition of risk, which focuses on the most likely and consequential impacts of climate change, is also reflected in the report on *Mitigation of Climate Change for the IPCC's Fifth Assessment Report* (2014),[57] although the authors of the latter report implicitly suggest that this definition should only be employed where there is a "detailed understanding of the characteristics of a specific event", so that there is agreement regarding estimates of the likelihood of its occurrence and its resulting consequences.[58]

The probabilistic analysis implied by these IPCC definitions has been the subject of criticism, including by Aven and Renn (2015).[59] In particular, they argued that the IPCC's conception of risk in the IPCC's Fifth Assessment Report as a function of probability and consequences was too narrow and, rather, new approaches to the definition of risk, which clearly distinguish between risk and uncertainty, provide a "stronger and more appropriate basis for climate change analysis".[60] In fact, the distinction drawn by Aven and Renn between risk and uncertainty is not new. Frank Knight drew this distinction in his seminal work undertaken in 1921. According to Knight:[61]

> Uncertainty must be taken in a sense radically distinct from the familiar notion of risk, from which it has never been properly separated. … The

55 IPCC, *Guidance Note for Lead Authors of the IPCC Fifth Assessment Report on Consistent Treatment of Uncertainties. Intergovernmental Panel on Climate Change* (IPCC, 2010), p. 1. The guidance note was prepared to assist lead authors of the IPCC Fifth Assessment Report (AR5) in the consistent treatment of uncertainties.

56 IPCC, "Summary for policymakers", in *Climate Change 2014: Impacts, Adaptation, and Vulnerability. Part A: Global and Sectoral Aspects. Contribution of Working Group II to the Fifth Assessment Report of the Intergovernmental Panel on Climate Change* (Cambridge University Press, 2014), pp. 1–32, at 5.

57 IPCC, "Integrated risk and uncertainty assessment of climate change response policies" in *Climate Change 2014: Mitigation of Climate Change. Contribution of Working Group III to the Fifth Assessment Report of the Intergovernmental Panel on Climate Change* (Cambridge University Press, 2014), p. 155.

58 Ibid. p. 155.

59 T. Aven & O. Renn, "An evaluation of the treatment of risk and uncertainties in the IPCC Reports on Climate Change" (2015) 35(4) *Risk Analysis*, pp. 701–12. As an aside, the authors of this article have both been involved in the development of the IRGC Framework.

60 Ibid. pp. 702, 703.

61 F.H. Knight, *Risk, Uncertainty and Profit* (Hart, 1921), pp. 19–20.

> essential fact is that "risk" means in some cases a quantity susceptible of measurement, while at other times it is something distinctly not of this character; and there are far-reaching and crucial differences in the bearings of the phenomena depending on which of the two is really present and operating. … It will appear that a measurable uncertainty, or "risk" proper, as we shall use the term, is so far different from an un-measurable one that it is not in effect an uncertainty at all.

There has been considerable debate about the meaning and significance of Knight's comments since they were first expressed.[62] The general consensus appears to be that the so-called "Knightian" distinction means that risk is measurable, whereas uncertainty is not.[63] In more practical terms, risk may be said to exist where sufficient information is available to make a probabilistic assessment. In contrast, uncertainty exists where there are indications that harm may occur but there is insufficient data to definitively confirm this using probabilistic methods.[64]

Notably, in its 2020 guidance for the Sixth Assessment Report, the IPCC revised its definition of risk to move away from a probabilistic approach towards one that is based more on uncertainty. In particular, risk has been defined as follows:[65]

> The potential for adverse consequences for human or ecological systems, recognising the diversity of values and objectives associated with such systems. In the context of climate change, risks can arise from potential impacts of climate change as well as human responses to climate change. Relevant adverse consequences include those on lives, livelihoods, health and wellbeing, economic, social and cultural assets and investments, infrastructure, services (including ecosystem services), ecosystems and species.

The IPCC clarifies that:

> [t]he word "potential" makes clear that uncertainty, or more broadly, incomplete knowledge … is a key element of the concept of risk. This uncertainty does not necessarily have to be quantified, but authors need to provide some sense of the nature and degree of uncertainty to allow a meaningful risk assessment and risk management responses to be undertaken.[66]

62 See, for example, J. Runde, "Clarifying Frank Knight's discussion of the meaning of risk and uncertainty" (1998) 22(5) *Cambridge Journal of Economics*, pp. 539–46.

63 J. Runde, ibid. p. 539; S.F. LeRoy & L.D. Singell, "Knight on risk and uncertainty" (1987) 95(2) *Journal of Political Economy*, pp. 394–406, at 394–5.

64 J. Runde, n. 62 above, pp. 539, 540.

65 IPCC, *The concept of risk in the IPCC Sixth Assessment Report: A Summary of Cross-Working Group Discussions. Guidance for IPCC authors* (IPCC, 2020), p. 4.

66 Ibid. p. 5.

The question arises as to what this shift in risk definition by the IPCC from one that is based on probability to another founded on uncertainty is likely to mean in practice.

While Knight's distinction between risk and uncertainty may appear conceptually clear and rational, as noted by O'Malley (2004), in practice, risk and uncertainty "blur, converge and overlap".[67] Surely, the dividing line between risk and uncertainty is not fixed and definitive. Rather, a situation of risk and one characterised by uncertainty are merely descriptions of opposite ends of a spectrum to capture the relative degree of understanding about an adverse event or outcome. Understanding is complete at one end of the spectrum (and, therefore, very little uncertainty exists) whereas there is limited understanding at the other end of the spectrum (with high degrees of uncertainty). When the distinction between risk and uncertainty is cast in these terms, it serves to highlight that the manner in which risk is defined – and, therefore, assessed and managed – will intrinsically be linked to how well it is understood, particularly the degree of uncertainty surrounding risk. So, where does this lead the debate regarding the appropriate definition for the regulation of systemic risks arising from climate change?

Deep uncertainty, which is a key characteristic of systemic risks including those caused by climate change, implies that, rather than being clear and unambiguous, many systemic risks are likely to be nebulous and ill-defined. This observation is important because it confronts the reality that regulating systemic risks is challenging and implies that the boundary of a regime to regulate this type of risk may well be "fuzzy" or, at worst, incapable of specific definition, rather than being characterised by clear, bright lines. This is not heartening news for policy- and law-makers tasked with designing regulatory regimes to address systemic risks, nor for regulators expected to apply those regimes. Nonetheless, it does stress the importance of coming to terms with the distinction between risk and uncertainty when regulating systemic risks and understanding what this distinction is likely to mean in practice.

2 Risk deconstruction

The profound uncertainty associated with many systemic risks is inextricably linked to the complexity of the systems and stressors giving rise to those risks. The systems themselves are typically characterised by tightly coupled, interdependent and dynamic components. The stressors to which a system might be subject are varied and their impacts on the system are likely to be unpredictable.

67 P. O'Malley, *Risk, Uncertainty and Government* (GlassHouse, 2004), p. 18.

Kessler (2008) suggests that "unstructured uncertainty needs to be absorbed and (re)structured in order to be subjected to risk decisions".[68] This observation reflects a recognition that a means to navigate uncertainty, including the uncertainty that characterises systemic risks, is essential to render uncertainty susceptible to management, including in a regulatory context. Systems thinking offers a way forward by effectively "deconstructing risk"[69] – that is, seeking to better understand the system at risk, including its main features and boundaries, as well as the stressors that place the system at risk, by decomposing risk into its constituent parts.

This is the focus of the second proposed Building Block 2: Risk Deconstruction for the development of an ideal to regulate systemic risks, which is highlighted in Figure 7.3. It involves understanding the system at risk (that is, the "risk absorbing system") as well as the process giving rise to risk (that is, the "risk agent"). By deconstructing systemic risks, this building block helps unpack uncertainty associated with these risks and, in turn, assists in determining the appropriate aspect of the risk that should be the focus of regulatory control.

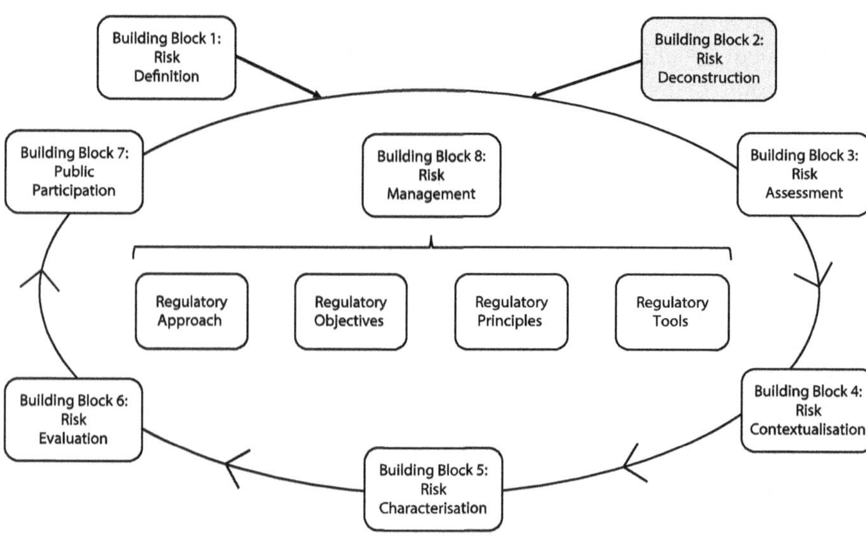

Figure 7.3 Building Block 2: Risk Deconstruction

68 O. Kessler, "Is risk changing the politics of legal argumentation?" (2008) 21(4) *Leiden Journal of International Law*, pp. 863–84, at 868.

69 A common meaning of "deconstruction" is the act of breaking something down into its separate parts in order to better understand that thing, especially when this understanding is different from the way the thing was previously understood.

Mapping complex "risk absorbing systems"

Modern society is characterised by a broad range of "risk absorbing systems", including ecosystems, social systems, financial systems and systems of infrastructure. Such systems are typically complex and are usually dynamic because the constituent parts are constantly changing.[70] A key challenge in determining how risks to these systems should be regulated is understanding the system itself.

In its 2011 report, *Future Global Shocks: Improving Risk Governance*, the OECD suggests that maps of complex systems could provide insights regarding the risks to which such systems are likely to be subject and, by implication, these maps could assist in responding to global shocks in the future:[71] "A major challenge in analysing future global shocks is the need to understand the way in which any system's various components are interrelated. Maps help identify and quantify the relationship of a system's components."

Maps of complex systems of the kind referred to by the OECD are not new. Indeed, they are not uncommon in areas where the application of systems theory has some history – such as, to design, better understand or manage information systems,[72] biological systems,[73] ecosystems[74] and geographical systems.[75] The OECD suggests that maps of complex systems may take the form of physical maps that capture spatial relationships between elements of the system, conceptual maps that reflect the system's intangible components, or process maps that provide insights about the structure or order of the system.[76] Nonetheless, the OECD concedes that mapping of complex systems is likely to be challenging:[77]

> Maps of complex systems are often hard to come by for three principle [sic.] reasons. First, the complexity of the systems makes mapping difficult. This is sometimes due to the sheer number of components and interconnections, but can also result from a lack of knowledge about key components and connections. Second, understanding a complex system

70 See Chapter 2 where the distinctive features of systemic risks are discussed.

71 OECD, *Future Global Shocks: Improving Risk Governance* (OECD, 2011), p. 58.

72 See, for example, S. Natarajan, G.M. Krishna, C. Subhrojyoti, P. Patwari, & A. Banerjee, "System maps: Integrating knowledge system models", in *Proceedings of the 9th India Software Engineering Conference* (ACM, 2016), pp. 146–50.

73 See, for example, E. Bullmore & O. Sporns, "Complex brain networks: Graph theoretical analysis of structural and functional systems" (2009) 10(3) *Nature Reviews Neuroscience*, pp. 186–98.

74 See, for example, F. Santos-Martin, A. Viinikka, L. Mononen, L. Brander, P. Vihervaara, I. Liekens, & M. Potschin-Young, "Creating an operational database for Ecosystems Services Mapping and Assessment Methods" (2018) 3 *One Ecosystem*, e26719.

75 See, for example, P.J. Phipps, "Terrain systems mapping" (2001) 18(1) *Geological Society, London, Engineering Geology Special Publications*, pp. 59–61.

76 OECD, n. 71 above, p. 58.

77 Ibid. p. 58.

> requires detailed knowledge of its component's functions and their interrelations. As each component on its own could be complex, and understanding the various components could invariably require specialised skills, developing a map often requires a large interdisciplinary effort. Finally, mapping complex systems requires a sustained effort since they evolve over time. Maps need to be revisited and revised to provide an accurate image of the system.

It would be fair to say that systems mapping is unusual in a regulatory context. One possible explanation is that legal scholarship has traditionally focused on the logic of legal cause and effect and this has led to a detailed "nano-scale" focus on individual elements of a system when designing regulatory regimes, rather than a high-level overview and understanding of entire systems, particularly when complex systems are involved. Arguably, this is evident in the traditional approach to the regulation of risk described earlier in this book.[78]

At a minimum, the design of regulatory regimes that seek to address systemic risks, like those arising from climate change, should make reference to maps of the system at risk, noting that the system and the risks to which it could be subject are likely to change over time. This exercise will help contextualise, in very practical terms, how risk is likely to affect the system and, in turn, provide useful insights about where regulatory intervention might be most effective. This is no easy task, however, because a crucial balance needs to be achieved between on the one hand reducing complexity so that one can comprehend the structure and ordering of the system, and on the other hand, ensuring that enough detail is at hand so that potential systemic weaknesses and vulnerabilities can be identified. Failure to achieve this balance could, in turn, compromise the capacity of the regulatory response to effectively address the systemic implications of risk – at worst, this could be collapse of the system.

Decomposition of the process giving rise to the "risk agent"

In addition to understanding the "risk absorbing system", gaining insights into the "risk agent" – particularly, the process giving rise to the risk – is also important. Breaking down or dissecting the risk process can clarify the sources of, context for, and possible consequences associated with the risk agent. In turn, this can assist policy-makers to determine the stage or stages in the risk process where regulatory intervention is likely to be most effective and the type of intervention that might best address risks at each stage.

The specific process that leads to the creation or exacerbation of risk is likely to be different for each type of risk. In some cases, the process will be relatively easy to identify and understand – such as the combustion of materials or discharge of effluent from an industrial facility leading to localised smog,

78 See Chapter 5 for an explanation of a traditional approach to the regulation of risk.

degradation of local soil quality, and depletion of fish stock in a neighbouring river. However, in other cases, the process will be less clear, such as cases where high levels of harmful radiation are detected in a building when there is no obvious source. Despite these differences, it is possible, to conceptualise the process giving rise to risk that has been triggered by human activities in a high-level, standardised way, as illustrated in Figure 7.4. This conceptualisation is helpful to understand and articulate the broad options that may be available to mitigate risk.

Four key stages are depicted in the typical risk process triggered by human activities – specifically, in Stage 1, human activities occur. In Stage 2, the human activities result in a loading or initiator – that is, some force or pressure that could precipitate adverse consequences. During Stage 3, the loading or initiator results in exposure or vulnerability. In Stage 4, effects or consequences are produced from the exposure or vulnerability.

Illustration of a risk process in this way is highly simplified and, by no means, revolutionary. Nevertheless, it helps crystallise and take stock of the main

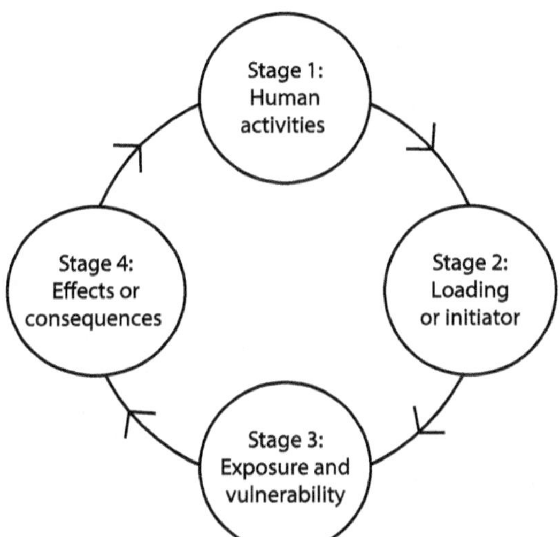

Figure 7.4 Illustration of risk process triggered by human activities[79]

79 The author's conceptualisation of a risk process triggered by human activity in Figure 7.4 is derived from modelling of risk processes undertaken by Morgan et al. (2000) for cases where regulatory intervention has been employed to protect health, safety or the environment: M.G. Morgan, H.K. Florig, M.L. DeKay, & P. Fischbeck, "Categorizing risks for risk ranking" (2000) 20(1) *Risk Analysis*, pp. 49–58, at 52. The authors of that article used this process of risk modelling in the context of air pollution, but explain that it can be utilised in relation to any risk domain.

choices faced by decision-makers regarding the need for and function of regulatory intervention to respond to risk – in the case of climate change risks, the main choice is whether an emphasis should be placed on anticipation and/or on reaction.[80] An anticipatory approach involves identification of risks and efforts to prevent the occurrence of adverse effects or to reduce their impact by focusing on the cause of the risk. This approach implies that attention will be focused on the early stages of the risk process.[81] In contrast, a reactive approach is aimed at dealing with the consequences once the risks have materialised, which implies that attention will be focused on the later stages of the risk process.[82]

In fact, regulatory intervention has a role to play at all stages of the climate change risk process, although the form and function of regulation will differ. Moreover, the relative effectiveness and directness of each form of regulatory intervention in addressing the risks arising from climate change are likely to differ. In particular, regulatory intervention is likely to be most effective and direct during the early stages of the climate process because, conceivably, such intervention could prevent risk from materialising, thereby obviating or, at least, reducing the need to focus on addressing exposure or vulnerability or the effects or consequences during later stages. However, in reality, the options for intervention that are available to policy- and law-makers at each stage of the risk process will be limited by a broad range of factors, including politics, resources and societal perception of risk.

Interaction between the risk-absorbing system and the risk agent

Breaking down the risk-absorbing system and the risk agent goes some way to enhancing the understanding of a system and its stressors. However, regulatory intervention to address systemic risks is best guided by an understanding of how a system and relevant risk agents are likely to interact in practice. Given the dynamic nature of systems and the broad range of stressors to which they may be subject, a static analysis that is based on understanding at a particular point in time will probably be insufficient to address future risk. Regulatory responses may need to be modified, augmented or abandoned as a system and stressors mutate with time.

This is easier said than done. In fact, the greater the degree of uncertainty, complexity and ambiguity associated with the risks to which a system may be

80 Baldwin and others characterise the choice differently as one between "anticipation" and "resilience": R. Baldwin, M. Cave, & M. Lodge, *Understanding Regulation: Theory, Strategy, and Practice* (Oxford University Press, 2012), p. 94.

81 An anticipatory approach would appear to be particularly relevant during Stage 1 – "human activities", Stage 2 – "loading/initiator" or Stage 3 – "exposure/vulnerability".

82 A reactive approach would be particularly relevant during Stage 4 – "effects or consequences".

subject, the more difficult it will likely be to determine how to respond. Nonetheless, in designing regimes for the regulation of systemic risks, it will be important to take stock of this difficulty and, to the extent possible, provide a mechanism within the regime for it to be addressed. Ideally, the regime would embed sufficient flexibility so that the changes in the likely interaction between the relevant system and associated stressors are accounted for on a continuous and dynamic basis.

The building blocks considered so far provide some insights regarding the system at risk, the risks to which it is likely to be subject, and areas of particular weakness and vulnerability for the system. They, therefore, enable – even if somewhat tentatively – the boundary of the regulatory regime to be drawn in order to reflect the system and to capture those risks which may pose the greatest risk to the system. The context for the regime, including the manner in which the risks that fall within the boundary are assessed, are considered in the next part.

III Context for an ideal regime to regulate systemic risks

Understanding the context for systemic risks is a critically important component of a regulatory regime. Most regulatory regimes embed mechanisms to provide some context for application of the regime, such as mechanisms to obtain information and make assessments about entities or activities covered by the regime.[83] These mechanisms can play a role in informing whether and how the regime should be applied in practice. Building blocks to provide context for an ideal regime to regulate systemic risks – specifically, risk assessment, risk contextualisation, risk characterisation, risk evaluation and public participation – are considered in turn below.

1 Risk assessment

The assessment of risk is the foundation for any response to risk, whether in a regulatory or non-regulatory context. In essence, risk assessment involves a determination of the relative importance, scale or seriousness of a risk which, in turn, assists in making decisions about whether a response is needed and, if so, what the response should be. The question of how risk should be assessed is related to but distinct from the debate about how risk should be defined, which was discussed earlier.[84] The latter essentially relates to the frame of reference for the assessment of risk, whereas the former relates to how risk is actually measured or quantified within that frame of reference.

83 See, for example EU Directive 2001/18 (amended by Directive 2018/35 in March 2018) which requires an environmental risk assessment (ERA) of Genetically Modified Organisms (GMOs) that are intended to be released into the environment.

84 See the discussion of Building Block 1: Risk Definition earlier in this chapter.

Notably, there is no single approach for the assessment of risk, but important implications may flow from the particular approach adopted, such as the nature of the assessment and the person or entity that is best placed to undertake the assessment. As noted by Baldwin, Cave and Lodge (2012):[85]

> [A] host of different approaches to the definition and measurement of risks can be taken. These very differences raise issues about regulatory responses to risks and the ways in which risk regulation regimes can be justified or legitimated. Thus, high confidence in technical approaches to risk might be expected to lead to an emphasis on leaving risk regulation to experts .... In contrast, a strong belief that risks are socially constructed might be taken to suggest that regulatory priorities and policies cannot be left to the "objective" evaluation of experts, but have to emerge from democratically legitimate processes of debate and consultation.

The third of the proposed building blocks of an ideal to regulate systemic risks – Building Block 3: Risk Assessment – deals with how systemic risks should be assessed and is highlighted in Figure 7.5. It forms a critical component of the context for the regulatory regime.

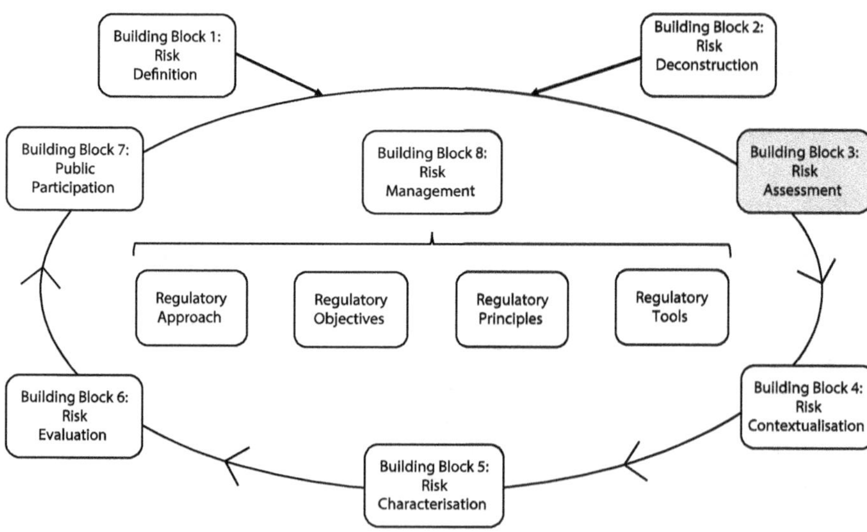

Figure 7.5 Building Block 3: Risk Assessment

85 R. Baldwin, M. Cave, & M. Lodge, n. 80 above, pp. 92–3.

This section involves an exploration of different approaches to the assessment of risk, including a consideration of the advantages and disadvantages that are likely to be particularly relevant for the regulation of systemic risks caused by climate change.

The battle between objectivity and subjectivity

A number of distinct approaches to the conception of risk for the purposes of risk assessment exist, largely grounded in the academic disciplines from which the respective approaches have originated.[86] Two conceptual models are dominant,[87] which differ not only in the definitions of risk they employ, but also in terms of their philosophical underpinnings.[88] In essence, the dichotomous models reflect a conflict between expert and lay conceptions of risk. Slovic, Fischhoff and Lichtenstein (1979) explain that the public ranks risks differently compared to scientific experts.[89] Vassie and colleagues (2005) argue that this is linked to divergences between the parameters considered by these respective groups when evaluating risk.[90] In general terms, the expert model adopts an objective, probabilistic approach to the assessment of risk, whereas the lay model is based upon social perceptions and involves a subjective approach. The two conceptions, which are generally presented as diametric, incompatible opposites, represent extreme ends of a continuum,[91] and are discussed in more detail below.

OBJECTIVE CONCEPTION OF RISK

The underlying claim of the objective conceptual model[92] is that the assessment of risk is based on fact and is value-neutral. More specifically, risk is determined by quantifying probability of an adverse event or outcome using data and statistics from past events and by employing "formulas, estimates, correlations, and predictions".[93]

86 O. Renn, "Concepts of risk: A classification" in Krimsky, S. and Golding, D. (eds.) *Social Theories of Risk* (Praeger, 1992), p. 56.
87 A. Ben-Ari & K. Or-Chen, "Integrating competing conceptions of risk: A call for future direction of research" (2009) 12(6) *Journal of Risk Research*, pp. 865–77, at 872.
88 Ibid. p. 867.
89 P. Slovic, B. Fischhoff, & S. Lichtenstein, "Rating the risks" (1979) 21(3) *Environment: Science and Policy for Sustainable Development*, pp. 14–39.
90 L. Vassie, P. Slovic, B. Fischhoff, & S. Lichtenstein, "Facts and fears: Understanding perceived risk" (2005) 3(Supplement 1) *Policy and Practice in Health and Safety*, pp. 65–102.
91 A. Ben-Ari & K. Or-Chen, n. 87 above, p. 872; S. Hornig, "Reading risk: Public response to print media accounts of technological risk" (1993) 2(2) *Public Understanding of Science*, pp. 95–109, at 95–6.
92 This model has been variously referred to as the objective, probabilistic or positivist approach to risk assessment.
93 A. Ben-Ari & K. Or-Chen, n. 87 above, p. 869; R. Baldwin, M. Cave, & M. Lodge, n. 80 above, p. 88.

The assumption underlying this approach is that decisions regarding responses to risks can be made "on the basis of objective evidence that can be treated quantitatively to produce a numerical result".[94] The methodology also claims to "assure equal treatment for all risks under consideration",[95] because the same quantitative approach is applied to assess all risks. In the context of this conceptual model, values and normative judgements are excluded from consideration[96] because risk is objectively determined, independently of subjective perception or evaluation.[97]

Such an approach can work well in areas such as food safety, where extensive and reliable historical data exists.[98] However, Aven (2009) comments that reliance on data alone to assess risk can lead to "poor predictions" if the past is unlikely to be representative of the future.[99] In addition, sample sizes may be too small, extrapolations may be ill-founded, and valuable sources of information may be excluded because they are not regarded as scientifically valid.[100]

The objective conception of risk has also been criticised by Ben-Ari and Or-Chen (2009) on the basis that expert perceptions of risk "do not reflect dimensions of risk which appear to be significant for the public".[101] More specifically, the expert, objective approach is blind to the social, political and cultural contexts in which risks arise.[102] Moreover, Black (2010) suggests that, in making their assessments, "experts are just as prey to cognitive biases" as lay people.[103] Black also considers that many government bodies are gradually recognising that basing decisions on scientific evidence alone may be politically unpalatable, even if expanding risk assessment beyond a technical, objective approach is "not accepted as being normatively desirable".[104]

94 R. Baldwin, M. Cave, & M. Lodge, n. 80 above, p. 88.
95 O. Renn, "Three decades of risk research: Accomplishments and new challenges", (1998) 1(1) *Journal of Risk Research*, pp. 49–71, at 64 referring to M. Merkhofer, "Comparative analysis of formal decision-making approaches" in V.T. Covello, J. Menkes and J. Mumpower (eds.), *Risk Evaluation and Management* (Plenum, M. 1984), pp. 183–220.
96 E.A. Rosa, n. 47 above, pp. 2–1.
97 P.B. Thompson & W. Dean, "Competing conceptions of risk" (1996) 7(4) *Risk: Health, Safety & Environment*, p. 361.
98 For example, Food Standards Australia New Zealand (FSANZ) is a statutory authority that develops food standards that are based on the "best available scientific evidence". These standards are reflected in the Australian New Zealand Food Standards Code, which is referred to in a range of legislative instruments that apply food standards – for example, to control the supply and sale of foods that pose a risk to human health. See the FSANZ website at: https://www.foodstandards.gov.au/Pages/default.aspx.
99 T. Aven, n. 47 above, pp. 64–5.
100 J. Black, "The role of risk in regulatory processes", in R. Baldwin, M. Cave, & M. Lodge (eds.), *Oxford Handbook of Regulation* (Oxford University Press, 2010), pp. 301–48, at 315.
101 A. Ben-Ari & K. Or-Chen, n. 87 above, p. 867.
102 E.A. Rosa, n. 47 above, p. 21.
103 J. Black, n. 100 above, p. 315.
104 Ibid. p. 317.

SUBJECTIVE CONCEPTION OF RISK

The emphasis in the alternative model[105] is the subjective reaction of people to risk, as well as the context within which the phenomenon of risk is situated. This approach is based on social science research and assumes that the notion of risk is culturally embedded and, therefore, its meaning varies from one social group to another. Accordingly, assessment of risk is necessarily a social, cultural and political exercise,[106] which involves studying particular social contexts to identify and assess risk in that context.[107] Under this model, risk is equivalent to the way risk is perceived, rather than how it is quantified.

It has been suggested by Ben-Ari and Or-Chen (2009) that a subjective approach to the assessment of risk allows a qualitative and complex evaluation, including consideration of uncertainty, catastrophic potential, equity and danger to future generations,[108] which may not be practically possible if a traditional, objective, quantitative approach is employed. Moreover, it has been suggested that perceptions of risk, which are at the heart of the subjective conceptual approach for the assessment of risk, are "formed by common sense reasoning, personal experience, social communication and cultural traditions".[109] Moreover, people do not use completely irrational strategies to assess information, but will generally "follow relatively consistent patterns of creating images of risks and evaluating them", which are related to "evolutionary bases of coping with dangerous situations".[110] Morgan (1993) goes further, arguing that lay assessment of risk may be more rational, and justifiable, than an expert one:[111]

> Laypeople have different broader definitions of risk, which in important respects can be more rational than the narrow ones used by experts. Furthermore, risk management is, fundamentally, a question of values. In a democratic society, there is no acceptable way to make these choices without involving the citizens who will be affected by them.

A broad range of factors have been said to affect perception of risks, including catastrophic potential, degree of personal control over the size or

105 This model has been variously referred to as the subjective, contextualist or constructivist approach to risk assessment.
106 E.A. Rosa, n. 47 above, p. 28; S. Jasanoff, "The songlines of risk" (1999) 8(2) *Environmental Values*, pp. 135–52, at 150.
107 E.A. Rosa, n. 47 above, p. 21.
108 A. Ben-Ari & K. Or-Chen, n. 87 above, p. 867.
109 IRGC, n. 9 above, p. 31. referring to, amongst others, N.F. Pidgeon, C.C. Hood, D.K.C. Jones, B.A. Turner, and R. Gibson, "Risk perception" in Royal Society Study Group (eds.), *Risk Analysis, Perception and Management* (The Royal Society, 1992), pp. 89–134.
110 IRGC, ibid. p. 31.
111 M.G. Morgan, "Risk analysis and management" (1993) 269(1) *Scientific American*, pp. 32–41, at 32.

probability of the risk, and familiarity with the risk.[112] In essence, risk perception is based on the availability of information regarding the risk, how that information is communicated, earlier experience of the risk within the cultural context, and the psychological mechanisms for processing uncertainty.[113]

Rosa (1998) suggests that a possible deficiency associated with the subjective conceptual model is that it could result in bias.[114] Research has shown that, where empirical data are missing, there are persistent biases in people's risk perceptions.[115] Similarly, when there is a series of extreme events, public preferences are likely to shift towards increased protection, even if there is a lack of clear scientific evidence to support such an approach.[116]

Klinke and Renn (2002) comment that, if risk assessments are nothing but social constructions, they lack normative validity for guiding regulatory action.[117] Indeed, basing risk assessments and regulatory responses on risk preferences of public interest groups and citizens at large could lead to irrational results. In this regard, Brandt (2014) suggests that there is even a risk that "special interests attempt to exploit extreme events to progress their particular agenda, which can have either positive or negative feedback for the chosen policy after an extreme event".[118]

A hybrid approach in the climate change context

It is difficult to see how an objective, quantitative approach could be implemented for climate change risks given the significant degree of uncertainty associated with many climate change hazards and associated impacts. While certain climate change outcomes are relatively predictable, and reliable data or models for such outcomes exist,[119] such as increasing temperatures,[120] there are

112 R. Baldwin, M. Cave, & M. Lodge, n. 80 above, pp. 89–90.

113 O. Renn, "Perception of risks" (2004) 29(1) *The Geneva Papers on Risk and Insurance – Issues and Practice*, pp. 102–14, at 104.

114 E.A. Rosa, n. 47 above, pp. 23–4.

115 D. Etkin & E. Ho, "Climate change: Perceptions and discourses of risk" (2007) 10 (5) *Journal of Risk Research*, pp. 623–41, at 628.

116 U.S. Brandt, "The implication of extreme events on policy responses" (2014) 17 (2) *Journal of Risk Research*, pp. 221–40, at 230.

117 A. Klinke & O. Renn, "A new approach to risk evaluation and management: Risk-based, precaution-based, and discourse-based strategies" (2002) 22(6) *Risk Analysis*, pp. 1071–94, at 1073.

118 U.S. Brandt, n. 116 above, p. 236.

119 In the context of its Fifth Assessment Report, the IPCC noted that the "high or very high confidence" findings in IPCC reports indicate that "uncertainties can be quantified probabilistically": IPCC, "Technical summary", in *Climate Change 2013: The Physical Science Basis. Contribution of Working Group I to the Fifth Assessment Report of the Intergovernmental Panel on Climate Change*, (Cambridge University Press, 2013), p. 36.

120 The IPCC stated that there is a very high confidence that climate models to simulate surface temperature reproduce observed large-scale surface temperature patterns: ibid. p. 75.

many others that are far less foreseeable because of a lack of data or appropriate models that can address the complexity of the climatic, human and ecological systems involved, such as global trends in lack of rainfall and drought.[121] In fact, the IPCC (2014a) has effectively acknowledged that the occurrence of many climate change hazards cannot reasonably be reduced to a well-defined probability.[122] Applying the Knightian distinction between risk and uncertainty, it is arguable that the main challenge associated with addressing the consequences of climate change is the fact that we are typically dealing with uncertainty, rather than risk.

If an objective, positivist approach to the conception of risk is inappropriate for at least some aspects of climate change because the risks are inherently unknowable, the question then arises as to whether, and how, a subjective, contextualist approach should be employed instead. This would involve a lay, subjective assessment of risk based upon the context in which those risks arise. Unfortunately, climate change is a particularly challenging subject for the public to grasp for a variety of reasons. These include complexities in applying a systems-based approach,[123] difficulties in thinking in probabilistic terms,[124] a lack of relevant and compelling narratives within which the debate about climate change risks can be placed,[125] and frustration, apathy and denial as a result of the perceived lack of scientific and political consensus.[126] Moreover, the level of uncertainty associated with many climate change impacts is so high that this could lead to a broad array of subjective perceptions of risk, some of which may be contradictory and irreconcilable when responses to risk are being devised.[127]

121 For example, the IPCC found that there is a low confidence in a global-scale observed trend in drought or dryness (lack of rainfall), owing to, among other things, lack of direct observations and geographical inconsistencies in trends: ibid. p. 50.

122 The IPCC has noted the "large number of key uncertainties with respect to the climate system" and "even greater uncertainties with respect to the impacts of changes in the climate system on humans and the ecological system as well as their costs to society". IPCC, "Integrated Risk and Uncertainty Assessment of Climate Change Response Policies" in *Climate Change 2014: Mitigation of Climate Change. Contribution of Working Group III to the Fifth Assessment Report of the Intergovernmental Panel on Climate Change* (Cambridge University Press, 2014), p. 158.

123 J.D. Sterman & L.B. Sweeney, "Cloudy skies: Assessing public understanding of global warming" (2002) 18(2) *System Dynamics Review*, pp. 207–40, at 208.

124 A.G. Patt & D.P. Schrag, "Using specific language to describe risk and probability" (2003) 61 *Climatic Change*, pp. 17–30, at 17.

125 K. McComas & J. Shanahan, "Telling stories about global climate change: Measuring the impact of narratives on issue cycles" (1999) 26(1) *Communication Research*, pp. 30–57. The authors state at p. 54 that "further attention to the narrative and dramatic mechanics of media coverage of environmental issues can result in a better understanding of the results for public awareness, knowledge and action".

126 S. Seacrest, R. Kuzelka, & R. Leonard, "Global climate change and public perception: The challenge of translation" (2000) 36(2) *Journal of the American Water Resources Association*, pp. 253–63, at 253–4.

127 D. Etkin & E. Ho, n. 115 above, p. 627.

Leiserowitz (2005, 2006), in a US national study examining risk perceptions associated with global warming, found that although most Americans displayed a high awareness, belief and concern about climate change, they perceived it as a "moderate risk" that will predominantly impact "people and places that are geographically and temporally distant".[128] Taylor and colleagues (2014) noted that, in the UK, the way residents assess and perceive climate change risks often diverges from the approach of experts and residents typically conflate climate change with other environmental problems.[129] Prokopy and colleagues (2015) considered risk perceptions of farmers in high-income countries, including the UK, the US, Australia and New Zealand, and found that a majority of farmers in each of these jurisdictions did not believe that climate change is a threat to local agriculture, even though they believed that climate change is occurring.[130]

The foregoing considerations regarding the implications of the objective and subjective approaches to the assessment of risk in the climate change context tend to indicate that a hybrid approach is appropriate. Support for this view is provided by Ben-Ari and Or-Chen (2009), who suggest that the two competing conceptual models for risk assessment are "too reductionist".[131] Likewise, Thompson and Dean (1996) argue that the objective and subjective approaches represent opposite ends of a continuum for conceptualising risk and practical conceptions of risk are likely to "fall somewhere along the axis between the extremes".[132]

Indeed, growing recognition of the respective shortcomings of the two competing approaches leads inexorably to the view that an integrated approach is needed – an approach that incorporates objective as well as subjective components and offers a "mutual enhancement between technical risk assessment and intuitive risk perception".[133] Such a hybrid approach recognises the "dual nature of risk" – that is, risk is both fact- and value-laden, and contains both objective and subjective components.[134]

128 A.A. Leiserowitz, "American risk perceptions: Is climate change dangerous?" (2005) 25(6) *Risk Analysis*, pp. 1433–42, at 1438; A. Leiserowitz, "Climate change risk perception and policy preferences: The role of affect, imagery, and values" (2006) 77 *Climatic Change*, pp. 45–72, at 62.

129 A.L. Taylor, S. Dessai, & W. Bruine de Bruin, "Public perception of climate risk and adaptation in the UK: A review of the literature" (2014) 4–5 *Climate Risk Management*, pp. 1–16, at 11–12.

130 L.S. Prokopy, J.G. Arbuckle, A.P. Barnes, V.R. Haden, A. Hogan, M.T. Niles, & J. Tyndall, "Farmers and climate change: A cross-national comparison of beliefs and risk perceptions in high-income countries" (2015) 56(2) *Environmental Management*, pp. 492–504, at 492, 501.

131 A. Ben-Ari & K. Or-Chen, n. 87 above, p. 872 expressing agreement with E.A. Rosa, n. 47 above.

132 P.B. Thompson & W. Dean, n. 97 above, pp. 383–4.

133 O. Renn, n 95 above, pp. 65–6; O. Renn, n. 113 above, p. 111.

134 S.O. Hansson, "Risk: objective or subjective, facts or values" (2010) 13(2) *Journal of Risk Research*, pp. 231–8, at 237.

This duality is recognised in the context of the "risk appraisal" element of the IRGC Framework by distinguishing between a "technical risk assessment" and a "concern assessment". It is also reflected in the IPCC's Fifth Assessment Report regarding climate change adaptation. In particular, in considering the "Foundations of Decision-making" for adaptation in the *Report on Impacts, Adaptation and Vulnerability for the IPCC's Fifth Assessment Report* (2014), the IPCC commented:[135]

> In complex situations, sociocultural and cognitive-behavioral contexts become central to decision making. This requires combining the scientific understanding of risk with how risks are framed and perceived by individuals, organizations, and institutions ... For that reason, formal risk assessment is moving from a largely technocratic exercise carried out by experts to a more participatory process of decision support, ... although this process is proceeding slowly.

The IPCC report includes the rationale for this integrated, hybrid approach. In particular, the IPCC explains that "[a]cceptance of the science behind controversial risks is strongly influenced by social and cultural values and beliefs". Moreover, the IPCC notes that:

> [s]cience is most suited to calculating risk in areas where it has predictive skill and will provide better estimates than may be obtained through more informal methods, ... but an assessment of what is at risk generally needs to be accepted by stakeholders. Therefore, the science always sits within a broader social setting.[136]

So, the IPCC's position appears to be that a subjective, contextualist approach should be integrated with one that is objective and quantitative when assessing climate change risks.

Recognising the duality of systemic risks in this way helps ensure a comprehensive assessment of risk, rather than one that is unidimensional and likely to be rejected by one stakeholder group or another. In terms of how an objective and subjective approach can practically be combined in the context of adaptation to climate change, a 2004 report prepared by the United Nations Development Programme (UNDP) on "Adaptation Policy Frameworks for Climate Change" noted that both quantitative and qualitative methods could be used to assess risk, depending on the quality of information needed by

135 IPCC, "Foundations for decision-making" in *Climate Change 2014: Impacts, Adaptation, and Vulnerability. Part A: Global and Sectoral Aspects. Contribution of Working Group II to the Fifth Assessment Report of the Intergovernmental Panel on Climate Change* (Cambridge University Press, 2014), p. 201.

136 Ibid. p. 202.

stakeholders and the data and knowledge available to provide that information.[137] This implicitly suggests that the relative importance of objective and subjective analyses when assessing the risks associated with climate change is likely to be linked to the relative degree of uncertainty associated with those risks. In practice, objective, quantitative assessments could be undertaken where possible, but these would need to be supplemented by subjective, qualitative assessments that are "anchored in political, cultural and social contexts",[138] particularly to help address deep uncertainty when it exists.

Traditionally, regulation – its design, implementation and application – has been the primary domain of government bureaucrats with lay stakeholders comparatively playing a limited role.[139] Reflecting a dualist approach to the assessment of risk in a regulatory regime necessarily implies taking regulation into a multidisciplinary domain, where law crosses paths with science, economics, psychology and sociology. This is a new frontier that must be advanced if systemic risks, like those associated with climate change, are to be regulated effectively.

2 Risk contextualisation

In common discourse, risk is often referred to in a generic manner as if it were a homogeneous phenomenon, regardless of the specific context in which a risk has arisen or is likely to materialise. For example, it is often stated that it is riskier to drive a car than fly in an aeroplane. Treating risk in this way has its advantages. It facilitates comparison of risks, particularly comparison between well-known risks and those that are less well-known, and allows intuitive conclusions to be drawn about the relative magnitude of various risks, even though they might actually be quite disparate. However, such an approach fails to account for the fact that each risk is actually a product of the particular context in which it arises.

Indeed, in the same way that a regulatory regime requires context to inform its content,[140] context is also necessary at a more granular level in relation to the assessment of risk that is governed by the regime. The context helps to convert risk from an abstract concept into something that is grounded in a realistic and practical setting. With all risks – but, particularly with systemic risks – the need to comprehensively understand the context for the assessment of risks is critical to ensure a deep understanding of the risk itself and this, in

137 R. Jones and R. Boer, "Technical Paper 4: Assessing Current Climate Risks" in B. Lim, E. Spanger-Siegfried, (eds.) and I. Burton, E. Malone, S. Huq (co-authors), *Adaptation Policy Frameworks for Climate Change: Developing Strategies, Policies and Measures* (UNDP, Cambridge University Press, Cambridge, 2004), p. 96.

138 A. Ben-Ari & K. Or-Chen, n. 87 above, p. 873.

139 The issue of "public participation" in regulation is discussed later in this chapter.

140 See the discussion of the architecture of a regulatory regime, including its content and context, in Chapter 4.

turn, can enhance effectiveness of the regulatory response. Accordingly, the fourth proposed building block of an ideal to regulate systemic risks is Building Block 4: Risk Contextualisation and is illustrated in Figure 7.6. This building block seeks to ensure that the context for the assessment of systemic risk is appropriately accounted for.

The importance of contextualising the assessment of risk

While driving a car and flying in an aeroplane both involve transportation between two locations, the risks associated with each mode of transportation are very different. These differences are associated with the distinctive contexts for each mode. In particular, the risks associated with driving a car will be linked to a broad range of factors, but the speed at which the driver and other proximate motorists are travelling is likely to be one of the most significant contextual factors. There will also be a diverse range of factors linked to the risks associated with flying in an aeroplane, although the mechanical integrity of the aeroplane and prevailing weather conditions are likely to be particularly important risk factors. These factors, which form the context for the assessment of risks associated with driving a car and flying in an aeroplane respectively, will play an essential role in determining how risk can be effectively mitigated. Specifically, they assist in identifying targets to reduce risk.

In general terms, the context for risk assessment comprises the main drivers of risk – namely, the hazard giving rise to risk, as well as the exposure and vulnerability of those people or things that might be at risk. The hazard is the

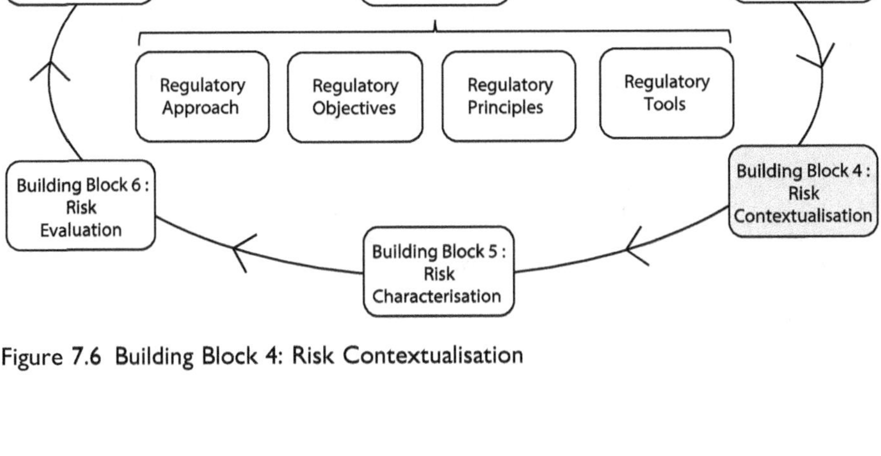

Figure 7.6 Building Block 4: Risk Contextualisation

"risk agent", exposure refers to the contact of the risk agent with the "risk-absorbing system", and vulnerability describes the degree of susceptibility of the risk-absorbing system to harm or damage that could arise from the risk agent. Accounting for these risk drivers helps to ensure that the future is anticipated on a specific and contextualised basis, rather than one that is generic and detached from the practical, on-the-ground reality.

Contextualising assessment of climate change risks

The IPCC has stated that adaptation to climate change is "place and context-specific" and that there is no single, one-size-fits all approach for reducing risks that is relevant in all settings.[141] This follows from the wide variation locally, regionally and globally in the nature, location, frequency and severity of climate hazards. Associated variation exists in relation to the consequences of these hazards for individuals, businesses, communities and society at large. To appropriately account for this variation, the assessment of climate change risk needs to be tailored to the relevant context in which it arises, which will help ensure that responses to risk are fit-for-purpose and capable of making a tangible and positive difference in the particular situation in which those responses will be applied.

In the domain of climate change, there has traditionally been a focus on natural hazards in assessing risk, to the exclusion of the other risk drivers – namely, vulnerability and exposure.[142] However, more recently, the IPCC (2012) has explicitly recognised that hazards are not the sole driver of climate change risk and that the relative levels of adverse consequences arising from climate hazards are determined in large part by the vulnerability and exposure of societies and socio-ecological systems.[143]

According to the IPCC (2014b), the risks posed by climate change result from the interaction of climate-related hazards with the vulnerability and exposure of human and natural systems.[144] Moreover, unlike hazards, which are a product of complex climatic processes that are typically difficult to control through human intervention, vulnerability and exposure could be more easily influenced by adaptive actions.[145] Consequently, these drivers of climate change risks are potentially better targets for regulatory intervention.

Gaining a practical understanding of the context for the assessment of climate change risks is challenging because the elements comprising that context – particularly, vulnerability and exposure – are complex and ambiguous concepts

141 IPCC, n. 56 above, p. 25.
142 IPCC, "Determinants of risk: Exposure and vulnerability" in IPCC, *Managing the Risks of Extreme Events and Disasters to Advance Climate Change Adaption: Special Report of the Intergovernmental Panel on Climate Change* (Cambridge University Press, 2012), p. 72.
143 Ibid. p. 69.
144 IPCC, n. 56 above, p. 7.
145 IPCC, n. 142 above, p. 72.

in themselves. They are uncertain,[146] dynamic, and vary across temporal and spatial scales.[147] In addition, individuals and communities will be differentially exposed and vulnerable based on an array of factors, such as wealth, education, race, ethnicity, religion, gender, age, class, caste, disability and health status.[148] Moreover, the relative importance of hazard, vulnerability and exposure in explaining and assessing risk may vary depending upon the particular event producing the risk and how the various contextual factors interact.[149]

The challenges associated with defining the context for the assessment of climate change risks, including addressing the ambiguity and dynamic nature of each of the contextual factors and the likely complex interaction between them, raise the question of whether it is feasible to contextualise these risks in a regulatory setting. As it turns out, vulnerability assessments are not uncommon in regulatory regimes. For example, a "risk and vulnerability assessment" is a pre-condition for development under the Norwegian land-use planning regime.[150] Such assessments are used to identify and assess the range of factors that may lead to greater exposure and vulnerability, which may warrant further stakeholder engagement and the implementation of risk mitigation mechanisms under the relevant regulatory regime.

The existence and extent of vulnerability is specific to the system that may be subject to a particular hazard and a system can be vulnerable to certain disturbances, but not to others.[151] The system-specific nature of vulnerability leads to a number of observations about vulnerability for the purposes of regulating systemic risks. First, vulnerability is not static but, rather, is likely to be constantly changing in line with changes to the system.[152] This implies that any assessment of vulnerability should be regularly reviewed to ensure that changes to the system are accurately captured in the assessment. Second, critical levels of vulnerability will differ depending upon the particular system at risk.[153] Establishing vulnerability thresholds, above which action might need to be taken to

146 The IPCC notes that "[u]ncertainties about future vulnerability, exposure, and responses of interlinked human and natural systems are large". IPCC, n. 56 above, p. 11.

147 IPCC, n. 142 above, pp. 69–72.

148 Ibid. pp. 76–89.

149 Ibid. FAQ 1.1, p. 33.

150 E. Junker, "Legal requirements for risk and vulnerability assessments in Norwegian land-use planning" (2015) 20(4) *Local Environment*, pp. 474–88.

151 G.C. Gallopín, "Linkages between vulnerability, resilience, and adaptive capacity" (2006) 16(3) *Global Environmental Change*, pp. 293–303, at 294.

152 R.W. Scholz, Y.B. Blumer, & F.S. Brand, "Risk, vulnerability, robustness, and resilience from a decision-theoretic perspective" (2012) 15(3) *Journal of Risk Research*, pp. 313–30, at 318.

153 K.S. Vatsa, "Risk, vulnerability, and asset-based approach to disaster risk management" (2004) 24(10/11) *International Journal of Sociology and Social Policy*, pp. 1–48, at 10, citing T.E. Downing and K. Bakker, "Drought discourse and vulnerability" in D. A. Wilhite (ed.) *Drought: A Global Assessment*, Vol. 2, 213–30 (Routledge, 2000).

mitigate risk will be a challenging exercise that may require input from a broad range of stakeholders and may need to be regularly revisited. Finally, the vulnerability of a system will be affected by the inherent resilience of the system itself but will also be linked to the individuals, groups and communities that depend upon the system.[154] This interlinkage should ideally be accounted for in undertaking a vulnerability assessment for systemic risks.

There has been some debate as to whether or not exposure is part of the vulnerability assessment.[155] As explained by the IPCC (2012), it is possible to be exposed but not vulnerable – for example, by living in a floodplain but having sufficient means to modify building structure and behaviour to mitigate potential loss.[156] This warrants treatment of exposure as a separate driver of climate change risk. In terms of how exposure can be addressed in a regulatory setting, maps of hazard-prone areas have been used in regimes such as land-use planning and building. These maps provide an effective visual representation of areas that could be exposed to a particular hazard and, by implication, the systems and communities that may be at risk. When employed in the context of a regulatory regime, such maps identify geographical areas where special regulatory requirements may apply to mitigate risk where the relative degree of exposure is high.[157] The preparation of such maps typically involves significant resources and depends upon a vast swathe of technical data to ensure they are as comprehensive and accurate as possible.

While vulnerability assessments and exposure maps are clearly useful context for the assessment of risk, it is the interaction between hazards, exposure and vulnerability that is most important. Scenario or stress testing is a means by which all these risk drivers can be accounted for collectively. While efforts have been made to undertake such testing for climate change risks in relation to a variety of systems – including health systems,[158] ecosystems,[159] financial systems,[160] and

154 IPCC, "Climate-resilient pathways: adaptation, mitigation, and sustainable development" in *Climate Change 2014: Impacts, Adaptation, and Vulnerability. Part A: Global and Sectoral Aspects. Contribution of Working Group II to the Fifth Assessment Report of the Intergovernmental Panel on Climate Change* (Cambridge University Press), p. 1116.

155 G.C. Gallopín, n. 151 above, p. 297.

156 IPCC, n. 142 above, p. 69.

157 For example, in Victoria, areas that face the highest bushfire risk are identified in maps of "Bushfire Prone Areas" and are subjected to special regulatory requirements. These requirements are considered in the case study on the regulation of bushfire risks to infrastructure in Victoria, Australia in Chapter 11.

158 K. Ebi, P. Berry, K. Hayes, C. Boyer, S. Sellers, P. Enright, & J. Hess, "Stress testing the capacity of health systems to manage climate change-related shocks and stresses" (2018) 15(11) *International Journal of Environmental Research and Public Health*, p. 2370.

159 M. Scholze, W. Knorr, N.W. Arnell, & I.C. Prentice, "A climate-change risk analysis for world ecosystems" (2006) 103(35) *Proceedings of the National Academy of Sciences*, pp. 13116–20.

160 H. Jung, R.F. Engle, & R. Berner, "Climate stress testing" (2021) 977 *Federal Reserve Bank of New York Staff Reports*.

systems of infrastructure[161] – the extent to which such testing is utilised in a regulatory context to manage climate change risks to these systems is unclear. Nonetheless, mechanisms to practically combine the drivers of risk are critically important to ensure a contextualised assessment of systemic risks and failure to do so could undermine regulatory efforts to manage such risks.

3 Risk characterisation

Even though all risks share a common feature – namely, uncertainty about the future – risks are typically quite different, and of unequal prevalence, scale and consequence. As a result, different risks are likely to warrant different treatment. Indeed, it is important to understand the inherent characteristics of a risk, so that it is managed appropriately and proportionately. Mischaracterising risk could result in over- or under-estimation of its relative importance, and result in a response that is more or less than is required. At worst, mischaracterising systemic risks could compromise the survival of the system at risk and have catastrophic consequences for the system, as well as those that depend upon it.

"Risk characterisation" is used to describe a risk's characteristics and is a means to create a bridge between risk assessment and risk management.[162] More specifically, it is a summary of information about a risk following its assessment, capturing what is known about the risk and what remains unknown.[163] This consolidation of information can guide the response to risk by accounting for the nature and completeness of knowledge that is available about the risk.[164] Risk characterisation, which is the basis for proposed Building Block 5 of an ideal to regulate systemic risks, forms a component of the regime's context and is highlighted in Figure 7.7.

The extent to which risk characterisation is likely to be helpful in developing a response to risk will depend upon its capacity to synthesise relevant information about risk in an understandable, accessible and meaningful way that provides decision-makers with relevant risk insights. Risk characterisation's potential to deliver this outcome will, in turn, depend upon the existence of apposite risk categories. Indeed, without meaningful risk categories, decision-makers could be faced with a list of risk characteristics that do not provide clear and coherent guidance regarding the appropriate response to risk.

Semantically, risk characterisation and risk categorisation may appear synonymous. Nevertheless, they are practically distinct, although inter-related. Whereas risk characterisation is the means to summarise and communicate the outcome of the risk assessment by identifying relevant characteristics of a risk

161 T. Comes, V. Bertsch, & Simon French, "Designing dynamic stress tests for improved critical infrastructure resilience" (2013) *Proceedings of the 10th International ISCRAM Conference.*

162 IRGC, n. 11 above, p. 17.

163 IRGC, n. 9 above, p. 39; IRGC, n. 11 above, p. 8.

164 IRGC, n. 9 above, p. 39.

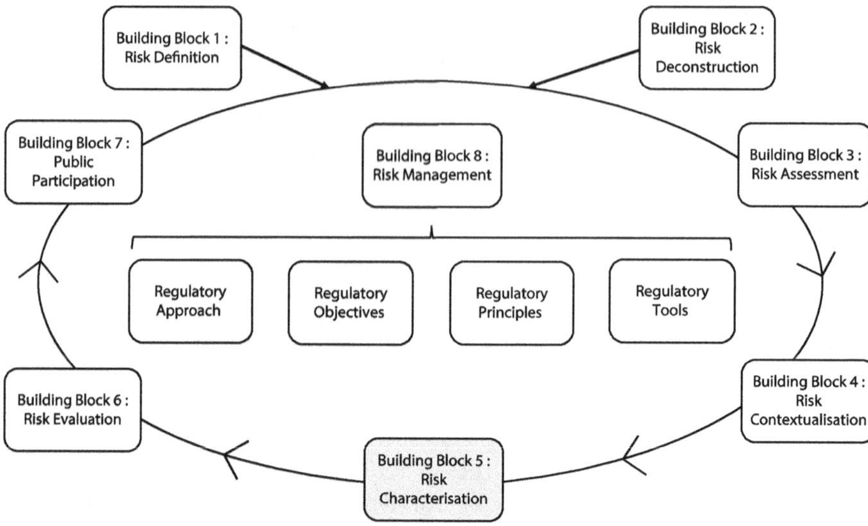

Figure 7.7 Building Block 5: Risk Characterisation

that has been assessed, risk categorisation involves grouping together risks that share common characteristics. In other words, risk categories represent groups of risks with similar characteristics.[165]

Benefits

It is worthwhile taking stock of the important practical benefits that risk categorisation and, by extension, risk characterisation can yield. Sokal (1974) suggests that a framework that allows identification and differentiation of specific categories or classes of risks can serve as a heuristic with associated hypotheses,[166] which may help to inform and guide the development of the response to risk. Cvetkovic and Earle (1985) further argue that the classification of risks into discrete categories can facilitate decision-making about which risks might be worthy of a response and, if so, what the response should be.[167]

At a more granular level, risk categorisation can be used for information management, to help simplify and order diverse and complex information

165 Some of the relevant literature refers to "risk classification", rather than "risk categorisation". These terms essentially relate to the same exercise – namely, grouping together risks that share common characteristics. Accordingly, these terms are used interchangeably in this book.

166 R.R. Sokal, "Classification: Purposes, principles, progress, prospects" (1974) 185 (4157) *Science*, pp. 1115–23, at 1117.

167 G. Cvetkovich & T.C. Earle, "Classifying hazardous events" (1985) 5(1) *Journal of Environmental Psychology*, pp. 5–35, at 8, 22.

about a range of risks.[168] In particular, categorisation simplifies the phenomena under consideration, thereby obviating the need to deal with a "world of single cases".[169] In addition, categorisation can aid and enhance the effectiveness of communication about different types of risks.[170] The description of each category can convey essential risk characteristics that define each category, thereby imparting important information for the purposes of the risk management process.[171] The simplification of risk information may also facilitate stakeholder engagement, particularly when public input into the establishment of risk management priorities is required.[172]

Above all, categorisation of risks can serve as an analytical and decision-making tool,[173] providing an important theoretical frame of reference for the management of risk. In particular, risk categories can provide a snapshot of the types of risks faced by society and what they are likely to mean in practical terms. A framework for the categorisation of risk can facilitate a systematic and focused approach for the management of risks within each risk category,[174] which is tailored to the features that epitomise risks within a category. Moreover, categorisation could enable the identification of generic responses that are likely to be appropriate and effective for particular risk categories.[175]

Challenges

Established, formal systems to categorise items exist in a variety of contexts, including to describe the natural world,[176] goods and services[177] and medical

168 Ibid. p. 8.
169 R.R. Sokal, n. 166 above, pp. 1116–17; G. Cvetkovich & T.C. Earle, n. 167 above, p. 8.
170 G. Cvetkovich & T.C. Earle, n. 167 above, p. 8.
171 M.G. Morgan, H.K. Florig, M.L. DeKay, & P. Fischbeck, n. 79 above, p. 49.
172 H.K. Florig, M.G. Morgan, K.M. Morgan, K.E. Jenni, B. Fischhoff, P.S. Fischbeck, & M.L. DeKay, "A deliberative method for ranking risks (I): Overview and test bed development" (2001) 21(5) *Risk Analysis*, pp. 913–913, at 920.
173 G. Cvetkovich & T.C. Earle, n. 167 above, p. 8.
174 Y. Zhang, S. Wu, E. Dai, D. Liu, & Y. Yin, "Identification and categorization of climate change risks" (2008) 18(3) *Chinese Geographical Science*, pp. 268–75, at 268.
175 G. Cvetkovich & T.C. Earle, n. 167 above, p. 8.
176 A modern taxonomy exists to classify groups of biological organisms (including both animals and plants) on the basis of shared characteristics, which is derived from a system for classifying living things developed by Carl Linnaeus in the 18th century known as the "Linnaean classification system".
177 The Harmonized Commodity Description and Coding System (generally referred to as the "Harmonized System") is an international nomenclature for goods developed by the World Customs Organization. The Harmonized System is used to classify internationally traded goods on the basis of factors including the product's composition, form and function. In relation to services, the World Trade Organization has developed a Services Sectoral Classification List, which is used as the basis for classification of internationally traded services.

conditions.[178] In summary, these systems convey important and, at times, complex information about the items governed by each system and practically enable items that share common characteristics to be grouped together to help systematise their analysis and management.

In comparison, a unique system for the categorisation of risks does not exist.[179] Rather, to date, categories of risk have been developed based on the particular context in which risk is sought to be mitigated.[180] Variations in policy and regulatory contexts as well as divergences between concerns, interests and values held by countries and groups across the world may well produce different approaches to the categorisation of risks.[181] Moreover, different systems for the categorisation of risk will likely serve different objectives.[182]

In practice, the appropriate system for risk categorisation will not necessarily be self-evident and, in some cases, development of such a system may be as complex and challenging as the actual assessment and management of those risks. Morgan and colleagues (2000) note that "[e]ven within a specific domain of risk such as air pollution or airline safety, the world is filled with numerous known and potential hazards."[183] The US National Research Council's Committee on Risk Characterization (1996) further argues that the challenges

178 The International Classification of Diseases, developed by the World Health Organization in collaboration with various international medical bodies, is an international system used, among other things, to help classify diseases.

179 M.G. Morgan, H.K. Florig, M.L. DeKay, & P. Fischbeck, n. 79 above, p. 50. Morgan and colleagues do, however, identify "desirable attributes of an ideal risk categorization system for risk ranking" at p. 54.

180 For example, categories of risk that could arise in the context of the development of disruptive technology are identified in A. Ganguly, R. Nilchiani, and J.V. Farr, "Identification, classification, and prioritization of risks associated with a disruptive technology process", (2011) 8(2) *International Journal of Innovation and Technology Management*, 273–293. Issues regarding risk characterisation (and, implicitly, risk categorisation) in the context of the use of nanotechnology are considered in R.A. Williams, K.M. Kulinowski, R. White, and G. Louis, "Risk characterization for nanotechnology", G. (2010) 30(11) *Risk Analysis*, 1671–1679. Risk classification of livestock farms was considered in the context of mitigation of animal disease in J.K. Niemi, T. Lyytikäinen, L. Sahlström, T. Virtanen, T. and H. Lehtonen, "Risk classification in animal disease prevention: Who benefits from differentiated policy?", *Paper presented at the 2009 Agricultural & Applied Economic Association (AAEA) and American Council on Consumer Interests (ACCI) Joint Annual Meeting*. A classification of risks that could arise in relation to food chains was proposed in C. Amesder, R. Haas, M. Fritz and G. Schiefer "Risk analysis in selected European and international food chains", *Proceedings of the 3rd International European Forum on System Dynamics and Innovation in Food Networks*, organised by the International Center for Food Chain and Network Research, University of Bonn, Germany (2009).

181 A. Klinke & O. Renn, n. 117 above, p. 1071; M.G. Morgan, H.K. Florig, M.L. DeKay, & P. Fischbeck, n. 79 above, pp. 50, 52.

182 M.G. Morgan, H.K. Florig, M.L. DeKay, & P. Fischbeck, n. 79 above, p. 50.

183 Ibid. p. 57.

associated with the categorisation of risks are particularly significant when the risks are characterised by high degrees of uncertainty:[184]

> It is difficult to characterize what is known about uncertainty without making the risk appear either larger or smaller than analysts believe it to be. … Characterizing uncertainty analytically puts risk analysts on the horns of a dilemma: simple characterizations are likely to give an erroneous impression of the extent of uncertainty, but more careful and elaborate characterizations may be incomprehensible to nonspecialists and so unusable by decision makers and some other participants.

While this quote refers to the challenges associated with risk characterisation, by corollary, risk categorisation is also likely to be especially challenging for systemic risks, particularly the risks associated with climate change, which are typically deeply uncertain. Striking the appropriate balance between coarse and more refined risk categories to ensure that the categories assist with communication, analysis and response to risk is unlikely to be easy for systemic risks in light of this uncertainty.

Categorising systemic risks caused by climate change

The question arises as to how the risks associated with climate change should be characterised and, potentially, categorised. More specifically, what are the key characteristics that should be accounted for when determining whether and how to respond to climate change risks? Moreover, based on these key characteristics, can climate change risks be grouped into meaningful categories that assist with the determination of the appropriate response to mitigate such risks?

In this regard, Zhang and colleagues (2008) argue that the risks associated with climate change are "messy" and, consequently, "it is not easy to package those risks neatly into distinct categories".[185] Indeed, as yet, an established framework for the categorisation and characterisation of climate change risks does not exist.[186] A possible approach for the characterisation and categorisation of such risks is considered below.

184 P.C. Stern, H.V. Fineberg, & National Research Council (U.S.) (eds.), *Understanding Risk: Informing Decisions in a Democratic Society* (National Academy Press, 1996), p. 67.

185 Y. Zhang, S. Wu, E. Dai, D. Liu, & Y. Yin, n. 174 above, p. 269.

186 It is worth noting that much effort has been invested in characterising the level of uncertainty associated with scientific information and findings contained in IPCC reports about climate change and associated risks. See, for example, IPCC, *Guidance Note for Lead Authors of the IPCC Fifth Assessment Report on Consistent Treatment of Uncertainties* (IPCC, 2010). However, this work does not extend to the categorisation and characterisation of the climate change risks themselves.

CHARACTERISTICS

A broad range of generic risk characteristics have been identified in risk and regulatory literature as a means of ordering the multitude of risks that pervade human existence. These characteristics are consolidated, organised and summarised in Table 7.2.

Table 7.2 Summary of risk characteristics[187]

Characteristic	*Description*
	Nature and Complexity of Risk
Catastrophic	Catastrophic risks are low probability events that can have major adverse consequences, including "eradication of humankind".[188]
Discrete	Discrete risks are highly identifiable threats and events of a precise, bounded nature.[189]
Pervasive	Pervasive risks are ubiquitous risks that exist everywhere, including in the context of the normal functioning of society.[190]
Persistent	Persistent risks are those for which loss or damage is likely to persist or endure over an extended period of time.[191]
Latent	Latent risks involve a delay between the risk-causing event or activity and the actual manifestation of loss or damage.[192]
Reversible	In the case of reversible risks, the original condition or state that existed before the risk materialised and loss or damage occurred can be restored.[193]
Non-reversible	In the case of non-reversible risks, the original condition or state that existed before the risk materialised and loss or damage occurred cannot be restored.[194]

187 This summary of risk characteristics has been derived from a review of relevant risk and regulatory literature cited in this table.

188 R. Baldwin, M. Cave, & M. Lodge, n. 80 above, p. 102 referring to R.E. Posner, *Catastrophe: Risk and Response* (Oxford University Press, 2004).

189 Ibid. p. 87.

190 A. Klinke & O. Renn, 'Precautionary principle and discursive strategies: classifying and managing risks' (2001) 4(2) *Journal of Risk Research*, pp. 159–73, at 162; R. Baldwin, M. Cave, & M. Lodge, n. 80 above, p. 87; M. Oppenheimer, 'Defining Dangerous Anthropogenic Interference: The Role of Science, the Limits of Science: Dangerous Anthropogenic Interference' (2005) 25(6) *Risk Analysis*, pp. 1399–407, at 1400.

191 A. Klinke & O. Renn, n. 190 above, p. 162.

192 Ibid. p. 162.

193 Ibid. p. 162; R. Baldwin, M. Cave, & M. Lodge, n. 80 above, p. 87; M. Oppenheimer, n. 190 above, p. 1400.

194 A. Klinke & O. Renn, n. 190 above, p. 162; R. Baldwin, M. Cave, & M. Lodge, n. 80 above, p. 87; M. Oppenheimer, n. 190 above, p. 1400.

Characteristic	Description
	Uncertainty about Risk
Probabilistic	Probabilistic risks are those that can be quantified based on available statistics concerning past events[195] and are characterised by the relative likelihood or frequency of events causing loss or damage.[196]
Uncertain	Uncertain risks are characterised by an inability to make reliable predictions about events causing loss or damage, primarily due to incomplete information or data.[197]
Unpredictable	Evidence of a causal connection between events and consequences may be weak and, consequently, difficult to establish or quantify for unpredictable risks.[198]
	Social Ambiguity about Risk
Inequitable	The consequences associated with inequitable risks are distributed unfairly across society, including on the basis of factors such as income and location.[199]
Contentious	Contentious risks are those that involve a violation of individual, social or cultural interests and values, resulting in social conflicts.[200]
Societally imposed	In the case of societally imposed risks, citizens have little choice as to exposure to those risks because control of the risks is in the hands of government.[201]
Voluntarily undertaken	Voluntarily undertaken risks are those where individuals or groups are aware of a risk but decide to take action that disregards the risk.[202]

Each characteristic set out in Table 7.2 conveys important information about a risk's attributes, including insights about whether a response to the risk is needed. For example, it could be considered rational to prioritise a response to a pervasive risk (such as the risk of increasing temperatures) over a latent risk (such as the risk of the collapse of an ecosystem, where critical tipping points for the ecosystem have not yet been reached). The risk characteristics also provide insights as to what the response to risk should be. So, for example, a pre-emptive response would be prudent for non-reversible risks (such as, the risk of extinction of a threatened species resulting from destruction of its local habitat), whereas a reactive response may be more

195 R. Baldwin, M. Cave, & M. Lodge, n. 80 above, pp. 86–7.
196 A. Klinke & O. Renn, n. 190 above, p. 161.
197 Ibid. pp. 161–2.
198 R. Baldwin, M. Cave, & M. Lodge, n. 80 above, pp. 86–7.
199 M. Oppenheimer, n. 190 above, p. 1400.
200 A. Klinke & O. Renn, n. 190 above, p. 162.
201 R. Baldwin, M. Cave, & M. Lodge, n. 80 above, p. 87.
202 Ibid. p. 87.

appropriate for certain reversible risks (such as the risks to ecosystems posed by invasive plants or animals caused by increasing temperatures that could be eradicated when detected).

Notwithstanding the insights provided by these risk characteristics, using them in practice to facilitate decision-making about the response to risk is not without challenge. Challenges arise from the fact that a particular risk may possess more than one characteristic, but the response indicated by each characteristic may be different. For example, investment in companies that are likely to be adversely affected by the uptake of net zero policies around the world (such as companies operating in carbon-intensive sectors) is a risk that is voluntarily undertaken, which tends to indicate that a government response to manage the risk is not appropriate because those exposed to the risk have chosen to engage in risky behaviour despite knowledge about the possible negative consequences. However, the same risk could also have widespread and cascading catastrophic consequences if, for example, there is a rapid and dramatic transition to net zero policies across the globe, which results in devaluation in the value of these companies across multiple sectors. In this scenario, a government response may be needed, particularly to limit the impact of the risks for those who have not engaged in the risky behaviour but could, nevertheless, be vulnerable due to the cascading effects of the risk if the entire financial system becomes destabilised. Determining which of these divergent responses that flow from different risk characteristics should prevail will not always be self-evident. Given that risks are typically multi-faceted, considering one characteristic in isolation and to the exclusion of others may lead to a response to risk that is inappropriate or ineffective. Grouping characteristics into meaningful and useful categories helps to overcome at least some of these challenges. Categories that may be particularly relevant in the context of climate change risks are considered below.

CATEGORIES

In 1985, Cvetkovic and Earle recognised the need to categorise risks associated with hazardous events. They noted that efforts had been made to:

> move beyond purely descriptive accounts of hazards, to develop more general comparisons and to classify events according to identified general characteristics precisely because of manifest information advantages. For the most part classification efforts have remained separate, often exploring different subsets of hazardous events … [A] next important step in the conceptualisation of hazards is the integration of the diverse proposed classifications into a single framework.[203]

203 G. Cvetkovich & T.C. Earle, n. 167 above, p. 6.

This begs the question of which "single framework" could capture the range of diverse systemic risks, including those caused by climate change, while remaining meaningful and useful? The answer could well lie in three core features that emerge from the risk characteristics identified in Table 7.2 above – namely, the nature and complexity of risk,[204] the relative degree of uncertainty about risk,[205] and the extent of social ambiguity about whether to respond to risk.[206] These core features cover the main characteristics of risk that have been identified to date and would, therefore, provide a rational basis for a risk categorisation system.

The IRGC Framework includes a categorisation system developed specifically for systemic risks which reflects these core features. In particular, the IRGC has identified four discrete categories of risk that together are intended to comprise a comprehensive categorisation framework for systemic risks – namely, "simple risks", "complex risks", "uncertain risks" and "ambiguous risks". The IRGC explains that these categories are distinguished based on the degree of difficulty of the cause–effect relationship between the risk agent and potential consequences on the risk-absorbing system (that is, the nature and complexity of the risk), the reliability of this relationship (which concerns the relative degree of uncertainty about the risk) and the degree of controversy with respect to what the risk means for those affected as well as the values to be applied when judging whether or not something needs to be done about the risk (which is relevant to the extent of social ambiguity about the risk).[207] These categories and the associated characterisation are summarised in Table 7.3, together with climate change examples for each category.

The primary purpose of the IRGC categorisation system is to demonstrate how each risk category calls for a specific set of strategies to respond to the key features that characterise each category. In general terms, the IRGC suggests that, given their relative regularity and predictability, simple risks could be managed by a specific and targeted approach. Complex risks are likely to be most effectively addressed through responses that are aimed at accessing and acting on the best available scientific information about the risk agent and its impact on the risk-absorbing system. Uncertain risks are likely to be best managed using a precautionary approach that is aimed at ensuring the reversibility of critical decisions. Ambiguous risks call for an approach that seeks to

204 The relevant risk characteristics from Table 7.2 are "catastrophic", "discrete", "pervasive", "persistent", "latent", "reversible" and "non-reversible".

205 The relevant risk characteristics from Table 7.2 are "probabilistic", "uncertain" and "unpredictable".

206 The relevant risk characteristics from Table 7.2 are "inequitable", "contentious", "societally imposed" and "voluntarily undertaken".

207 IRGC, n. 9 above, p. 12.

Table 7.3 Categories of systemic risks

Category	*Characterisation*	*Climate Change Example*
Simple	Simple risks display a relatively clear causal connection between a risk agent and risk-absorbing system. In addition, the potential negative consequences are obvious, the uncertainty is low and there is limited controversy regarding the interpretation of the risk. Simple risks are recurrent and not affected by ongoing or expected major changes. As a consequence, statistics are available and their use to assess future risks is meaningful.[208]	Sea level rise is an example of a simple risk. There is limited uncertainty concerning the occurrence of sea level rise. There is also a relatively clear connection between this type of climate change hazard and the things at risk as a result of coastal flooding and erosion caused by sea level rise. The consequences for affected systems, particularly local infrastructure, are obvious – namely, damage or destruction.
Complex	Complexity refers to the difficulty of identifying and quantifying causal links between a risk agent and specific adverse effects for a risk-absorbing system. Sophisticated models of probabilistic inferences are required because of this complexity.[209]	Bushfires caused by extreme heat can be regarded as an example of a complex risk. Bushfires are unpredictable acts of both nature and human intervention. Factors influencing fire ignition and progression are diverse, with the consequence that no two bushfires are exactly the same in terms of the triggers, location, nature and scale of impacts. The precise origin and timing of a particular bushfire are difficult to predict. Likewise, the progression of any specific fire, once started, is equally unforeseeable and unpredictable, because this depends on the prevailing weather, the available fuel to feed the fire and the terrain traversed by the fire. Moreover, it is likely to be difficult to identify causal links between, on the one hand, increasing and extreme temperatures caused by climate change and, on the other hand, the specific adverse effects associated with bushfires, including the location and scale of consequences.

208 Ibid. p. 45; IRGC, n. 11 above, p. 17; O. Renn, "White Paper on Risk Governance: Towards an Integrative Framework" in O. Renn & K.D. Walker (eds.), n. 7 above, pp. 37–8; O. Renn, A. Klinke, & M. van Asselt, n. 23 above p. 234.

209 IRGC, n. 9 above pp. 29–30, 45; IRGC, n. 11 above, p. 18; O. Renn, "White Paper on Risk Governance: Towards an Integrative Framework" in O. Renn & K.D. Walker (eds.), n. 7 above, p. 38; O. Renn, A. Klinke, & M. van Asselt, n. 23 above, p. 234.

Category	*Characterisation*	*Climate Change Example*
Uncertain	Uncertainty refers to the incompleteness of knowledge (particularly, scientific or technical data and information) about a risk agent and its interaction with a risk-absorbing system, and the available knowledge relies on uncertain assumptions, assertions and predictions. This makes it difficult to assess the probability of adverse outcomes for the risk-absorbing system.[210]	Floods are an example of an uncertain risk because historical data cannot be relied upon to predict floods, as various factors typically combine to produce extreme flood events. Floods can be influenced by both small- and large-scale weather patterns, modes of variability, thermodynamic processes, land-atmosphere feedbacks and antecedent conditions. Uncertainty regarding the interaction of these factors could lead to under-estimation of flood frequency and levels.
Ambiguous	In the case of ambiguous risks, there are different legitimate viewpoints about whether there are or could be adverse effects for a risk-absorbing system arising from a particular risk agent and whether these risks are acceptable, tolerable or intolerable. There are difficulties in agreeing on the appropriate values, priorities, assumptions or boundaries to be applied when determining whether and how to respond to such risks.[211]	Increasing temperatures can be considered as an example of an ambiguous risk. Compared to extreme events like bushfires and floods, increasing temperatures are unlikely to cause obvious, immediate or permanent harm or damage – at least, not in the short term. In addition, increasing temperatures have varied and diffuse implications for a broad range of sectors, that may be difficult to identify and disaggregate from other stressors on those sectors.

ensure mutual understanding of conflicting views and values with a view to eventually reconciling them.[212]

These categories of systemic risks are illustrated in Figure 7.8. The diagram illustrates the relative dominance of the core risk features described above for each category of systemic risk – namely, the nature and complexity of the risk, the relative uncertainty of the risk and the social ambiguity about the response

210 IRGC, n. 9 above, p. 30; IRGC, n. 11 above, p. 18; O. Renn, "White Paper on Risk Governance: Towards an Integrative Framework" in O. Renn & K.D. Walker (eds.), n. 7 above, p. 39; O. Renn, A. Klinke, & M. van Asselt, n. 23 above, p. 235.

211 IRGC, n. 9 above, pp. 30–1; IRGC, n. 11 above, p. 18; O. Renn, "White Paper on Risk Governance: Towards an Integrative Framework" in O. Renn & K.D. Walker (eds.), n. 7 above, pp. 39–40; O. Renn, A. Klinke, & M. van Asselt, n. 23 above, p. 235.

212 IRGC, n. 11 above, pp. 24–5; O. Renn, "White Paper on Risk Governance: Towards an Integrative Framework" in O. Renn & K.D. Walker (eds.), n. 7 above, pp. 36–40. The risk management strategies for each category of system risk are considered further below.

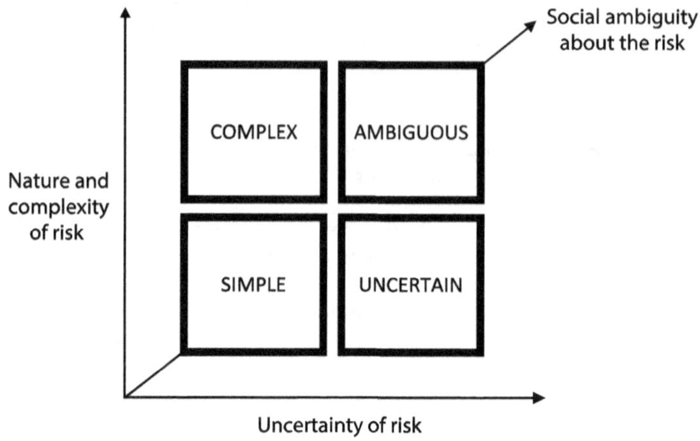

Figure 7.8 Illustration of systemic risk categories[213]

to risk. The greater the relative degree of complexity and uncertainty for a particular risk, the greater the overall ambiguity regarding the risk. By corollary, the regulatory challenges are likely to be particularly pronounced as a risk's complexity, uncertainty and ambiguity intensify.

The categories of systemic risks put forward by the IRGC share some similarities with a "typology of extreme climate events" suggested by the UNDP in 2004.[214] Moreover, the categories have been specifically applied in the climate change context. Zhang and colleagues (2008) used the IPCC's assessment of uncertainties associated with risks caused or exacerbated by climate change and found that most of the risks could be allocated into one of the IRGC's four systemic risk categories.[215] Based on the results of their analysis, the authors concluded that those categories "could greatly enhance the efficiency of risk management and might help advance scientific development in climate change research".[216]

Nevertheless, it is unlikely that the application of the IRGC systemic risk categories in the climate change context will be without difficulty, as is evident from a variety of comments that have been made about their application in practice. In particular, it has been suggested that the risk categories are neither clear-cut nor

213 This illustration is the author's conceptualisation of the relationships between the four categories of systemic risks put forward by the IRGC.

214 The UNDP's typology of extreme climate events includes the following categories: "Simple": individual local weather variables exceeding critical levels on a continuous scale, such as heavy rainfall. "Complex": severe weather associated with particular climatic phenomena, often requiring a critical combination of variables, such as droughts. "Unique or singular": a plausible future climatic state with potentially extreme large-scale or global outcomes, such as major circulation changes. R. Jones, and R. Boer, n. 137 above, p. 98.

215 Y. Zhang, S. Wu, E. Dai, D. Liu, & Y. Yin, n. 174 above, p. 273.

216 Ibid. p. 274.

mutually exclusive[217] and that the risk categories may simply describe different aspects of the same risk.[218] In fact, Van Asselt and Renn (2011) acknowledge that uncertainty often results from complexity and that, in turn, high degrees of complexity and uncertainty result in ambiguity.[219] In addition, the characteristics of a risk may shift over time,[220] and a risk may migrate between different categories.[221] Moreover, in practice, risks may be inappropriately classified because they may initially appear simple in nature, but may, in fact, be practically more complex, uncertain and ambiguous than originally thought.[222] This could result in an amplification or attenuation of the response to risk.

Despite the difficulties associated with the systemic risk categories, there are only a limited number of categories, which makes them manageable. The labels used to describe each category are relatively intuitive and easy to comprehend, enhancing the role of these criteria as information management tools. The categories are logically consistent in that they are based on the relative dominance of each of the core features that have been used to characterise risks. All in all, they offer a rational starting point for synthesising information about climate change risks, particularly in light of the fact that they capture the key features that have been generally used to characterise risks to date and they have been specifically designed to address systemic risks.

4 Risk evaluation

The ultimate function of establishing the context for application of a regulatory regime is to determine whether to respond to risk, but this objective cannot be achieved unless the relevant risk has been evaluated. "Risk evaluation" is a "deliberative effort to qualify risks in terms of social acceptability and tolerability"[223] and, implicitly, recognises that different levels of risk warrant

217 Ibid. p. 271.

218 K. Kuenzi and J. McNeely, "Nature-based tourism" in O. Renn & K.D. Walker (eds.), n. 7 above, p. 176.

219 M.B.A. van Asselt & O. Renn, "Risk governance" (2011) 14(4) *Journal of Risk Research*, pp. 431–49, at 438.

220 IRGC, n. 11 above, p. 17.

221 de Vries, G., Verhoeven, I. and Boeckhout, M. (2011) "Taming uncertainty: the WRR approach to risk governance" (2011) 14(4) *Journal of Risk Research*, pp. 485–499, at 485.

222 O. Renn, A. Klinke, & M. van Asselt, n. 23 above, pp. 235–6. Flood prevention provides an example. The authors note that, for a long time, flooding was presented as a "simple" risk problem that was well understood and deemed manageable with established prevention-oriented forms of risk management. However, today, because of climate change, flood prevention also entails "uncertain" and "ambiguous" risk problems. As a result, policy-makers have been forced to fundamentally adapt their strategies: G. de Vries, I. Verhoeven, & M. Boeckhout, "Taming uncertainty: The WRR approach to risk governance" (2011) 14(4) *Journal of Risk Research*, pp. 485–99, at 488.

223 O. Renn, A. Klinke, & M. van Asselt, n. 23 above, p. 240.

different reactions.[224] At one extreme, an "acceptable" risk requires no response. At the other extreme, an "intolerable" risk means that risk must be avoided. Between these two extremes, risk is "tolerable" but risk reduction may be necessary to render the risk more acceptable.[225]

Risk evaluation is both a process and an outcome. In terms of the process, it involves determining the relative significance of risks in practical terms by "making judgments and choices, which are often social, technical, economic, political or strategic".[226] Regarding the outcome, the process of evaluating risk yields a decision about whether a risk is acceptable, tolerable or intolerable and essentially enables risks to be ranked and prioritised for the purpose of responding to risk. In this regard, Linkov and colleagues (2011) note that "[a]ll risks are not created equal, and proper risk management recognizes that different levels of risk warrant different reactions".[227]

In a corporate setting, risk evaluation is a well-established practice that is undertaken in the context of a company's "enterprise risk management" plan.[228] Risks faced by a company (such as those associated with a proposed merger or acquisition) are evaluated based on the board's risk appetite and tolerance for risk. Risk appetite is the amount of risk exposure or potential adverse impact from an action or event the company is willing to accept while pursuing its objectives and strategies. The risk appetite of the company's board is then translated into practical and, at times, quantitative terms in the form of risk tolerances.

The practice of risk evaluation by government in a regulatory context, as opposed to an operational context, has comparatively had a shorter history.[229] This

224 I. Linkov, M. Bates, D. Loney, M. Sparrevik, & T. Bridges, "Risk management practices: Cross-agency comparisons and tolerable risk", in I. Linkov & T.S. Bridges (eds.), *Climate Change: Global Change and Local Adaptation* (Springer Netherlands, 2011), pp. 133–55, at 134.

225 IRGC, n. 9 above, pp. 36–7; IRGC, n. 11 above, p. 20.

226 IRGC, n. 11 above, p. 20.

227 I. Linkov, M. Bates, D. Loney, M. Sparrevik, & T. Bridges, n. 224 above, p. 134.

228 "Enterprise risk management" is broadly defined as a strategy designed to identify, assess and prepare for dangers, hazards and events that may affect an entity, and to manage risk to within the entity's risk appetite in order to provide reasonable assurance regarding achievement of objectives.

229 The reference here to risk evaluation by government in a *regulatory context* is distinct from risk evaluation by government in relation to its *operational context*, which is typically referred to as "enterprise risk". In Australia, management of enterprise risk is required by Commonwealth government agencies under the *Public Governance, Performance and Accountability Act 2013 (Cth)* and this obligation is supported by the Commonwealth Risk Management Policy accessible at: https://www.finance.gov.au/sites/default/files/2019-11/commonwealth-risk-management-policy_0.pdf. Similarly, Hardy suggests that, in relation to the US government, "[Enterprise Risk Management] is fast becoming an important activity for many agencies to undertake as a solution for bringing together various agency risk activities": K. Hardy, *Enterprise Risk Management: A Guide for Government Professionals* (Jossey-Bass, 2015), p. 29.

is likely linked to the traditional view of regulation as a means to secure compliance with rules, rather than a mechanism to manage risk. Indeed, Jordana and Levi-Faur (2004) note that "in its narrowest sense and simplest sense, 'regulation refers to the promulgation of an authoritative set of rules, accompanied by some mechanism, typically a public agency, for monitoring and promoting compliance with these rules'".[230] When viewed as a tool to secure compliance with rules, it is difficult to imagine how the concept of risk evaluation might readily apply to regulation because, at least in theory, compliance with all rules should be ensured. However, when considered through the lens of risk, the relevance becomes clearer – specifically, risks that are most likely to result in harm that is sought to be addressed under a regulatory regime ought to take priority over risks that are relatively less likely to produce harm because the former are less tolerable than the latter.

The process of risk evaluation is particularly important in the context of systemic risks, including those caused by climate change, where the harm that could eventuate if risks to the relevant system are not adequately managed could be catastrophic. Questions arise in relation to the criteria for establishing what is socially acceptable. In addition, how can these criteria be applied in practice? What does the outcome from the evaluation process mean for the management of risk in practical terms? In light of the importance of the evaluation of systemic risks, the sixth proposed building block of an ideal to regulate systemic risks is Building Block 6: Risk Evaluation, which is highlighted in Figure 7.9. It is a component of the context for the regulatory regime.

Exemplars for risk evaluation in a regulatory setting

Regulators from countries including the US, the UK and Australia have adopted a variety of so-called "regulatory strategies" that help ensure a rational basis for evaluating risk. This essentially involves relativising risk arising under a regulatory regime so that regulatory responses can be chosen in line with how risk is evaluated by the regulator.[231] The higher the relative risk, the more intensive, severe and intrusive the regulator's response. The most relevant of these strategies for the regulation of climate change risks are discussed below.

230 J. Jordana and D. Levi-Faur, D. "The politics of regulation in the age of governance" in J. Jordana & D. Levi-Faur (eds.), *The Politics of Regulation: Institutions and Regulatory Reforms for the Age of Governance* (Edward Elgar, 2004), at pp. 3–4 referring to R. Baldwin, C. Scott, and C. Hood, "Introduction", in R. Baldwin, C. Scott, and C. Hood (eds.), *A Reader on Regulation* (Oxford University Press, 1998) p. 3.

231 The strategies include advice and persuasion, deterrence, responsive regulation, risk-based regulation, smart regulation, metaregulation and criteria strategies: N. Gunningham, "Compliance, enforcement, and regulatory excellence", in C. Coglianese (ed.), *Achieving Regulatory Excellence* (Brookings Institution Press, 2016), p. 188, at 189–91.

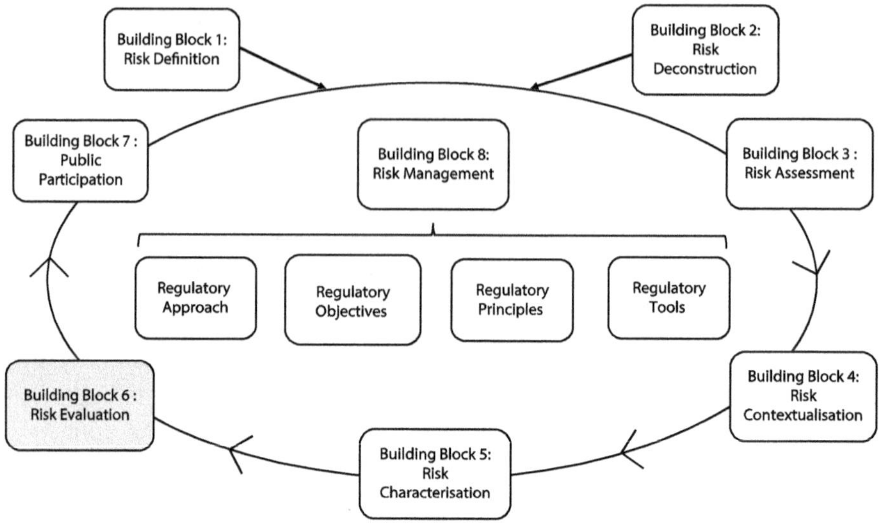

Figure 7.9 Building Block 6: Risk Evaluation

Responsive regulation is an approach that responds to the context in which regulation is applied. It counsels regulators to modulate regulatory intervention to a greater or lesser degree based on a contextual assessment,[232] which considers the behaviour of regulated entities – specifically, the disposition of regulated entities towards compliance with the regulatory regime. This strategy rests on the assumption that the best regulatory outcomes will be achieved if the regulatory response is adapted to the conduct of the regulated entity.[233]

The essence of responsive regulation can be depicted as a pyramid of regulatory enforcement measures – from a broad base of light-handed "education" for those entities that are still learning how to comply, escalating up to successively stronger and more severe legal measures leading ultimately to heavy-handed sanctions to address relatively more concerning attitudes towards compliance and more egregious instances of non-compliance. A simplified example of an enforcement pyramid is set out in Figure 7.10.

Baldwin and Black's model of "really responsive regulation" proposes a broader context for the determination of the regulatory response compared to its "responsive regulation" predecessor. It accounts not just for the conduct and compliance disposition of regulated entities, but also relevant aspects of the

232 I. Ayres & John Braithwaite, *Responsive Regulation: Transcending the Deregulation Debate* (Oxford University Press, 1992), pp. 4–5.

233 N. Gunningham, n. 231 above, p. 190.

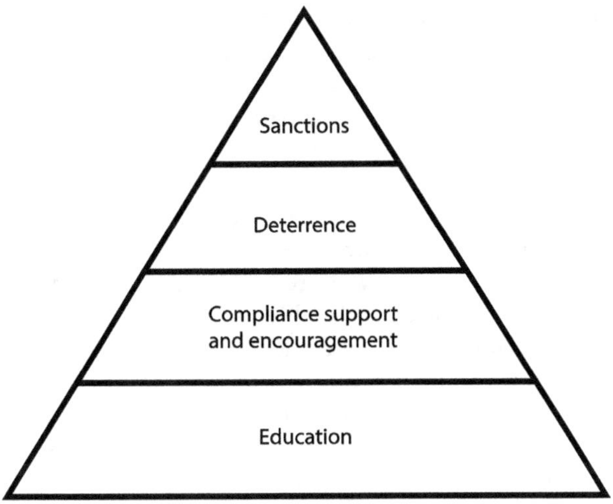

Figure 7.10 Example of a pyramid of enforcement[234]

regulatory regime.[235] Under this more expansive approach, regulators are encouraged to consider the attitude of the regulated entities towards compliance, as well as their operating and cognitive frameworks, the institutional environment of the regulatory regime, the different logics of regulatory tools and strategies, and changes in these elements.[236] Regulation is considered to be really responsive when the regulator "knows its regulatees and its institutional environment, when it is capable of deploying different regulatory logics coherently, when it is performance sensitive and when it grasps what its shifting challenges are".[237]

The above-mentioned strategies provide insights about the application of risk evaluation in a regulatory context. Indeed, implicitly, an enforcement pyramid (of the kind set out in Figure 7.10) embeds a regulator's risk appetite and risk tolerance. Each tier of the pyramid reflects different levels of the regulator's risk appetite and the interface between one tier and the next corresponds to the various risk tolerance levels established by the regulator. As a regulator moves up the pyramid, the relative risk progressively increases as do the intensity, severity and consequences associated with the regulatory response.

Risk-based regulation is another regulatory strategy that demonstrates how risk evaluation has been used in a regulatory setting. Risk-based regulation is commonly understood to comprise decision-making frameworks and procedures used

234 There are numerous examples of pyramids of enforcement employed by regulators around the world. This illustration is a simplified example of a typical pyramid of enforcement.

235 R. Baldwin & J. Black, "Really responsive regulation" (2008) 71(1) *The Modern Law Review*, pp. 59–94.

236 Ibid. p. 61.

237 Ibid. p. 61.

by regulators to prioritise regulatory activities and the deployment of resources according to an assessment of the risks that regulated entities pose to the regulatory regime's objectives.[238] The core premise of risk-based regulation is that regulators should "focus their efforts on the most serious risks that they face in achieving their objectives".[239] It requires the regulator to identify, assess and control risks.[240]

Risk-based regulation was originally introduced to provide a way of prioritising regulatory inspections based on the relative risk of non-compliance but it has become, for some regulators, a more holistic concept, encompassing wider strategic decision-making. Black and Baldwin (2010) argue that "in its idealized form, risk-based regulation offers an evidence-based means of targeting the use of resources and of prioritising attention to the highest risks in accordance with a transparent, systematic and defensible framework".[241] Peterson and Fensling (2011) further suggest that "[w]hile regulators have always made allocation choices, partly to manage limited resources, risk-based regulation aims to formalise this within a consistent structure for decision-making process".[242] Gunningham (2017) explains that:

> [l]ike responsive regulation, risk-based regulation provides a strategy for regulators to determine their response to compliance with the regulatory regime, but it does so on the basis of risk of harm to the regulatory objectives rather than on the basis of the regulatee's degree of cooperation or responsiveness to the regulator.[243]

Typically, under this approach, "risks are identified and assessed, a ranking or score is assigned on the basis of this assessment, and inspection and enforcement are undertaken on the basis of these scores".[244] In general terms, the higher the relative risk associated with non-compliance, the more intensive the response required to secure compliance.

238 R. Baldwin & J. Black, "Driving priorities in risk-based regulation: What's the problem?" (2016) 43(4) *Journal of Law and Society*, pp. 565–95, at 567. See also H. Rothstein, M. Huber, & G. Gaskell, "A theory of risk colonization: The spiralling regulatory logics of societal and institutional risk" (2006) 35(1) *Economy and Society*, pp. 91–112.

239 R. Baldwin & J. Black, n. 238 above, p. 566.

240 N. Gunningham, n. 231 above, p. 190.

241 J. Black & R. Baldwin, "Really responsive risk-based regulation" (2010) 32(2) *Law & Policy*, pp. 181–213, at 181.

242 D. Peterson and S. Fensling, "Risk-based regulation: Good practice and lessons for the Victorian context," *Conference Paper presented at the Victorian Competition and Efficiency Commission Regulatory Conference* (Melbourne, 2011), p. 7, referring to M. Sparrow, *The Regulatory Craft* (Brookings Institution Press, 2000).

243 N. Gunningham, n. 231 above, p. 192.

244 Ibid. p. 192.

Relevance of exemplars for the regulation of climate change risks

It is notable that the strategies of responsive regulation and a risk-based approach to regulation have been designed for and applied primarily in the compliance and enforcement context. They provide the basis for decision-making to deal with regulatory breaches when they occur. More specifically, they are applied to detect non-compliant behaviour and to take action to respond to non-compliance.[245] They assume that the underpinning frameworks regulate risk by proscribing, limiting or requiring certain behaviour or activity. The objective of the regulatory responses in the context of such frameworks is to secure compliance.

In contrast, as explained earlier in this book, the primary objective of responses in the context of climate change adaptation is not so much to secure compliance, but rather to facilitate change so as to minimise exposure and vulnerability.[246] Therefore, the purpose, form and targets of regulatory responses in relation to the regulation of climate change risks are likely to be starkly different to responses to which a responsive or risk-based approach to regulation is traditionally applied. Equally, the context within which risk is evaluated in relation to climate change risks is quite distinct, focusing on factors that make society and individuals vulnerable to harm, rather than factors influencing a propensity not to comply with proscriptive regulatory frameworks.

Moreover, the basis for risk evaluation in the context of a responsive or risk-based approach to regulation – that is, to determine whether risks of non-compliance are tolerable or acceptable – appears meaningless for purely natural hazards, like the hazards associated with climate change, since humans have limited, if any, choice in tolerating or accepting them. However, human activities can influence the impact of natural hazards – specifically, by changing the degree of vulnerability or exposure. Therefore, evaluation in the climate change context is more appropriately linked to the acceptability of the risks associated with not taking action compared to taking action, not with the acceptability of risks *per se*.

Nevertheless, despite these differences, responsive regulation and a risk-based approach to regulation still provide a useful reference point for the evaluation of systemic risks, including those caused by climate change. Both strategies involve determination and application of criteria to distinguish between different levels of risk, based on a regulator's risk appetite and tolerance. They use the context to modulate the regulatory response employed by the regulator based on relative risk. In other words, it is the context that informs how a particular risk arising under the regulatory regime is evaluated and, ultimately, this dictates the regulatory response to the risk. These features would appear to be equally relevant for the evaluation of systemic risks caused by climate change.

245 R. Baldwin & J. Black, n. 235 above, p. 68.

246 This issue was discussed in the context of an analysis of the relevance of traditional approach to the regulation of risk in Chapter 5.

A key practical issue in relation to the regulation of systemic risks is the analysis that should be employed to evaluate tolerance for risk. It is clear that the judgement depends upon employing a broad variety of sources of information, given that systemic risks are characterised by complexity, uncertainty and ambiguity. Indeed, as noted by the IRGC (2005), "[w]hat society is supposed to tolerate or accept can never be derived from looking at evidence alone".[247] Peterson and Fensling (2011) capture what this might mean for regulators well. They state that "[r]egulators face the challenge of determining the weight to be given to dealing with subjective perceptions of risk versus objective risks, which may be quite different" and that "data, public perceptions and political views may be contradictory".[248]

Clearly, the process of risk evaluation raises important questions, not least of which relates to the identification of stakeholders who should be involved in determining whether or not a risk is acceptable and from whose perspective the assessment is to be made. These considerations highlight the important role of the public in decision-making in the context of the regulation of systemic risks, which is discussed in the next section.

5 Public participation

Traditionally, government bodies and their expert advisors have tended to assume sole responsibility for regulating to resolve policy dilemmas, including those involving the management of significant risks facing society.[249] This so-called "technocratic approach" assumes that regulation should be based on the judgement of established "experts".[250] Meijer and colleagues (2013) suggest that this leads to an evidence-based approach to regulation.[251] Schelsky (1961, 1965), who was a supporter of such an approach, believed in "the superiority of

247 IRGC, n. 9 above, p. 37.

248 D. Peterson and S. Fensling, n. 242 above, pp. 12 and 13 referring to J. Black, "Risk-based regulation: Choices, practice and lessons being learnt" in OECD (ed.), *Risk and Regulatory Policy: Improving the Governance of Risk* (OECD, 2010) and to J. Black, "Risk-based Regulation," *Presentation to the Future of Legal Regulation Conference* (Birkbeck College, London, 2010).

249 G. Rowe, T. Horlick-Jones, J. Walls, W. Poortinga, & N.F. Pidgeon, "Analysis of a normative framework for evaluating public engagement exercises: Reliability, validity and limitations" (2008) 17(4) *Public Understanding of Science*, pp. 419–41, at 419 referring to S. Jasanoff, *The 5th Branch: Science Advisers as Policy-makers* (Harvard University Press, 1990).

250 R. Fischer, "European governance still technocratic? New modes of governance for food safety regulation in the European Union" (2008) 12(6) *European Integration Online Papers*, p. 4; A. Meijer, W. Boon, & E. Moors, "Stakeholder engagement in pharmaceutical regulation: Connecting technical expertise and lay knowledge in risk monitoring" (2013) 91(3) *Public Administration*, p. 696, at 697.

251 A. Meijer, W. Boon, & E. Moors, n. 250 above, p. 697.

scientific and expert knowledge" over what he labelled as "irrational, biased and value-laden policy-making".[252]

However, as explained in more detail in this section, the so-called technocratic approach has been challenged in preference for a more inclusive, participative approach. In essence, it represents a devolved form of administrative decision-making, where the public are as critical in the process of managing risks as experts. In an ideal regulatory context, such an approach can create a virtuous cycle that builds trust, strengthens democracy, enhances legitimacy and, ultimately, leads to better regulatory outcomes.[253] The OECD (2020) notes that, while this approach is not new, there has been a growing "wave" in the uptake of public participation in the context of public decision-making and that this could be seen as "the start of a period of transformation to adapt the architecture of representative democracy".[254]

The seventh proposed building block for an ideal to regulate systemic risks is Building Block 7, which deals with "public participation". This building block, highlighted in Figure 7.11, is an important component of a regime's context, and is explored in this section.

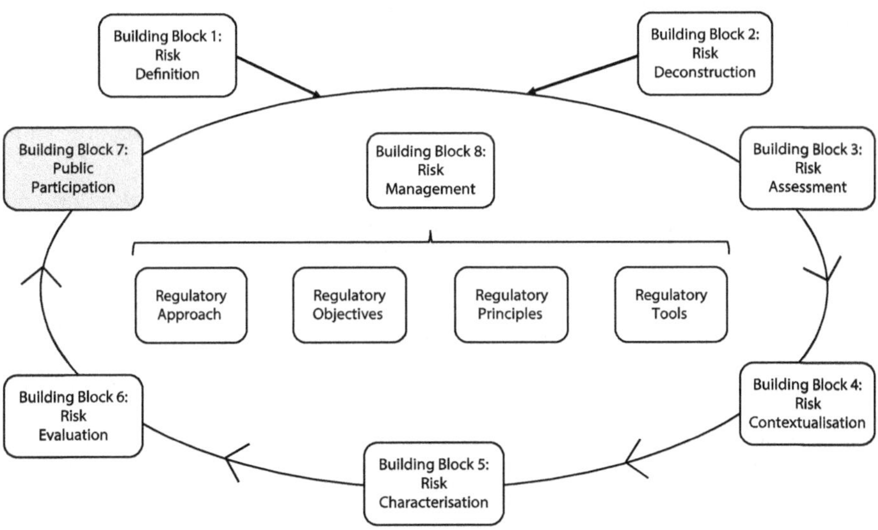

Figure 7.11 Building Block 7: Public Participation

252 R. Fischer, n. 250 above, p. 5 referring to H. Schelsky, *Der Mensch in der wissenschaftlichen Zivilisation* (Westdt. Verl., 1961) and H. Schelsky, *Auf der Suche nach Wirklichkeit. Gesammelte Aufsätze* (Eugen Diederichs Verlag, 1965).

253 OECD, *Innovative Citizen Participation and New Democratic Institutions: Catching the Deliberative Wave – Highlights 2020* (OECD, 2020), pp. 3, 6.

254 OECD, n. 253 above, pp. 3, 20.

The imperative for public participation in regulation

There are three intertwined factors that, together, create an irresistible imperative for greater public participation in the context of regulation, particularly the regulation of systemic risks.

The first factor is declining trust in government.[255] A trust deficit has arisen in the wake of a number of risk management disasters, such as the so-called "mad cow" crisis in the UK.[256] The regulatory response in place at the height of that crisis was considered to lack transparency and to be based on biased scientific opinions favouring vested interests in the agriculture sector, which led to ineffective regulation of the risks.[257] Regulatory crises such as this one have led to a general diminution of trust in government bodies and their advisors.[258] There is also specific evidence to indicate that such distrust exists in the context of the regulation of climate change risks. A study undertaken by Poortinga and Pidgeon (2003) examined how British people perceived their government and its policies toward risk regulation in various contexts, including climate change. It found that "in general, there is a tendency toward distrust in government".[259]

In turn, mistrust can compromise the ability of regulators to effectively manage risk. Indeed, research indicates that there is a positive relationship between trust in government and support for governmental activities.[260] In the field of risk research, there is now "general agreement" that trust in bodies tasked with risk management is an important factor in public perception and acceptance of risks.[261] Risk acceptance will be difficult to achieve if there is a misalignment between the government's assessment of risk on the one hand,

255 G. Stoker, M. Evans, & M. Halupka, *Trust and democracy in Australia: Democratic decline and renewal*, Report No. 1 (Democracy 2025, 2018).

256 BSE (bovine spongiform encephalopathy) was first detected in cattle in the UK in 1982. The European Commission accepted assurances from the British Ministry of Agriculture that it posed no danger to humans until 1990. While concern among the British public over health effects of eating meat from BSE-diagnosed cattle continued to grow throughout the 1990s, the British government denied the legitimacy of the public's concerns, and its position was accepted by the European Commission, which placed no restrictions on the sale of British beef. The EU's slowness to recognise health hazards to humans associated with British beef undermined trust in EU food safety regulation and the scientific expertise on which it was based.: D. Vogel, *The New Politics of Risk Regulation in Europe* (CARR, 2001), pp. 1, 11–12.

257 R. Fischer, n. 250 above, p. 3; J.E. Aldy, "Risk regulation lessons from mad cows" (2013) 8(4) *Foundations and Trends in Microeconomics*, pp. 231–313.

258 G. Rowe, T. Horlick-Jones, J. Walls, W. Poortinga, & N.F. Pidgeon, n. 249 above, p. 419.

259 W. Poortinga & N.F. Pidgeon, "Exploring the dimensionality of trust in risk regulation" (2003) 23(5) *Risk Analysis*, pp. 961–72, at 969.

260 M.N. Herian, N.C. Shank, & T.L. Abdel-Monem, "Trust in government and support for governmental regulation: The case of electronic health records" (2014) 17(6) *Health Expectations*, pp. 784–94, at 786.

261 W. Poortinga & N.F. Pidgeon, n. 259 above, p. 961.

and the public's perception of risk on the other. Such a misalignment can arise when trust is absent.

The second factor driving public participation in the context of regulation arises from democratic ideals underpinning many States around the world, which effectively assume active participation of members of the public in the business of government.[262] Holmes (2011) comments that "[p]articipation by citizens in the governance of their society is the bedrock of democracy".[263] Norris (2010) also considers that "[a]ctive citizenship is one of the key defining features of democratic governance".[264] However, she points to evidence that "in many countries today, satisfaction with the performance of democracy diverges from public aspirations" and links this divergence to, among other things, failing government performance.[265] A "democratic" approach contrasts with the "technocratic" approach, and assumes that regulation needs to be based on interactions not just among experts, but also between professional and lay stakeholders because, as noted by Windholz (2017), "[g]overnments and regulators, no matter how expert, do not have all the information and knowledge with which to diagnose issues and to identify, assess and compare all design options".[266]

The third factor in support of the incorporation of public participation into regulation is the emergence of intractable policy problems, like systemic risks. Stewart-Weeks (2011) argues that as "risks … become more complex and interconnected, government is discovering the limits of its reach".[267] Social trust is considered to be particularly critical where complex socio-technical systems generate risks that are remote from everyday experience.[268] Indeed, systemic risks have been said to be linked to "post-normal science," which requires public participation in problem-solving and decision-making to effectively address such risks.[269]

262 L. Diamond and L. Morlino, "The quality of democracy" in L. Diamond (ed.), *In Search of Democracy* (Routledge, 2016).
263 B. Holmes, *Citizens' Engagement in Policymaking and the Design of Public Services* (Parliamentary Library, 2011), p. 39.
264 P. Norris, *Democratic Deficit: Critical Citizens Revisited* (Cambridge University Press, 2010), p. 220.
265 Ibid. pp. 4, 5.
266 E.L. Windholz, *Governing through Regulation: Public Policy, Regulation and the Law* (Routledge, 2017), p. 88.
267 M. Stewart-Weeks, *The Future Course of Modern Government*, An IPAA Policy Paper (IPAA, 2011) accessible at: https://www.ipaa.org.au/wp-content/uploads/2019/06/The-Future-Course-of-Modern-Govenment-policy-paper.pdf.
268 W. Poortinga & N.F. Pidgeon, n. 259 above, p. 963 referring to T.C. Earle and G.T. Cvetkovich, "Social trust and culture in risk management" in G. T. Cvetkovich and R. E. Löfstedt (eds.), *Social Trust and the Management of Risk* (Earthscan, 1999).
269 Ravetz explains the concept of "post-normal science" as a form of "assistance" for experts. He argues that "[t]he world of 'normal science' in which they were trained has its place in any scientific study of the environment, but it needs to be

More than 20 years ago, one of the main proponents of post-normal science, Ravetz (1999)[270] advocated reliance on this approach for the management of modern, natural risks as follows:[271]

> In response to the new conditions of science in its social context, with increasing turbulence and uncertainty, the idea of "Post-Normal Science" has been developed. ... science derived from textbooks must be supplemented by other ways of knowing ... involving "extended peer communities" who deploy "extended facts" and take an active part in the solution of their problems.
>
> Of course, there have always been problems that science could not solve. But increasingly over recent generations, our civilisation has been able to tame nature in so many ways, so that for unprecedented numbers of people, life is more safe, convenient and comfortable than could ever have been imagined in earlier times. But now we are finding that the conquest of nature is not, and cannot be, complete. As we now confront nature in its disturbed and reactive state, we find extreme uncertainties in our understanding of its complex systems, uncertainties which will not be resolved by mere growth in our data bases or computing power. And since we are all involved with managing the natural world to our personal and sectional advantage, any policy for change is bound to affect our interests. Hence in any problem-solving strategy, the decision-stakes of the various stakeholders must also be reckoned with.

The relevance of these comments for the regulation of climate change risks is apparent. The disruption to natural processes caused by anthropogenic climate change has resulted in risks that are highly complex, deeply uncertain, and implicate a broad range of stakeholders. The confluence of declining trust, an evident democratic deficit, and the emergence of confounding policy problems like climate change paints a very clear picture. In particular, public participation is not only beneficial to enhance legitimacy and help build trust, it would also appear to be an indispensable means to help navigate systemic risks, particularly those caused by climate change.[272]

supplemented by awareness of the 'post-normal' nature of the problems we now confront": J.R. Ravetz, "What is post-normal science" (1999) 31(7) *Futures*, p. 647, at 653.

270 The need for "post-normal science", particularly in the context of global environmental risks, was proposed in S.O. Funtowicz & J.R. Ravetz, "Uncertainty, Complexity and Post-normal Science" (1994) 13(12) *Environmental Toxicology and Chemistry*, pp. 1881–5.

271 J.R. Ravetz, n. 269 above, pp. 647, 650.

272 S. Jodoin, S. Duyck, & K. Lofts, "Public participation and climate governance: An introduction" (2015) 24(2) *Review of European, Comparative & International Environmental Law*, pp. 117–22.

Yet, despite the relative simplicity of public participation as a concept and the theoretical benefits, Jones and Irwin (2010) question the extent to which it is practically possible to provide for meaningful interaction between technocrats and the public.[273] Indeed, as noted by Plough and Krimsky (1987), there is an underlying antagonism between rationality and democracy.[274] This begs the question of whether, in practice, public participation is appropriate for the regulation of systemic risks, including the risks associated with climate change and, more specifically, how this "antagonism" can be resolved, if at all.

Rationale for public participation in the regulation of climate change risks

The UNFCCC – the global treaty which seeks to address both climate change and the risks arising from it – specifically contemplates public participation. In particular, Article 6 of the UNFCCC outlines States' responsibilities to, amongst other things, promote and facilitate education and public awareness, public access to information, and public participation in addressing climate change and its effects.[275] The Parties to the UNFCCC have reiterated the importance of public participation and public access to information since inception of the UNFCCC.[276] However, the precept that the public should be engaged in the management of the risks arising from climate change is not just a legal one deriving from the UNFCCC. There is also a broader governance imperative, which is linked to the uncertainty, complexity and ambiguity surrounding these risks.

Etkin and Ho (2007) argue that the risks associated with climate change require public participation to help assess and, ultimately, respond to such risks:[277]

> [The] continuing lack of conceptual convergence about climate change raises the question of how knowable the risks might be. … [C]lasses of problems exist that are intrinsically unquantifiable. On the one hand, this may arise because the systems themselves are highly unstable and unpredictable (e.g. chaotic or complexly interactive systems); on the other hand it may occur because there is no means of achieving consensus about the values that need to

273 K.E. Jones & A. Irwin, "Creating space for engagement? Lay membership in contemporary risk governance", in B.M. Hutter (ed.), *Anticipating Risks and Organising Risk Regulation* (Cambridge University Press, 2010), pp. 185–207, at 185.

274 A. Plough & S. Krimsky, "The emergence of risk communication studies: Social and political context" (1987) 12(3/4) *Science, Technology, & Human Values*, pp. 4–10, at 7.

275 Article 6(a)(ii) and (iii) UNFCCCC.

276 Article 12 of the 2015 Paris Agreement, which builds upon the UNFCCC, provides that "Parties shall cooperate in taking measures, as appropriate, to enhance climate change education, training, public awareness, public participation and public access to information".

277 D. Etkin & E. Ho, n. 115 above, pp. 632–3.

> be applied in assessment due, for example, to deep-seated cultural or philosophical differences (e.g. religious or political convictions) between the stakeholders concerned. From an ecological and ethical perspective, the climate change issue … has elements of both of these problems.

They suggest that climate change is an example of a problem that "fits into post-normal science" because there is significant uncertainty regarding climate change risks, complexity arises from the feedbacks between the climate system, the human system and ecosystems and, consequently, a plurality of legitimate perspectives regarding the management of risks arising from climate change exist.[278] In their opinion, "[i]t is not surprising … that the public discourse on climate change [is] often ridden by misinformation, issue confusion, poor communication and unstated assumptions".[279] Indeed, as noted by Pidgeon and Fischhoff (2014), the complexity of the science underlying climate change makes it difficult to garner public support for "ambitious climate policies".[280]

Risk perception is complex, very diverse amongst different sectors and interest groups, and subject to inherent biases. In the climate change context, Poortinga and colleagues (2011) refer to research they say indicates awareness and knowledge of climate change and the associated risks have been rising steadily over the last two decades. However, they also point to more recent research that, in their view, suggests that scepticism and uncertainty about climate change has increased and could become a major barrier to effectively respond to climate change risks.[281] They note in this regard that "[i]t will be a difficult task to convince the public to make sacrifices in terms of their lifestyle … if they do not believe the climate is changing or will have a real impact on their lives".[282] Climate change scepticism appears to be associated, at least in part, with a "widespread perception among the public that climate change risks are spatially and temporally remote".[283]

Research also indicates that climate scepticism is often closely linked to people's core values and world views, which are difficult to alter.[284] In fact, there may be a variety of psychological reasons for not engaging with climate change, including disinterest, distrust, an unwillingness to change behaviour, or

278 Ibid. p. 627.
279 Ibid. p. 638.
280 N. Pidgeon and B. Fischhoff, "The role of social and decision sciences in communicating uncertain climate risks", in J. Árvai & L. Rivers (eds.), *Effective Risk Communication* (Routledge, 2014), p. 319.
281 W. Poortinga, A. Spence, L. Whitmarsh, S. Capstick, & N.F. Pidgeon, "Uncertain climate: An investigation into public scepticism about anthropogenic climate change" (2011) 21(3) *Global Environmental Change*, pp. 1015–24, at 1016.
282 Ibid. p. 1016.
283 Ibid. p. 1016. See also N. Pidgeon and B. Fischhoff, n. 280 above, p. 323 where the authors contend that most people find climate change "psychologically distant", irrespective of their beliefs.
284 W. Poortinga, A. Spence, L. Whitmarsh, S. Capstick, & N.F. Pidgeon, n. 281 above, p. 1022.

despondency resulting from feelings of helplessness and lack of control.[285] Accordingly, merely providing information about climate change risks, without appropriate engagement, is unlikely to be successful in enhancing appreciation of such risks because, in the absence of such engagement, people are likely to interpret such information in line with their existing attitudes and worldviews, rather than on the basis of a critical evaluation of the available evidence.[286]

Shifting perception of climate change risks through meaningful public participation will be critical to the effectiveness of responses to systemic risks to help overcome some of these psychological barriers. Understanding how the concept of public participation – encompassing both risk communication and stakeholder engagement – can be practically enlivened is an important first step in determining how public perceptions can be shifted in practice and is discussed below.

Public participation mechanisms

The field of *risk communication* initially developed as a means of exploring how risk information – particularly, risk assessments and risk management decisions – could be communicated to the public to help bridge the divide between expert judgements and public perceptions of risk.[287] Over time, however, there has been a growing realisation that the general public does not necessarily incorrectly assess science and expert opinions regarding risks.[288] Indeed, Frewer (2004) argues that technocrats have "underestimated the ability of non-experts to understand uncertainty … [thereby] increasing public distrust in the motives of regulators and scientific advisors".[289] Renn and colleagues (2011) further suggest that "risk communication … needs to accommodate far more than a simple one-way transfer of information … the mere provision of 'expert' information is unlikely to address public and stakeholder concerns or resolve any underlying societal issues".[290]

In fact, the specific form and content of communication may need to differ depending upon the context in which the communication occurs, including the political context, dominant social values among stakeholders, and the status of pre-existing trust relationships between the various actors in the risk governance process.[291] The main challenge appears to be to ensure that information provided to the public about risk is reliable, meaningful, useful, accessible and suited to the context.

There are a number of suggestions that rise to the challenge. Frewer (2004) advocates providing the public with a clear message regarding risks and associated uncertainties, including the source of uncertainty and the nature and

285 Ibid. p. 1022.
286 Ibid. p. 1022.
287 IRGC, n. 9 above, p. 54.
288 Ibid. p. 54.
289 L. Frewer, "The public and effective risk communication" (2004) 149(1–3) *Toxicology Letters*, pp. 391–7, at 394.
290 O. Renn, A. Klinke, & M. van Asselt, n. 23 above, p. 242.
291 Ibid. p. 243.

extent of disagreement between experts.[292] Leiss (2004) suggests that the focus of exchanges with relevant stakeholders should be on interpreting the results of scientific assessments for non-expert audiences, understanding the basis of public risk perceptions, and working with interested parties towards a shared understanding of risk factors.[293] Boholm and Corvellec (2014) further suggest that more attention needs to be paid to the public's values and preferences, in order to enhance the efficiency of risk communication.[294] These inter-related suggestions are all laudable and desirable in that they aim to ensure that risk information is tailored to the interests, concerns and needs of the recipients of that information. Nevertheless, they may well be difficult to implement in practice, particularly in as politicised an area as climate change, where information may be filtered, packaged and publicised in a manner to suit political objectives, rather than to genuinely enhance public understanding.

Public engagement goes beyond risk communication and implies a two-way interaction with the public. Such an approach recognises the importance of involving people who may be exposed or affected by risk to participate in the decision-making process to manage risk.[295] This can make the decision-making process more democratic and increases the legitimacy and public acceptance of the resulting decisions.[296] It can also inform regulatory decisions by providing regulators with access to the public's diverse perspectives, knowledge, information and expertise and, thereby, help to enhance the quality of regulatory decisions.[297] In summary, engagement can provide the public with a voice, as well as enable genuine consideration to be given to the public's issues, concerns and perspectives in the decision-making process.

As in the case of risk communication, engagement should ideally be tailored to the context in which it will be applied. In the climate change context, Jodoin and colleagues (2015) have noted that "creating meaningful opportunities for public participation may contribute to the emergence of the sort of productive, collaborative and inclusive policy process that the governance of climate change requires".[298]

292 L. Frewer, n. 289 above, pp. 395–6.

293 W. Leiss, "Effective risk communication practice" (2004) 149(1–3) *Toxicology Letters*, pp. 399–404, at 402.

294 Å. Boholm and H. Corvellec, "A relational theory of risk: Lessons for risk communication" in J. Árvai & L. Rivers (eds.), n. 280 above, p. 6.

295 O. Renn, "Three decades of risk research: Accomplishments and new challenges" (1998) 1(1) *Journal of Risk Research*, pp. 49–71, at 63; G. Rowe, T. Horlick-Jones, J. Walls, W. Poortinga, & N.F. Pidgeon, n. 249 above, p. 420.

296 G. Rowe, T. Horlick-Jones, J. Walls, W. Poortinga, & N.F. Pidgeon, n. 249 above, p. 420; OECD, *Emerging Systemic Risks in the 21st Century: An Agenda for Action* (OECD, 2003), p. 95.

297 J. Nash & D.E. Walters, "Public engagement and transparency in regulation: A field guide to regulatory excellence" (2015) *Penn Program on Regulation*, p. 49, at 29; E. Heims and M. Lodge, "Customer engagement: Towards a new era in economic regulation?" in M. Lodge (ed.), *Customer engagement in regulation, Discussion Paper No. 82* (CARR, 2016), p. 4.

298 S. Jodoin, S. Duyck, & K. Lofts, n. 272 above, p. 121.

Once again, this is likely to be easier said than done. A variety of mechanisms exist to provide for public engagement, which vary in the scope of participation, intensity of engagement, and extent of authority that regulators share with the public.[299] As discussed below, the success, or otherwise, of a mechanism to engage the public is likely to depend critically on how it is designed and implemented.

Designing public participation for regulatory responses to systemic risks

Despite the apparent benefits, effective public participation can be challenging and is likely to be particularly so in the context of the regulation of systemic risks caused by climate change. Among other things, there is a risk of inefficiency, inadequate or imbalanced public representation, regulatory capture and, at worst, compromised regulatory outcomes.[300] Moreover, Pidgeon (2014) comments that, as a society, "[w]e lack a fundamental strategic capacity not only in 'risk communication' …, but [also] in uncertainty assessment, in appropriate methods for situating 'values' in public and stakeholder engagement, and in fostering citizen deliberation for the wider public good".[301] Research and empirical evidence tends to bear out Pidgeon's concerns.

Nash and Walters (2015) consider that public participation may contribute to inferior regulatory outcomes resulting from regulatory capture associated with inexpert decision-making and domination of vested interests in the decision-making processes.[302] They note that these mechanisms may also result in burdensome, protracted and ineffectual regulatory processes.[303] Löfstedt and van Asselt (2008) argue that greater public participation may divert attention from the main task – that is, managing the risks at hand.[304]

The relationship between public participation and regulatory effectiveness has been tested in the context of governance of pharmaceutical risks. Meijer and colleagues (2013) surprisingly concluded that more lay stakeholder engagement does not result in a stronger process, nor better performance of the regulatory framework.[305] Nevertheless, the absence of public engagement was found to result in a decline in the long-term willingness to use pharmaceuticals if the public is not satisfied with the system of pharmaceutical regulation.[306]

299 Various public participation mechanisms are set out in J. Nash & D.E. Walters, n. 297 above, pp. 8–15 and P.C. Stern, H.V. Fineberg, & National Research Council (U.S.) (eds.), n. 184 above, Appendix B.

300 J. Nash & D.E. Walters, n. 297 above, pp. 5–6, 29.

301 N. Pidgeon, "Complexity, uncertainty and future risks" (2014) 17(10) *Journal of Risk Research*, pp. 1269–71, p. 1271.

302 J. Nash & D.E. Walters, n. 297 above, p. 5.

303 Ibid. p. 5.

304 R. Löfstedt and M. van Asselt, "A framework for risk governance revisited" in O. Renn & K.D. Walker (eds.), n. 7 above, p. 82.

305 A. Meijer, W. Boon, & E. Moors, n. 250 above, p. 709.

306 Ibid. p. 709.

Accordingly, the researchers concluded that public engagement leads to higher stakeholder satisfaction.[307]

These findings tend to confirm Dahl's (1994) "democratic dilemma" – namely, that democratic participation and system effectiveness are irreconcilable.[308] Meijer and colleagues (2013) argue that this dilemma "becomes poignant in the governance of topics that are complex and inherently uncertain".[309] In these cases, policy- and law-makers are faced with a difficult quandary. Public participation is needed to legitimise public decision-making, even in cases where technical expertise is critically important. However, as noted by Renn (2006), public perceptions can be based, at least in part, on biases, anecdotal evidence and false assumptions.[310]

The question arises as to how public participation can be undertaken in a way that balances the benefits against the possible costs and disadvantages, particularly diminution of regulatory effectiveness. Research suggests that there are no fixed formulas for success, as the results of public participation depend not just on the characteristics of the methods used, but also on factors that shape the context in which regulators act.[311]

The OECD (2020) has established a set of "Good Practice Principles for Deliberative Processes for Public Decision Making".[312] Similar guidance can be found in jurisdictions around the world, and at various levels of government. Such guidance is useful because it sets out the main principles and considerations to be accounted for when designing a mechanism to engage the public in a policy or regulatory context, but the guidance is necessarily high-level and abstract. The proponents of the IRGC Framework have gone one step further by proposing generic approaches to public participation, based on the relevant category of systemic risk.

For *simple risks*, the IRGC suggests that "a sophisticated approach to involve all potentially affected parties is not necessary. Most actors would not even seek to participate since the expected results are more or less obvious" and that, therefore, "relatively basic consultation with experts to decide which management option should be employed" is appropriate.[313] In relation to *complex risks*, the IRGC argues that decisions about the management of risk "may benefit from a wider dialogue amongst a broader range of experts and affected stakeholders".[314] It suggests that, for complex risks, "discourse should be inspired by different science

307 Ibid. p. 709.

308 R.A. Dahl, "A democratic dilemma: System effectiveness versus citizen participation" (1994) 109(1) *Political Science Quarterly*, p. 23.

309 A. Meijer, W. Boon, & E. Moors, n. 250 above, p. 696.

310 O. Renn, "Participatory processes for designing environmental policies" (2006) 23 (1) *Land Use Policy*, pp. 34–43, at 35.

311 J. Nash & D.E. Walters, n. 297 above, p. iii.

312 OECD, n. 253 above, pp. 9–11.

313 IRGC, n. 11 above, pp. 29, 30; O. Renn "White Paper on Risk Governance: Towards an Integrative Framework" in O. Renn & K.D. Walker (eds.), n. 7 above, p. 46.

314 IRGC, n. 11 above, p. 30.

camps and the participation of experts and knowledge carriers".[315] Regarding *uncertain risks*, the IRGC proposes that decisions about the management of risk "may benefit from a wider dialogue amongst a broader range of experts and affected stakeholders"[316] and recommends "collective reflection about balancing the possibilities for over- and under-protection".[317] With respect to *ambiguous risks*, the IRGC suggests that discourse is needed "where competing arguments, beliefs and values are openly discussed. The opportunity for resolving these conflicting expectations lies in the process of identifying common values, defining options … to find equitable and just distribution rules."[318] They further recommend that "involving civil society is recommended, in part to capture and reconcile the various existing perceptions of a risk and options for its management".[319]

In summary, a more intensive discourse with a broader range of stakeholders is advocated as the degree of uncertainty, complexity and ambiguity associated with the systemic risk that is sought to be managed increases. This sliding scale approach to public participation is not dissimilar to the "continuum" referred to by Windholz (2017) regarding both "who" to engage and "how" to engage[320] and is a variation of Arnstein's "Ladder of Citizen Participation", which involves a typology of escalating levels of public participation.[321]

From a purely theoretical perspective, a graduated spectrum of public participation approaches based on the applicable category of systemic risk seems sound and rational. Indeed, there is a compelling logic associated with modulating the scope of stakeholders and form of public participation based on the relative degree of uncertainty, complexity and ambiguity associated with systemic risks. Such an approach also appears to be in line with observations by the OECD (2020) that "the most appropriate deliberative model depends primarily on the policy problem. The more complex the question is and the wider its implications … the more elaborate deliberative process is applicable".[322]

However, it is unclear whether the application of this escalating approach could work as a matter of practice. Indeed, it implies that public participation would be characterised by the greatest number of stakeholders and the most inclusive and intensive engagement for risks that are at the extreme end of the spectrum – specifically, those that are the most complex, uncertain and ambiguous. Yet, the prospect of the process becoming derailed would likely be highest in this context given the expansive scope and nature of public participation that is contemplated. In fact, the risk of Dahl's dilemma playing out in

315 O. Renn, n. 313 above, p. 46.
316 IRGC, n. 11 above, p. 30.
317 O. Renn, n. 313 above, p. 46.
318 Ibid. p. 47.
319 IRGC, n. 11 above, p. 30.
320 E.L. Windholz, n. 266 above, p. 88.
321 S.R. Arnstein, "A ladder of citizen participation" (2019) 85(1) *Journal of the American Planning Association*, pp. 24–34.
322 OECD, n. 253 above, p. 16.

this scenario is real and could altogether defeat the purpose of providing for public participation. Moreover, while it makes rational sense for a largely technocratic approach at the opposite end of the spectrum, with limited or no public participation, because the relative degree of complexity, uncertainty and ambiguity is low, it conflicts with growing momentum to provide for public participation for the regulation of all types of risks. Indeed, public participation does not merely yield benefits for the government, but is also fundamental to restoring public trust, addressing the democratic deficit, and placing some degree of regulatory power in the hands of the public.

Overall, despite these misgivings, the systematic, structured and seemingly rational approach to establish the basic parameters for public participation based on the category of systemic risk is worth considering. At the least, this approach is a decent starting point in determining how to embed an effective public participation mechanism in a regime to regulate systemic risks caused by climate change and, thereby, address Dahl's democratic dilemma.

IV Content of an ideal regime to regulate systemic risks

A framework to manage risk is typically characterised by multiple stages to gather information, understand and analyse risk, and culminates in action that is designed to transform unacceptable risk so that it is deemed acceptable by relevant stakeholders or, at least, more tolerable. This "stagist" approach to risk management is aimed at building a robust foundation for an effective response to risk. As Windholz (2017) puts it, stagist approaches in a regulatory context "enable us to examine the regulatory endeavour without being overwhelmed by its complexity … and provide much needed structure and functional guidance for the policy-maker that alternative frameworks do not provide".[323]

The final stage of the risk management process involves deciding how to respond to risk with the primary objective of preventing or, at a minimum, reducing to an acceptable level harm to humans, including the communities to which they belong, and the things that they value. In relation to the design of a regulatory regime, the response to risk will involve choosing among the available regulatory options for inclusion in the regime in order to best mitigate risk. The decision about how to respond to risk during this final stage will be framed by the regulatory boundaries that have been established to manage risk and informed by the context for the regulatory regime, which were considered earlier in this chapter.

This part explores the eighth and final building block for the regulation of systemic risk – Building Block 8: Risk Management. This building block involves identification of the main regulatory options that should ideally be available under the regulatory regime to respond to systemic risk. As illustrated in Figure 7.12, there are several facets of a regulatory regime that could be used to craft the regulatory content to respond to risk – namely, the regulatory approach, objectives, principles and tools.

323 E.L. Windholz, n. 266 above, pp. 83–4.

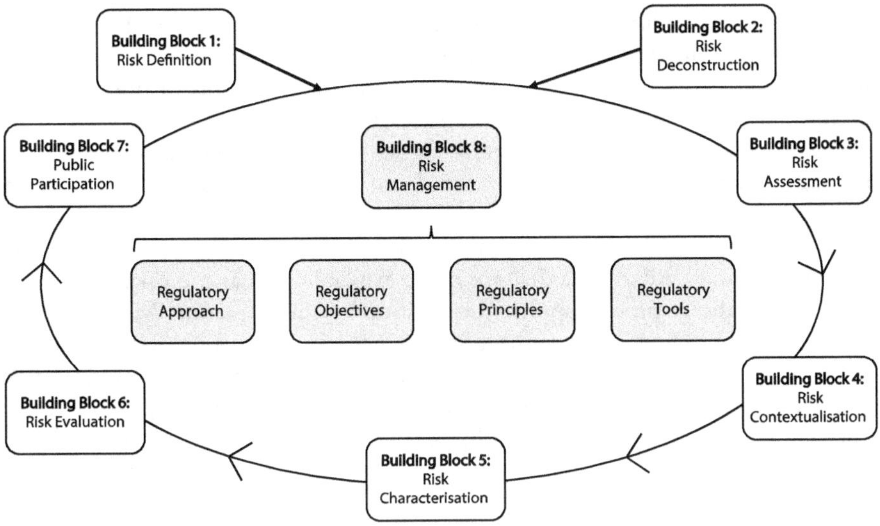

Figure 7.12 Building Block 8: Risk Management

I The facets of regulatory content

In simplistic terms, the content of a regulatory regime is embodied in the rules and requirements contained in the regime that can be relied upon to manage risk governed by the regime. As previously noted, Levi-Faur (2011) describes regime content as referring to the internal structure of the regime.[324] In more practical terms, it is a composite of various concrete elements that, together, are designed to manage risks covered by the regime. These include the regulatory approach, objectives, principles and tools, which are discussed below.

Regulatory approach

Each regulatory regime will be characterised by a particular regulatory approach, which will typically be reflected in the overall structure of the regime. There is a variety of regulatory approaches that have been employed to manage risks, ranging from traditional, prescriptive command and control regulatory instruments to those that are much more flexible and principle-based where entities that are the subject of regulation have considerable discretion in complying with the applicable regulatory requirements.[325]

As the name suggests, prescriptive regulation specifically proscribes activities that are prohibited or prescribes the activities that may be undertaken or how those activities must be undertaken. Such an approach assumes that the risks

324 D. Levi-Faur, "Regulation & regulatory governance" in D. Levi-Faur, *Handbook on the Politics of Regulation* (Edward Elgar Publishing, 2011), p. 14.

325 A. Freiberg, *Regulation in Australia* (Federation Press, 2017), Chapter 6.

governed by the regime are relatively clear and defined, which enables them to be specifically targeted by prescriptive rules. In contrast, principle-based regulation relies upon principles, which are overarching precepts of the regulatory regime that are used to drive achievement of regulatory outcomes.

The regulatory approach selected for a regulatory regime serves several important purposes. It signals the attitude of policy- and law-makers towards the risks that are sought to be regulated. The regulatory approach will typically dictate the specific manner in which risks are regulated under the regime, including the particular tools that are made available to manage risks. It will also inform how the regime is applied in practice by the regulator. Moreover, the regulatory approach can affect the regime's capacity to adapt to changes in the regime's context, including the augmentation and, at times, mutation of risks over time. In summary, the regulatory approach is likely to pervade every aspect of the regulatory regime, either explicitly or implicitly, emphasising the need for the implications of each approach to be fully appreciated before one is employed to manage risks.

Regulatory objectives

Regulatory objectives are at the core of every regulatory regime and are typically set out in the "objects" clause of the relevant regulatory framework. In essence, regulatory objectives represent the desired outcomes that a regulatory regime aspires to and is designed to deliver. These objectives will inform the interpretation of the regime's detailed requirements, but will also affect the regulatory tools made available to the regulator and the manner in which those tools are applied in practice.

In general terms, the objectives of many regulatory regimes focus on one or more drivers of risk, to avoid or, at least, mitigate risk. In the context of the regulation of the risks associated with climate change, the objectives may aim at removing or containing the source of *hazards* that cause harm. For example, a regime aimed at reducing the risks arising from bushfires may regulate the management of vegetation that could be fuel for such fires. The objectives may seek to isolate activities so that *exposure* to harm is prevented or reduced by, for example, prohibiting settlements in areas that are prone to bushfires. Alternatively or in addition, the objectives may be directed at making risk targets less *vulnerable* to potential harm, such as by requiring structures that could be in the pathway of a bushfire to be built to a particular standard that has been proven to provide resilience to bushfire attack.

Depending upon the system at risk from the effects of climate change, and the scope of risks to which the system may be subject, it may not be practically possible or cost-effective to pursue each of these objectives concurrently in a single regulatory regime; the process of deconstructing risk to assist in targeting regulatory intervention can be useful to clarify the most important regulatory objectives. The manner in which objectives for the regulation of systemic risks are formulated, particularly their capacity to encompass future contexts, will be critical in dealing with risks that could arise in the future.

Regulatory principles

Many regulatory regimes are underpinned by overarching principles, which serve a similar function to regulatory objectives. They express key ideals that are sought to be achieved or preserved by the regulatory regime. Black and colleagues (2007) note that principle-based rules "express fundamental obligations that all should observe".[326] More practically, principles guide interpretation of a regime and assist in determining when regulatory intervention might be needed – that is, when the principles are at risk of being breached.[327]

Principles are likely to be particularly important in the context of the regulation of systemic risks because of their generic, non-specific nature, which make them malleable to the context in which they are to be applied. More specifically, principles can be used to guide and "stabilise" the regulation of risk, but do not embed assumptions about the context in which risk has arisen or pre-emptively dictate or determine the response to risk. As noted by Black (2010), "[i]n order to provide some stability, some foundation on which to base risk regulation, governments and policy makers have tried to devise various 'rules' or principles for how to make decisions about risk".[328] In the specific context of climate change, Driessen and van Rijswick (2011) suggest that normative principles may "promote a more predictable and stable rule of climate law and policy", while avoiding what they refer to as "legal instrumentalism".[329]

Two principles that would appear to be particularly relevant in relation to the regulation of systemic risks caused by climate change are the "precautionary principle" and "resilience". In a regulatory context, the former principle is well-established in many legal systems and means that, when activities or omissions may lead to irreversible harm that is scientifically plausible but uncertain, regulatory action should be taken to avoid or diminish that harm before it materialises.[330] Regulatory protection can only be relaxed if "scientific evidence has progressed to a point that one would be able to translate former uncertainties in terms of risk and adverse effects to terms of defined, consensual levels of harm and damage".[331] In practice, the principle demands a proactive, precautionary approach for systemic risks, one which seeks to avoid such risks materialising, particularly where the risks to the system in question are likely to be irreversible.

As for resilience, it has been defined as the "ability to respond to external (or internal) disturbances while undergoing change so as to still preserve essentially the

326 J. Black, M. Hopper, & C. Band, "Making a success of principles-based regulation" (2007) 1(3) *Law and Financial Markets Review*, pp. 191–206, at 192.
327 Black and colleagues note that principles are "purposive, expressing the reason behind the rule": ibid. p. 192.
328 J. Black, n. 100 above, p. 309.
329 P.P.J. Driessen & H.F.M.W. van Rijswick, "Normative aspects of climate adaptation policies" (2011) 2(4) *Climate Law*, pp. 559–81, at 564.
330 R. von Schomberg, "The precautionary principle: Its use within hard and soft law" (2012) 3(2) *European Journal of Risk Regulation*, pp. 147–56, at 148.
331 Ibid. p. 156.

same functions".[332] In practice, resilience is the capacity of a system to "bounce back" from an adverse disruption or shock.[333] A resilience-based approach requires consideration of "future threats to system stability" and development of "countermeasures" to address those threats to allow a "soft landing" for the system when subjected to a significant external shock.[334] Such an approach accepts that it is not always possible to identify potential hazards to a system, nor to quantify the likelihood of occurrence. Even where this is possible, it may not be practically feasible to reduce the chance that the hazard will occur. Moreover, resilience objectives focus less on preserving the status quo and more on determining how to achieve a necessary level of performance – for example, in the case of infrastructure systems, this may involve continuing to deliver essential services.[335] Despite the clear relevance of resilience to risk management, it does not have the same status in legal systems as the precautionary principle.

Regulatory tools

One of the core facets of a regulatory regime is the suite of regulatory tools that equip the regulator to achieve the outcomes and objectives underlying the regime. Without these tools, the regulator is largely powerless to respond to risks governed by the regulatory regime. Freiberg (2017) observes that the concept of a "regulatory tool" is a broad one and covers the spectrum of action that may be employed to address a public problem, including regulatory interventions aimed at inducing individuals and groups to make decisions and take actions compatible with public policies.[336] It is obvious from this description that regulatory tools are numerous and that they may take a broad variety of forms.[337] This book concerns regulatory tools

332 D.J. Yu, P.S.C. Rao, C.J. Klinkhamer, E.H. Krueger, N. Sangwan and K. Sung, "Aligning different schools of thought on resilience of complex systems and networks" in M.V. Florin and I. Linkov (eds.), *IRGC Resource Guide on Resilience (Volume 1)* (EPFL International Risk Governance Center, 2016), p. 268. See also OECD, *Boosting Resilience through Innovative Risk Governance* (OECD, 2014), at p. 23.

333 The principle of resilience in a regulatory context – particularly in relation to climate change adaptation law – is closely linked to the concept of adaptive capacity of socioecological systems. Kundis Craig suggests that "[r]esilience reflects a system's ability to absorb impacts and continue to function, while adaptive capacity refers to a system's ability to change to adjust to new conditions": R.K. Craig, "'Stationarity is dead' – long live transformation: Five principles for climate change adaptation law" (2010) 34(1) *The Harvard Environmental Law Review*, p. 9, at 22.

334 I. Linkov, B.D. Trump and C. Fox-Lent, "Resilience: Approaches to risk analysis and governance" in M.V. Florin and I. Linkov (eds.), n. 332 above, p. 4.

335 E.D. Vugrin, "Critical infrastructure resilience" in M.V. Florin and I. Linkov (eds.), n. 332 above, p. 237.

336 A. Freiberg, n. 325 above, pp. 199–200, citing L.M. Salamon, *The Tools of Government: A Guide to the New Governance* (Oxford University Press, 2002) and R. Landry and F. Varone, "The choice of policy instruments: Confronting the deductive and interactive approaches" in R. Eliadis, M.M. Hill and M. Howlett (eds.), *Designing Government: From Instruments to Governance* (McGill-Queen's University Press, 2005).

337 OECD, *Risk and Regulatory Policy: Improving the Governance of Risk* (OECD Publishing, 2010), pp. 148–9.

in the form of government, rule-based intervention through regulatory regimes, while acknowledging that a raft of administrative and policy tools exist outside the context of a regulatory regime, as well as court-based tools such as the institution of proceedings for damages or civil liability. As a general rule, regulatory tools are not automatically prescribed for a particular regime; rather, it is ultimately for policy- and law-makers to decide which tools to include in a regulatory regime and up to the relevant regulator to decide which of those tools to employ when applying the regime in practice on a case-by-case basis. This is not an easy task.

2 *Matching regulatory content to policy problems*

Freiberg (2017) argues that the task of matching policy problems to regulatory options is difficult and seldom perfect.[338] In practical terms, the exercise involves identifying the appropriate mix of regulatory approach, objectives, principles and tools to address a particular problem and the associated risks. Howlett and Mukherjee (2018) note "the many dilemmas that can appear in the path of the development of effective policy tool or 'toolkit' designs as well as the realities of policy implementation that they face on the ground."[339] The challenges associated with determining how regulatory content should be designed are likely to be particularly marked in the context of regulation of systemic risks, like those caused by climate change. Howlett and Mukherjee note that "better-designed policies are more likely to correctly identify or solve the problems they are expected to address … while poorly or non-designed ones are more likely to fail for the same reason".[340]

In summary, when designing the content of a regulatory regime, the orchestration of the various available regulatory options is likely to be challenging. A key issue for the regulation of systemic risks – as with many other risks – is to determine how best to combine the available options for the regulatory approach, objectives, principles and tools in order to maximise the chance of delivering an effective regulatory response to risk when the regime is applied in practice.

The IRGC has suggested generic strategies for the management of systemic risks. The notion that risks could be tackled using regulatory strategies is very appealing. It implies that the emphasis in developing the content of a regulatory regime is on strategic goals and outcomes underlying a regulatory regime, rather than the "nuts and bolts". In practical terms, such strategies establish a blueprint of the fundamental aspects of the regime's content. Provided those fundamentals are present, theoretically, they should lead to an appropriate regulatory response when the regime is applied in practice.

It would be fair to say that this method of designing regulation sounds more aspirational, rather than realistic. Surely, the typically painstaking process of

338 A. Freiberg, n. 325 above, p. 199.
339 M. Howlett and I. Mukherjee, "The importance of policy design: Effective processes, tools and outcomes" in M. Howlett & I. Mukherjee (eds.), *Routledge Handbook of Policy Design* (Routledge, 2018), p. 6.
340 Ibid. p. 4.

developing the detailed provisions of a regulatory regime is indispensable to the regime's effectiveness. Nonetheless, given the challenges presented by systemic risks and the associated challenges of designing regimes to regulate them, it is worth exploring the potential of the IRGC strategies to overcome at least some of those challenges. The generic risk management strategies that have been put forward in the IRGC Framework are outlined below, together with a possible explanation of how those strategies could be translated into regulatory form.

A routine approach for simple risks

Generally speaking, "simple risks" – like those associated with sea-level rise[341] – are relatively well-known and understood. The potential negative consequences associated with this category of risks are obvious and clearly causally connected to the hazard, uncertainty regarding the probability and consequences associated with these risks is relatively low, and there is limited controversy regarding the interpretation of the risk.[342] In light of these core characteristics, the proponents of the IRGC Framework have suggested a strategy for simple risks that is "routine", with a focus on risk reduction and the use of "traditional decision-making".[343] They further suggest that readily available data can be utilised to undertake a statistical analysis and the results of this analysis can be used to determine the response to risk.[344]

The concept of a routine approach to risk that is characterised by traditional decision-making suggests that the process to address risk follows a predictable, well-traversed pattern and could be interpreted as implying a standardised, linear response to risk, for which the bells and whistles of risk governance would not be required. Rather, a prescriptive regulatory regime, with the primary objective to reduce risk that is well-known and understood, would be appropriate, particularly to specify activities that are prohibited or authorised, the circumstances when authorised activities can be undertaken, and/or how those activities must be undertaken. Readily available data could be used to undertake a cost-benefit analysis, which would help identify specific regulatory tools to be employed to reduce risk in a particular case.

Improving robustness for complex risks

In the case of "complex risks" – such as, the bushfire risks[345] – identifying and quantifying causal links between a multitude of potential causal agents and specific

341 As suggested above in the section on "Risk Characterisation" in this chapter, sea-level rise is arguably an example of a "simple risk".

342 IRGC, n. 9 above, p. 45; IRGC, n. 11 above, p. 17.

343 IRGC, n. 9 above, p. 45; IRGC, n. 11 above, p. 24.

344 O. Renn, "White Paper on Risk Governance: Towards an Integrative Framework" in O. Renn & K.D. Walker (eds.), n. 7 above, p. 37.

345 The section on "Risk Characterisation" above suggests that the risks associated with bushfires may be categorised as "complex".

observed effects is extremely difficult because of the inherent complexity associated with this category of systemic risk.[346] There is typically scientific dissent regarding causal links between actions or events and adverse consequences.[347] Consequently, it is difficult to determine the extent of investment in resilience required to effectively mitigate the consequences when they arise.[348] The proponents of the IRGC Framework have suggested a multi-faceted response to complex risks. First, they say that the response should focus on being "risk-informed" regarding the risk agent (i.e. the hazard) by seeking to access and act on the best available scientific information and expertise about the risk agent.[349] Second, the response should improve the "robustness" of the risk-absorbing system by aiming to reduce the system's vulnerability to known hazards and threats.[350] They suggest that sophisticated models of probabilistic inferences may be required for this category of risks to address the significant degree of causal complexity.[351]

The suggested risk management strategy indicates that the primary objective of the regulatory regime for complex risks would be on reducing vulnerability and the principle of resilience would be critical to achieving this objective. Moreover, the regulatory form of this strategy would involve a multi-pronged approach. One facet of the regime would focus on enhancing information about the likely behaviour of the risk agent and the other facet would seek to reduce vulnerability of the risk-absorbing system to risk by enhancing resilience.

A precautionary approach for uncertain risks

"Uncertain risks" – including flood risks[352] – are characterised by incompleteness of knowledge, particularly, scientific knowledge. The knowledge that does exist relies on uncertain assumptions and predictions. This makes it difficult to assess the probability and possible outcomes of adverse consequences.[353] The proponents of the IRGC Framework have suggested that, rather than investing significant effort to obtain more knowledge about uncertain risks, which may or may not enhance the effectiveness of the response to such risks, another option is to develop better ways to co-exist with uncertainty regarding the risk agent through a "precautionary" approach.[354] They also suggest that uncertain risks are best managed using a "resilience-focused" strategy regarding the risk-absorbing system, with the intention of

346 IRGC, n. 9 above, pp. 29–30, 45; IRGC, n. 11 above, p. 18.
347 IRGC, n. 9 above, p. 45.
348 O. Renn, A. Klinke, & M. van Asselt, n. 23 above, p. 241.
349 IRGC, n. 9 above, pp. 15, 45; IRGC, n. 11 above, p. 24.
350 IRGC, n. 9 above, pp. 15, 45; IRGC, n. 11 above, p. 24.
351 A. Klinke & O. Renn, "Adaptive and integrative governance on risk and uncertainty" (2012) 15(3) *Journal of Risk Research*, pp. 273–92, at 275.
352 In the section on "Risk Characterisation" above, it was suggested that flood risks can be categorised as "uncertain".
353 IRGC, n. 9 above, p. 30; IRGC, n. 11 above, p. 18.
354 O. Renn, A. Klinke, & M. van Asselt, n. 23 above, p. 241.

increasing a system's coping capacity to the point where it can withstand surprises.[355]

Based on the suggested risk management strategy for uncertain risks, a principle-based approach would appear to be particularly appropriate in the context of a regime to regulate this category of systemic risk. The core principles of the regime would include the precautionary principle to ensure that the characteristics of the risk agent are properly accounted for in responding to risk, as well as the principle of resilience to reduce the vulnerability of the risk-absorbing system to risk.

Discourse for ambiguous risks

"Ambiguous risks" – such as increasing temperatures[356] – are characterised by different viewpoints regarding the interpretation of observations or data. Differences in views may also exist in relation to whether or not risks are tolerable as well as the appropriate risk management actions.[357] In the case of this category of risks, the proponents of the IRGC Framework recommend a "discourse" approach, which involves a deliberative style and participative process, involving discourse between all relevant stakeholders. Such a process is aimed at producing a collective understanding among all stakeholders on how to interpret the risks, particularly the acceptability or tolerability of risks and the appropriate risk management actions.[358] A co-operative regulatory approach would appear to be particularly appropriate for this category of risk, to help facilitate discourse and drive consensus, particularly regarding the objectives of and mechanisms to respond to risk under the regulatory regime.

3 A strategic approach to managing systemic risks using regulation

There are several important facets of the content of a regulatory regime – that is, the regulatory approach, objectives, principles and tools. Each facet has a different yet complementary and mutually reinforcing role to play in managing risks that are sought to be addressed under the regime. When designing a regulatory regime, and deciding upon its composition, there are various options available to policy- and law-makers for each facet. As explained, choosing between these options can be challenging, particularly for systemic risks for which there are no existing regulatory paradigms. Nonetheless, the choice is likely to have an important impact on the regime's capacity to effectively mitigate risk.

The use of the IRGC risk management strategies to determine how to regulate particular categories of systemic risks is appealing and what is produced from that exercise appears to make intuitive sense. Specifically, as the level of complexity,

355 IRGC, n. 11 above, p. 24.

356 In the section on "Risk Characterisation" above, it was suggested that increasing temperatures can be categorised as "ambiguous".

357 IRGC, n. 9 above, pp. 30–1; O. Renn, A. Klinke, & M. van Asselt, n. 23 above, p. 242.

358 IRGC, n. 9 above, p. 30; IRGC, n. 11 above, p. 25.

uncertainty and ambiguity of the risk being regulated intensifies, the regulatory regime should ideally focus less on technocracy, objectivity and prescription and more on democracy, subjectivity and flexibility. At the very least, the strategies provide a starting point to guide policy- and law-makers through the challenging task of matching regulatory options to risk. Regulatory strategies of this kind are not intended to be comprehensive nor prescriptive but, rather, indicate the regulatory options that could be applied for each category of systemic risk. Indeed, in the absence of detailed guidance from government about how risks should be managed – other than the standard exhortation to choose options that offer the greatest net benefit – risk management strategies can help crystallise choices about the appropriate regulatory options to manage a particular risk or set of risks.

What follows in the next four chapters is a consideration of the design of regulatory regimes to address systemic risks caused by climate change from across the world to determine the extent to which they are aligned with the ideal for the regulation of systemic risks outlined in this chapter or, at least, evidence a risk governance approach. The case studies cover regulatory regimes to address climate change risks to ecosystems in the EU, the UK health system, the US financial system, and systems of infrastructure in Australia.

Bibliography

Aldy, J.E., & Kip Viscusi, W. (2013). "Risk regulation lessons from mad cows." *Foundations and Trends in Microeconomics*, 8(4), 231–313. https://doi.org/10.1561/0700000046.

Arnstein, S.R. (2019). "A ladder of citizen participation." *Journal of the American Planning Association*, 85(1), 24–34. https://doi.org/10.1080/01944363.2018.1559388.

Árvai, J., & Rivers, L. (eds.). (2014). *Effective Risk Communication*. Routledge.

Aven, T. (2009). "Risk analysis and management. Basic concepts and principles." *Reliability: Theory & Applications*, 4(1(12)), 57–73.

Aven, T. (2013a). "On the meaning and use of the risk appetite concept." *Risk Analysis*, 33(3), 462–468. https://doi.org/10.1111/j.1539-6924.2012.01887.x.

Aven, T. (2013b). "The concepts of risk and probability: An editorial." *Health, Risk & Society*, 15(2), 117–122. https://doi.org/10.1080/13698575.2013.775729.

Aven, T. (2020). "Three influential risk foundation papers from the '80s and '90s: Are they still state-of-the-art?" *Reliability Engineering & System Safety*, 193, 106680. https://doi.org/10.1016/j.ress.2019.106680.

Aven, T., & Renn, O. (2009). "On risk defined as an event where the outcome is uncertain." *Journal of Risk Research*, 12(1), 1–11. https://doi.org/10.1080/13669870802488883.

Aven, T., & Renn, O. (2015). "An evaluation of the treatment of risk and uncertainties in the IPCC Reports on Climate Change." *Risk Analysis*, 35(4), 701–712. https://doi.org/10.1111/risa.12298.

Aven, T., Renn, O., & Rosa, E.A. (2011). "On the ontological status of the concept of risk." *Safety Science*, 49(8–9), 1074–1079. https://doi.org/10.1016/j.ssci.2011.04.015.

Ayres, I., & Braithwaite, J. (1992). *Responsive Regulation: Transcending the Deregulation Debate*. Oxford University Press.

Baldwin, R., & Black, J. (2008). "Really responsive regulation." *The Modern Law Review*, 71(1), 59–94. https://doi.org/10.1111/j.1468-2230.2008.00681.x.

Baldwin, R., & Black, J. (2016). "Driving priorities in risk-based regulation: What's the problem?" *Journal of Law and Society*, 43(4), 565–595. https://doi.org/10.1111/jols.12003.

Baldwin, R., Cave, M., & Lodge, M. (2012). *Understanding Regulation: Theory, Strategy, and Practice*. Oxford University Press. https://doi.org/10.1093/acprof:osobl/9780199576081.001.0001.

Baldwin, R., Scott, C., & Hood, C. (1998). "Introduction." In R. Baldwin, C. Scott, & C. Hood (eds.), *A Reader on Regulation*. Oxford University Press.

Ben-Ari, A., & Or-Chen, K. (2009). "Integrating competing conceptions of risk: A call for future direction of research." *Journal of Risk Research*, 12(6), 865–877. https://doi.org/10.1080/13669870902899674.

Black, J. (2010a). "Risk-based regulation: Choices, practice and lessons being learnt." In OECD (ed.), *Risk and Regulatory Policy: Improving the Governance of Risk*. Organisation for Economic Co-operation and Development.

Black, J. (2010b). "Risk-based regulation." *Presentation to theFuture of Legal Regulation Conference*, Birkbeck College, London.

Black, J. (2010c). "The role of risk in regulatory processes." In R. Baldwin, M. Cave, & M. Lodge (eds.), *The Oxford Handbook of Regulation*. Oxford University Press. https://doi.org/10.1093/oxfordhb/9780199560219.003.0014.

Black, J., & Baldwin, R. (2010). "Really responsive risk-based regulation." *Law & Policy*, 32(2), 181–213. https://doi.org/10.1111/j.1467-9930.2010.00318.x.

Black, J., Hopper, M., & Band, C. (2007). "Making a success of principles-based regulation." *Law and Financial Markets Review*, 1(3), 191–206. https://doi.org/10.1080/17521440.2007.11427879.

Boholm, Å., & Corvellec, H. (2014). "A relational theory of risk: Lessons for risk communication." In J. Árvai & L. Rivers (eds.), *Effective Risk Communication*. Routledge.

Brandt, U.S. (2014). "The implication of extreme events on policy responses." *Journal of Risk Research*, 17(2), 221–240. https://doi.org/10.1080/13669877.2013.794151.

Brooks, N. (2003). "Vulnerability, risk and adaptation: A conceptual framework." *Working Paper* 38. Tyndall Centre for Climate Change Research.

Bullmore, E., & Sporns, O. (2009). "Complex brain networks: Graph theoretical analysis of structural and functional systems." *Nature Reviews Neuroscience*, 10(3), 186–198. https://doi.org/10.1038/nrn2575.

Comes, T., Bertsch, V., & French, S. (2013). "Designing dynamic stress tests for improved critical infrastructure resilience." *Proceedings of the 10th International ISCRAM Conference*. ISCRAM.

Craig, R.K. (2010). "'Stationarity is dead' – long live transformation: Five principles for climate change adaptation law." *The Harvard Environmental Law Review*, 34(1), 9.

Cvetkovich, G., & Earle, T.C. (1985). "Classifying hazardous events." *Journal of Environmental Psychology*, 5(1), 5–35. https://doi.org/10.1016/S0272-4944(85)80036–0

Dahl, R.A. (1994). "A democratic dilemma: System effectiveness versus citizen participation." *Political Science Quarterly*, 109(1), 23. https://doi.org/10.2307/2151659.

Davis, D.W. (2004). Contemporary Risk Analysis as the Foundation of a New National Flood Standard. *Reducing Flood Losses: Is the 1% Chance (100-Year) Flood Standard Sufficient?* https://www.nrcs.usda.gov/Internet/FSE_DOCUMENTS/16/nrcs143_009401.pdf.

de Vries, G., Verhoeven, I., & Boeckhout, M. (2011). "Taming uncertainty: The WRR approach to risk governance." *Journal of Risk Research*, 14(4), 485–499. https://doi.org/10.1080/13669877.2011.553728.

Dimasi, J. (2015). "Rethinking utility regulation in Australia." *Monash Business Policy Forum*. Monash Business School. https://www.monash.edu/__data/assets/pdf_file/0005/906971/rethinking-utility-regulation-in-australia-final.pdf.

Downing, T.E., & Bakker, K. (2000). "Drought discourse and vulnerability." In D.A. Wilhite (ed.), *Drought: A Global Assessment, Vol.2*, pp. 213–230. Routledge.

Driessen, P.P.J., & van Rijswick, H.F.M.W. (2011). "Normative aspects of climate adaptation policies." *Climate Law*, 2(4), 559–581. https://doi.org/10.3233/CL-2011-051.

Earle, T.C., & Cvetkovich, G.T. (1999). "Social trust and culture in risk management." In G. T. Cvetkovich and R.E. Löfstedt (eds.), *Social Trust and the Management of Risk*. Earthscan.

Ebi, K., Berry, P., Hayes, K., Boyer, C., Sellers, S., Enright, P., & Hess, J. (2018). "Stress testing the capacity of health systems to manage climate change-related shocks and stresses." *International Journal of Environmental Research and Public Health*, 15(11), 2370. https://doi.org/10.3390/ijerph15112370.

Etkin, D., & Ho, E. (2007). "Climate change: Perceptions and discourses of risk." *Journal of Risk Research*, 10(5), 623–641. https://doi.org/10.1080/13669870701281462.

Falkner, R., Stephan, H., & Vogler, J. (2010). "International climate policy after Copenhagen: Towards a 'building blocks' approach." *Global Policy*, 1(3), 252–262. https://doi.org/10.1111/j.1758-5899.2010.00045.x.

Fischer, R. (2008). "European governance still technocratic? New modes of governance for food safety regulation in the European Union." *European Integration Online Papers*, 12(6).

Fischhoff, B., Watson, S.R., & Hope, C. (1984). "Defining risk." *Policy Sciences*, 17(2), 123–139. https://doi.org/10.1007/BF00146924.

Florig, H.K., Morgan, M.G., Morgan, K.M., Jenni, K.E., Fischhoff, B., Fischbeck, P.S., & DeKay, M.L. (2001). "A deliberative method for ranking risks (I): Overview and test bed development." *Risk Analysis*, 21(5), 913–921. https://doi.org/10.1111/0272-4332.215161.

Freiberg, A. (2017). *Regulation in Australia*. Federation Press.

Frewer, L. (2004). "The public and effective risk communication." *Toxicology Letters*, 149(1–3), 391–397. https://doi.org/10.1016/j.toxlet.2003.12.049.

Funtowicz, S.O., & Ravetz, J.R. (1994). "Uncertainty, complexity and post-normal science." *Environmental Toxicology and Chemistry*, 13(12), 1881–1885. https://doi.org/10.1002/etc.5620131203.

Gallopín, G.C. (2006). "Linkages between vulnerability, resilience, and adaptive capacity." *Global Environmental Change*, 16(3), 293–303. https://doi.org/10.1016/j.gloenvcha.2006.02.004.

Gunningham N. (2017). "Compliance, enforcement, and regulatory excellence." In C. Coglianese (ed.), *Achieving Regulatory Excellence*. Brookings Institution Press. https://doi.org/10.7864/j.ctt1hfr226.15.

Hansson, S.O. (2010). "Risk: Objective or subjective, facts or values." *Journal of Risk Research*, 13(2), 231–238. https://doi.org/10.1080/13669870903126226.

Heims, E. & Lodge, M. (2016). "Customer engagement: Towards a new era in economic regulation?" In M. Lodge (ed.), *Customer Engagement in Regulation*, CARR Discussion Paper No. 82.

Herian, M.N., Shank, N.C., & Abdel-Monem, T.L. (2014). "Trust in government and support for governmental regulation: The case of electronic health records." *Health Expectations*, 17(6), 784–794. https://doi.org/10.1111/j.1369-7625.2012.00803.x.

Holmes, B. (2011). *Citizens' Engagement in Policymaking and the Design of Public Services*. Parliamentary Library.

Hornig, S. (1993). "Reading risk: Public response to print media accounts of technological risk." *Public Understanding of Science*, 2(2), 95–109. https://doi.org/10.1088/0963-6625/2/2/001.

Howlett, M., & Mukherjee, I. (eds.). (2018). *Routledge Handbook of Policy Design*. Routledge. https://doi.org/10.4324/9781351252928.

IPCC (2010). *Guidance Note for Lead Authors of the IPCC Fifth Assessment Report on Consistent Treatment of Uncertainties*. Intergovernmental Panel on Climate Change.

IPCC. (2012). *Managing the Risks of Extreme Events and Disasters to Advance Climate Change Adaption: Special Report of the Intergovernmental Panel on Climate Change*. Cambridge University Press.

IPCC. (2013). "Technical summary." In *Climate Change 2013: The Physical Science Basis. Contribution of Working Group I to the Fifth Assessment Report of the Intergovernmental Panel on Climate Change*. Cambridge University Press.

IPCC. (2014a). *Climate Change 2014: Impacts, Adaptation, and Vulnerability. Part A: Global and Sectoral Aspects. Contribution of Working Group II to the Fifth Assessment Report of the Intergovernmental Panel on Climate Change*. Cambridge University Press.

IPCC. (2014b). "Summary for policymakers." In *Climate Change 2014: Impacts, Adaptation, and Vulnerability. Part A: Global and Sectoral Aspects. Contribution of Working Group II to the Fifth Assessment Report of the Intergovernmental Panel on Climate Change*. Cambridge University Press.

IPCC. (2020). "The concept of risk in the IPCC Sixth Assessment Report: A Summary of Cross-Working Group Discussions." *Guidance for IPCC authors*.

IRGC. (2005). *White Paper on Risk Governance: Toward an Integrative Framework*. International Risk Governance Council.

IRGC. (2017). *Introduction to the IRGC Risk Governance Framework, Revised Version*. EPFL International Risk Governance Center. https://doi.org/10.5075/EPFL-IRGC-233739.

Jasanoff, S. (1990). *The 5th Branch: Science Advisers as Policy-makers*. Harvard University Press.

Jasanoff, S. (1999). "The songlines of risk." *Environmental Values*, 8(2), 135–152. https://doi.org/10.3197/096327199129341761.

Jodoin, S., Duyck, S., & Lofts, K. (2015). "Public participation and climate governance: An introduction." *Review of European, Comparative & International Environmental Law*, 24(2), 117–122. https://doi.org/10.1111/reel.12126.

Jones, K.E., & Irwin, A. (2010). "Creating space for engagement? Lay membership in contemporary risk governance." In B.M. Hutter (ed.), *Anticipating Risks and Organising Risk Regulation*. Cambridge University Press. https://doi.org/10.1017/CBO9780511761553.010.

Jones, R., & Boer, R. (2004). "Technical Paper 4: Assessing Current Climate Risks." In Lim, B., Spanger-Siegfried, E. (eds.) and Burton, I., Malone, E. & Huq, S. (co-authors), *Adaptation Policy Frameworks for Climate Change: Developing Strategies, Policies and Measures*. UNDP.

Jordana, J., & Levi-Faur, D. (eds.) (2004). *The Politics of Regulation: Institutions and Regulatory Reforms for the Age of Governance*. Edward Elgar. http://www.elgaronline.com/view/1843764644.xml.

Jung, H., Engle, R.F., & Berner, R. (2021). "Climate Stress Testing." *Federal Reserve Bank of New York Staff Reports*, 977. https://doi.org/10.2139/ssrn.3931516.

Junker, E. (2015). "Legal requirements for risk and vulnerability assessments in Norwegian land-use planning." *Local Environment*, 20(4), 474–488. https://doi.org/10.1080/13549839.2013.852164.

Kaplan, S., & Garrick, B.J. (1981). "On the quantitative definition of risk." *Risk Analysis*, 1(1), 11–27. https://doi.org/10.1111/j.1539-6924.1981.tb01350.x.
Kelman, I. (2003). "Defining risk." *FloodRiskNet Newsletter*, 2 Winter, 7.
Kessler, O. (2008). "Is risk changing the politics of legal argumentation?" *Leiden Journal of International Law*, 21(4), 863–884. https://doi.org/10.1017/S0922156508005426.
Klinke, A., & Renn, O. (2001). "Precautionary principle and discursive strategies: Classifying and managing risks." *Journal of Risk Research*, 4(2), 159–173. https://doi.org/10.1080/13669870175012810 5.
Klinke, A., & Renn, O. (2002). "A new approach to risk evaluation and management: Risk-based, precaution-based, and discourse-based strategies." *Risk Analysis*, 22(6), 1071–1094. https://doi.org/10.1111/1539-6924.00274.
Klinke, A., & Renn, O. (2012). "Adaptive and integrative governance on risk and uncertainty." *Journal of Risk Research*, 15(3), 273–292. https://doi.org/10.1080/13669877.2011.636838.
Landry, R. & Varone, F. (2005). "The choice of policy instruments: Confronting the deductive and interactive approaches." In R. Eliadis, M.M. Hill and M. Howlett (eds.), *Designing Government: From Instruments to Governance*. McGill-Queen's University Press.
Leiserowitz, A. (2006). "Climate change risk perception and policy preferences: The role of affect, imagery, and values." *Climatic Change*, 77, 45–72. https://doi.org/10.1007/s10584-006-9059-9.
Leiserowitz, A.A. (2005). "American risk perceptions: Is climate change dangerous?" *Risk Analysis*, 25(6), 1433–1442. https://doi.org/10.1111/j.1540-6261.2005.00690.x.
Leiss, W. (2004). "Effective risk communication practice." *Toxicology Letters*, 149(1–3), 399–404. https://doi.org/10.1016/j.toxlet.2003.12.050.
LeRoy, S.F., & Singell, L.D. (1987). "Knight on risk and uncertainty." *Journal of Political Economy*, 95(2), 394–406. https://doi.org/10.1086/261461.
Levi-Faur, D. (2011). *Handbook on the Politics of Regulation*. Edward Elgar Publishing. https://doi.org/10.4337/9780857936110.
Linkov, I., Bates, M., Loney, D., Sparrevik, M., & Bridges, T. (2011). "Risk management practices: Cross-agency comparisons and tolerable risk." In I. Linkov & T. S. Bridges (eds.), *Climate: Global Change and Local Adaptation* (pp. 133–155). Springer Netherlands. https://doi.org/10.1007/978-94-007-1770-1_8.
Löfstedt, R., & van Asselt, M. (2008). "A framework for risk governance revisited." In O. Renn & K.D. Walker (eds.), *Global Risk Governance: Concept and Practice Using the IRGC Framework*. Springer Netherlands.
McComas, K., & Shanahan, J. (1999). "Telling stories about global climate change: Measuring the impact of narratives on issue cycles." *Communication Research*, 26(1), 30–57. https://doi.org/10.1177/009365099026001003.
Meijer, A., Boon, W., & Moors, E. (2013). "Stakeholder engagement in pharmaceutical regulation: Connecting technical expertise and lay knowledge in risk monitoring." *Public Administration*, 91(3), 696.
Morgan, M.G. (1993). "Risk analysis and management." *Scientific American*, 269(1), 32–41. https://doi.org/10.1038/scientificamerican0793-32.
Morgan, M.G., Florig, H.K., DeKay, M.L., & Fischbeck, P. (2000). "Categorizing risks for risk ranking." *Risk Analysis*, 20(1), 49–58. https://doi.org/10.1111/0272-4332.00005.
Nash, J., & Walters, D.E. (2015). "Public engagement and transparency in regulation: A field guide to regulatory excellence." *Penn Program on Regulation*. https://www.law.upenn.edu/live/files/4709-nashwalters-ppr-researchpaper062015.pdf.

Natarajan, S., Krishna, G.M., Subhrojyoti, C., Patwari, P., & Banerjee, A. (2016). "System Maps: Integrating Knowledge System Models." *Proceedings of the 9th India Software Engineering Conference*. https://doi.org/10.1145/2856636.2856647.

Norris, P. (2010). *Democratic Deficit: Critical Citizens Revisited*. Cambridge University Press.

NSW Regulation Review Committee. (2001). *Proceedings of the International Conference on Regulation Reform Management and Scrutiny of Legislation, Parliament House, Sydney, 9–13 July 2001*. Parliament of New South Wales. https://www.parliament.nsw.gov.au/ladocs/inquiries/2361/Conference%20Proceedings%2001%20September%202001%20-%20Inquiry.pdf.

OECD. (2003). *Emerging Systemic Risks in the 21st Century: An Agenda for Action*. Organisation for Economic Co-operation and Development. https://doi.org/10.1787/9789264101234-fr.

OECD. (2010). *Risk and Regulatory Policy: Improving the Governance of Risk*. OECD Publishing.

OECD. (2011). *Future Global Shocks: Improving Risk Governance*. Organisation for Economic Co-operation and Development. https://doi.org/10.1787/9789264114586-en.

OECD. (2014). *Boosting Resilience through Innovative Risk Governance*. Organisation for Economic Co-operation and Development.

OECD. (2020). *Innovative Citizen Participation and New Democratic Institutions: Catching the Deliberative Wave – Highlights 2020*. Organisation for Economic Co-operation and Development. https://doi.org/10.1787/339306da-en.

O'Malley, P. (2004). *Risk, Uncertainty and Government*. GlassHouse. http://site.ebrary.com/id/10398778.

Oppenheimer, M. (2005). "Defining dangerous anthropogenic interference: The role of science, the limits of science." *Risk Analysis*, 25(6), 1399–1407. https://doi.org/10.1111/j.1539-6924.2005.00687.x.

Patt, A. G., & Schrag, D. P. (2003). "Using specific language to describe risk and probability." *Climatic Change*, 61, 17–30. https://doi.org/10.1023/A:1026314523443.

Perhac̀, R.M. (1996). "Defining risk: Normative considerations." *Human and Ecological Risk Assessment: An International Journal*, 2(2), 381–392. https://doi.org/10.1080/10807039609383615.

Peterson, D. & Fensling, S. (2011). "Risk-based regulation: Good practice and lessons for the Victorian context." *Conference Paper presented at the Victorian Competition and Efficiency Commission Regulatory Conference*.

Phipps, P.J. (2001). "Terrain systems mapping." *Geological Society, London, Engineering Geology Special Publications*, 18(1), 59–61. https://doi.org/10.1144/GSL.ENG.2001.018.01.10.

Pidgeon, N. (2014). "Complexity, uncertainty and future risks." *Journal of Risk Research*, 17(10), 1269–1271. https://doi.org/10.1080/13669877.2014.940599.

Pidgeon, N., & Fischhoff, B. (2014). "The role of social and decision sciences in communicating uncertain climate risks." In J. Árvai & L. Rivers (eds.), *Effective Risk Communication*. Routledge.

Plough, A., & Krimsky, S. (1987). "The emergence of risk communication studies: Social and political context." *Science, Technology, & Human Values*, 12(3/4), 4–10.

Poortinga, W., & Pidgeon, N.F. (2003). "Exploring the dimensionality of trust in risk regulation." *Risk Analysis*, 23(5), 961–972. https://doi.org/10.1111/1539-6924.00373.

Poortinga, W., Spence, A., Whitmarsh, L., Capstick, S., & Pidgeon, N.F. (2011). "Uncertain climate: An investigation into public scepticism about anthropogenic climate change." *Global Environmental Change*, 21(3), 1015–1024. https://doi.org/10.1016/j.gloenvcha.2011.03.001.

Productivity Commission. (2006). *Rethinking regulation, Report of the Taskforce on Reducing Regulatory Burdens on Business.* Commonwealth of Australia. https://www.pc.gov.au/research/supporting/regulation-taskforce/report.

Prokopy, L.S., Arbuckle, J.G., Barnes, A.P., Haden, V.R., Hogan, A., Niles, M.T., & Tyndall, J. (2015). "Farmers and climate change: A cross-national comparison of beliefs and risk perceptions in high-income countries." *Environmental Management*, 56 (2), 492–504. https://doi.org/10.1007/s00267-015-0504-2.

Ravetz, J.R. (1999). "What is post-normal science." *Futures*, 31(7), 647.

Renn, O. (1998). "Three decades of risk research: Accomplishments and new challenges." *Journal of Risk Research*, 1(1), 49–71. https://doi.org/10.1080/136698798377321.

Renn, O. (2004). "Perception of risks." *The Geneva Papers on Risk and Insurance – Issues and Practice*, 29(1), 102–114. https://doi.org/10.1111/j.1468-0440.2004.00275.x.

Renn, O. (2006). "Participatory processes for designing environmental policies." *Land Use Policy*, 23(1), 34–43. https://doi.org/10.1016/j.landusepol.2004.08.005.

Renn, O., Klinke, A., & van Asselt, M. (2011). "Coping with complexity, uncertainty and ambiguity in risk governance: A synthesis." *AMBIO*, 40(2), 231–246. https://doi.org/10.1007/s13280-010-0134-0.

Renn, O., & Walker, K.D. (eds.). (2008). *Global Risk Governance: Concept and Practice Using the IRGC Framework.* Springer Netherlands. https://doi.org/10.1007/978-1-4020-6799-0.

Rosa, E.A. (1998). "Metatheoretical foundations for post-normal risk." *Journal of Risk Research*, 1(1), 15–44. https://doi.org/10.1080/136698798377303.

Rothstein, H., Huber, M., & Gaskell, G. (2006). "A theory of risk colonization: The spiralling regulatory logics of societal and institutional risk." *Economy and Society*, 35 (1), 91–112. https://doi.org/10.1080/03085140500465865.

Rowe, G., Horlick-Jones, T., Walls, J., Poortinga, W., & Pidgeon, N.F. (2008). "Analysis of a normative framework for evaluating public engagement exercises: Reliability, validity and limitations." *Public Understanding of Science*, 17(4), 419–441. https://doi.org/10.1177/0963662506075351.

Runde, J. (1998). "Clarifying Frank Knight's discussion of the meaning of risk and uncertainty." *Cambridge Journal of Economics*, 22(5), 539–546. https://doi.org/10.1093/cje/22.5.539.

Santos-Martin, F., Viinikka, A., Mononen, L., Brander, L., Vihervaara, P., Liekens, I., & Potschin-Young, M. (2018). "Creating an operational database for Ecosystems Services Mapping and Assessment Methods." *One Ecosystem*, 3, e26719. https://doi.org/10.3897/oneeco.3.e26719.

Scholz, R.W., Blumer, Y.B., & Brand, F.S. (2012). "Risk, vulnerability, robustness, and resilience from a decision-theoretic perspective." *Journal of Risk Research*, 15(3), 313–330. https://doi.org/10.1080/13669877.2011.634522.

Scholze, M., Knorr, W., Arnell, N.W., & Prentice, I.C. (2006). "A climate-change risk analysis for world ecosystems." *Proceedings of the National Academy of Sciences*, 103(35), 13116–13120. https://doi.org/10.1073/pnas.0601816103.

Seacrest, S., Kuzelka, R., & Leonard, R. (2000). "Global climate change and public perception: The challenge of translation." *Journal of the American Water Resources Association*, 36(2), 253–263. https://doi.org/10.1111/j.1752-1688.2000.tb04265.x.

Slovic, P. (1999). "Trust, emotion, sex, politics, and science: Surveying the risk-assessment battlefield." *Risk Analysis*, 19(4), 689–701. https://doi.org/10.1111/j.1539-6924.1999.tb00439.x.

Slovic, P., Fischhoff, B., & Lichtenstein, S. (1979). "Rating the risks." *Environment: Science and Policy for Sustainable Development*, 21(3), 14–39. https://doi.org/10.1080/00139157.1979.9933091.

Sokal, R.R. (1974). "Classification: Purposes, principles, progress, prospects." *Science*, 185(4157), 1115–1123.

Sparrow, M. (2000). *The Regulatory Craft*. Brookings Institution Press.

Sterman, J.D., & Sweeney, L.B. (2002). "Cloudy skies: Assessing public understanding of global warming." *System Dynamics Review*, 18(2), 207–240. https://doi.org/10.1002/sdr.242.

Stern, P.C., Fineberg, H.V., & National Research Council (U.S.) (eds.). (1996). *Understanding Risk: Informing Decisions in a Democratic Society*. National Academy Press.

Stewart, R.B., Oppenheimer, M., & Rudyk, B. (2013). "Building blocks for global climate protection." *Stanford Environmental Law Journal*, 32(2).

Stewart-Weeks, M. (2011). "The future course of modern government." *An IPAA Policy Paper* accessible at: https://www.ipaa.org.au/wp-content/uploads/2019/06/The-Future-Course-of-Modern-Govenment-policy-paper.pdf.

Stoker, G., Evans, M. & Halupka, M. (2018). "Trust and democracy in Australia: Democratic decline and renewal." *Report No. 1*. Democracy 2025. https://apo.org.au/node/208536.

Taylor, A.L., Dessai, S., & Bruine de Bruin, W. (2014). "Public perception of climate risk and adaptation in the UK: A review of the literature." *Climate Risk Management*, 4–5, 1–16. https://doi.org/10.1016/j.crm.2014.09.001.

Thompson, P.B., & Dean, W. (1996). "Competing conceptions of risk." *Risk: Health, Safety & Environment*, 7(4), 361.

van Asselt, M.B.A., & Renn, O. (2011). "Risk governance." *Journal of Risk Research*, 14 (4), 431–449. https://doi.org/10.1080/13669877.2011.553730.

Vassie, L., Slovic, P., Fischhoff, B., & Lichtenstein, S. (2005). "Facts and fears: Understanding perceived risk." *Policy and Practice in Health and Safety*, 3 (Supplement 1), 65–102. https://doi.org/10.1080/14774003.2005.11667668.

Vatsa, K. S. (2004). "Risk, vulnerability, and asset-based approach to disaster risk management." *International Journal of Sociology and Social Policy*, 24(10/11), 1–48. https://doi.org/10.1108/01443330410791055.

von Schomberg, R. (2012). "The precautionary principle: Its use within hard and soft law." *European Journal of Risk Regulation*, 3(2), 147–156. https://doi.org/10.1017/S1867299X00001987.

Waters, E. (2017). "Why the world should act like children: Using the building blocks method to combat climate change, beginning with methane." *University of Richmond Law Review*, 51(4), S27–42.

Windholz, E.L. (2017). *Governing through Regulation: Public Policy, Regulation and the Law*. Routledge. https://doi.org/10.4324/9781315677286.

Zhang, Y., Wu, S., Dai, E., Liu, D., & Yin, Y. (2008). "Identification and categorization of climate change risks." *Chinese Geographical Science*, 18(3), 268–275. https://doi.org/10.1007/s11769-008-0268-1.

Chapter 8

Regulating Climate Change Risks to EU Ecosystems

The view that humans are threatening the Earth and its ecosystems is not new and dates back many centuries.[1] However, there is evidence to indicate that the impact of human activity on the Earth and its natural systems has escalated since the 1970s, in line with growing populations and increasing consumption, resulting in unprecedented declines in the extent and integrity of ecosystems.[2] Moreover, some believe that we are now in a new epoch in the Earth's history known as the "Anthropocene", where "human enterprise and its overwhelming footprint on the Earth and its systems are rivaling the great forces of nature that are gradually pushing the Earth and its systems, including its ecosystems, into a highly unstable state that is much less biologically diverse."[3]

Biological diversity – otherwise referred to as "biodiversity" – refers to the variability among living organisms, including plants, animals and micro-organisms, as well as the broader ecosystems of which they respectively are a part.[4] Biodiversity can help ensure the stability and resilience of ecosystems. Yet, some models indicate dramatic, if not alarming, consequences of climate change for biodiversity with worst-case scenarios suggesting that it could lead to the next mass extinction event in the Earth's history.[5] This case study considers aspects of the regulation of

1 A. Goudie, *The Human Impact on the Natural Environment: Past, Present, and Future* (Wiley, 2013), p. 1.
2 S. Díaz et al., "Pervasive human-driven decline of life on Earth points to the need for transformative change" (2019) 366(1327) *Science*.
3 L.J. Kotzé & T. Marauhn, "A new paradigm for biodiversity governance in a transnational world", p. 3 (referring to W. Steffen, P.J. Crutzen, & J.R. McNeill, "The anthropocene: Are humans now overwhelming the great forces of nature?" (2007) 36(8) *Ambio*, pp. 614–21) in L.J. Kotzé & T. Marauhn (eds.), *Transboundary Governance of Biodiversity* (Brill, 2014).
4 Article 2 of the Convention on Biological Diversity.
5 C. Bellard, C. Bertelsmeier, P. Leadley, W. Thuiller, & F. Courchamp, "Impacts of climate change on the future of biodiversity: Biodiversity and climate change" (2012) 15(4) *Ecology Letters*, pp. 365–77; G. Ceballos, P.R. Ehrlich, & P.H. Raven, "Vertebrates on the brink as indicators of biological annihilation and the sixth mass extinction" (2020) 117(24) *Proceedings of the National Academy of Sciences*, pp. 13596–602. Five mass extinction events have occurred so far in the Earth's history,

DOI: 10.4324/9781003253464-8

biodiversity in the EU to determine how risks posed by climate change to ecosystems are managed, particularly to assess the extent to which the EU regulatory regime reflects an ideal for the regulation of systemic risks based on a risk governance approach described earlier in this book.

I The role and importance of ecosystems

An ecosystem is a community of living organisms and their physical environment.[6] There are different types of ecosystems on earth, including terrestrial ecosystems,[7] marine ecosystems[8] and aquatic ecosystems.[9] The location, physical context, structure, components and characteristics vary between ecosystems. Nonetheless, they each deliver essential services for life on Earth, like providing sources of food and fresh water, enhancing air quality, and regulating the climate.

Biodiversity is critical to the optimal functioning of ecosystems. The Convention on Biological Diversity (CBD), which is an international legal instrument for the conservation of biological diversity, has been described as a global legal framework to safeguard the valuable services provided by ecosystems.[10] The Secretariat for the CBD explains the importance of biodiversity as follows:[11]

> The biodiversity we see today is the fruit of billions of years of evolution, shaped by natural processes and, increasingly, by the influence of humans. It forms the web of life of which we are an integral part and upon which we so fully depend.

As the local environment fluctuates over time, different species and their habitats will respond differently. For example, whereas one species may decrease in abundance in response to an environmental event such as drought, another species may be affected less dramatically, which helps to compensate for the

which have arisen from major environmental changes. It has been suggested by some scientists that the sixth mass extinction event caused by human activity is already underway and poses a threat to human civilization because changes to the Earth and its natural systems are irreversible.

6 Article 2 of the Convention on Biological Diversity defines an ecosystem as "a dynamic complex of plant, animal and micro-organism communities and their non-living environment interacting as a functional unit".

7 Terrestrial ecosystems are land-based ecological communities in places like the Arctic tundra, deserts, forests and grasslands.

8 Marine ecosystems include the oceans, sandy shores, estuaries, mud flats, Antarctic waters, salt marshes and coral reefs.

9 Aquatic ecosystems include lakes, rivers, ponds and wetlands.

10 C. Prip, "The Convention on Biological Diversity as a legal framework for safeguarding ecosystem services" (2018) 29 *Ecosystem Services*, pp. 199–204.

11 Secretariat of the Convention on Biological Diversity, *Sustaining Life on Earth: How the Convention on Biological Diversity Promotes Nature and Human Well-being* (Secretariat of the Convention on Biological Diversity, 2000), p. 2.

impact of the event on the first species. The more species there are in an ecosystem – that is, the more biodiversity – and the more asynchronous their responses to disturbances, the greater the stability of the ecosystem as a whole.[12] In fact, there appears to be broad acceptance that biodiversity enhances the stability of ecosystems,[13] including ecosystems with multiple stable states.[14]

Stability enables ecosystems to provide critically important "ecosystem services".[15] The 2005 UN Millennium Ecosystem Assessment, which assessed the consequences of ecosystem change for human well-being, categorised ecosystem services into four categories.[16] "Provisioning services" cover the products provided by well-functioning ecosystems such as food, fresh water and genetic resources. "Regulating services" are the regulating benefits obtained from ecosystems, including water purification, climate regulation and natural hazard protection. "Cultural services" are non-material benefits from ecosystems, such as aesthetic enjoyment, recreation and tourism and spiritual benefits. "Supporting services" relate to the services necessary for all other ecosystem services and cover soil formation, photosynthesis and nutrient cycling. These ecosystem services have been characterised as the "bridge" between the natural and social worlds.[17]

An ecosystem's state of equilibrium may be disrupted by natural disturbances, such as volcanoes or the arrival of invasive species within a local environment. Human disturbances may also cause disruption, including through land clearing for livestock, overfishing, discharge of toxic waste and pollution, and anthropogenic climate change. An ecosystem will be resilient in the face of some of these disturbances – that is, it is capable of absorbing the disturbance and retaining its identity, including its structure, function and feedbacks.[18] However, there are limits to an ecosystem's resilience. Disturbances that took place

12 M. Loreau, *From Populations to Ecosystems: Theoretical Foundations for a New Ecological Synthesis* (Princeton University Press, 2010), Chapter 5 – Stability and Complexity of Ecosystems: New Perspectives on an Old Debate.

13 B.J. Cardinale et al., "Biodiversity loss and its impact on humanity" (2012) 486 (7401) *Nature*, pp. 59–67.

14 J. N. Griffin et al., "Biodiversity and the stability of ecosystem functioning", p. 91 in S. Naeem, D.E. Bunker, A. Hector, M. Loreau, & C. Perrings, *Biodiversity* (eds.), *Ecosystem Functioning, and Human Wellbeing* (Oxford University Press, 2009), Chapter 6.

15 S. Knapp, "The link between diversity, ecosystem functions, and ecosystem services" in C. Baessler, A. Bonn, S. Klotz, M. Schröter, & R. Seppelt (eds.), *Atlas of Ecosystem Services: Drivers, Risks, and Societal Responses* (Springer International Publishing, 2019).

16 Millennium Ecosystem Assessment Board, *Ecosystems and Human Well-being: Synthesis* (Island Press, 2005).

17 M. Ibrahim, H. Karimi, Y. Mustafa, & M. Hassan, "Mapping ecosystem service: Challenges and solutions" (2021) 2(1) *Journal of Geoinformatics & Environmental Research*, pp. 62–77, at 62.

18 E.J. Sterling, A. Gómez, & A.L. Porzecanski, "A systemic view of biodiversity and its conservation: Processes, interrelationships, and human culture" (2010) 32(12) *BioEssays*, pp. 1090–8, at 1094.

more than a millennium ago can continue to have an impact on ecosystem dynamics today and a disturbance that is severe enough could lead to permanent alteration, if not collapse, of an entire ecosystem.[19]

In a report prepared in 2019 for the G7 Environment Ministers' Meeting, the OECD noted the economic costs associated with the adverse impact of human activity on ecosystems and the services they provide. For example, historically, changes in land-cover have affected biodiversity and, in turn, led to reduced crop yields and increased risk of natural hazard disasters, such as flooding.[20] The World Bank (2021) has predicted that ecosystem collapse will continue to have a significant economic impact in the future and that low-income countries are particularly vulnerable to the loss of valuable ecosystem services that this will entail.[21]

Clearly, there is a delicate balance to be struck between, on the one hand, facilitating human exploitation of ecosystems so that the benefits of ecosystem services can be enjoyed and, on the other, exceeding the limits of an ecosystem's resilience that could lead to system collapse.[22] Finding this balance is particularly challenging given ecosystem complexity. In fact, ecosystems are among the most complex systems known to science. They contain many different components that interact directly and indirectly in highly interconnected networks.[23] Complexity exists at all levels of an ecosystem from molecular processes to the dynamic interactions among species,[24] which vary in time and across space.[25] Moreover, ecosystems rarely have discrete boundaries because of "leakage" of individuals, species, nutrients and energy.[26] This intrinsic complexity is compounded by the fact that ecosystems have a substantial random or stochastic component,[27] which creates uncertainty about ecosystem behaviour.[28]

19 A.S. Mori, "Ecosystem management based on natural disturbances: hierarchical context and non-equilibrium paradigm" (2011) 48 *Journal of Applied Ecology*, pp. 280–92, at 285.

20 OECD, *Biodiversity: Finance and the Economic and Business Case for Action*, Report Environment Ministers' Meetings, 5–6 May 2019, p. 12.

21 J.A. Johnson et al., *The Economic Case for Nature* (World Bank Group, 2021), p. vi.

22 R.T. Lackey, "Seven pillars of ecosystem management" (1998) 40(1–3) *Landscape and Urban Planning*, pp. 21–30, at 26.

23 O.J. Schmitz, *Resolving Ecosystem Complexity* (Princeton University Press, 2010), Chapter 1 – Introduction.

24 E.J. Sterling, A. Gómez, & A.L. Porzecanski, n. 18 above, p. 1095.

25 D. Moreno-Mateos, A. Alberdi, E. Morriën, W.H. van der Putten, A. Roíguez-Uña, & D. Montoya, "The long-term restoration of ecosystem complexity" (2020) 4 *Nature Ecology & Evolution*, pp. 676–85, at 678.

26 P.F. Brussard, J.M. Reed, & C.R. Tracy, "Ecosystem management: What is it really?" (1998) 40 *Landscape and Urban Planning*, pp. 9–20, at 11.

27 M. Anand, A. Gonzalez, F. Guichard, J. Kolasa, & L. Parrott, "Ecological systems as complex systems: Challenges for an emerging science" (2010) 2 *Diversity*, pp. 395–410, at 404.

28 K. Norris, "Biodiversity in the context of ecosystem services: The applied need for systems approaches" (2012) 367(1586) *Philosophical Transactions. Biological Sciences*, pp. 191–9, at 196.

As noted by Schmitz (2010), "it is not easy to sort through this 'tangled-bank' to arrive at an … understanding of whole-system function".[29] Norris (2012) considers that the only way to understand the future impact of biodiversity changes within an ecosystem on the delivery of ecosystem systems is through a systems approach.[30] However, according to Norris, biodiversity science has instead, to a large extent, adopted a retrospective approach that focuses on the drivers for past biodiversity loss.[31] A backward-looking approach is plainly inappropriate for systemic risks to ecosystems caused by climate change, which are complex, uncertain and ambiguous and are explored in the next section.

II The impact of climate change on ecosystems

Ecosystems face an array of threats associated with a range of human activity, including anthropogenic climate change. It is well-established that climate change is causing increasingly significant adverse consequences for ecosystems and biological diversity across the globe. The IPCC (2014a) has found that many terrestrial, freshwater and marine species have shifted their geographic ranges, seasonal activities, migration patterns, abundances and species interactions in response to climate change.[32] These impacts pose important systemic risks – namely, the loss of biodiversity, ecosystems and associated ecosystem services.[33]

More specifically, the IPCC has found that a large fraction of terrestrial and freshwater species face increased extinction risk under projected climate change, especially as climate change interacts with other pressures, such as habitat modification, over-exploitation, pollution and invasive species.[34] Climate change is also projected to significantly reduce surface water and groundwater resources in certain regions, which will intensify competition for water, including among ecosystems.[35] In addition, climate change will alter processes and feedbacks for microbes that play key roles in marine ecosystems,[36] which will increasingly put

29 O.J. Schmitz, n. 23 above, Chapter 2 – Conceptualizing Ecosystem Structure.
30 K. Norris, n. 28 above, p. 197.
31 Ibid. p. 197.
32 IPCC, "Summary for Policymakers", in *Climate Change 2014: Impacts, Adaptation, and Vulnerability. Part A: Global and Sectoral Aspects. Contribution of Working Group II to the Fifth Assessment Report of the Intergovernmental Panel on Climate Change* (Cambridge University Press, 2014), pp. 1–32, at 4.
33 Ibid. p. 13.
34 IPCC, "Terrestrial and inland water systems", in *Climate Change 2014: Impacts, Adaptation, and Vulnerability. Part A: Global and Sectoral Aspects. Contribution of Working Group II to the Fifth Assessment Report of the Intergovernmental Panel on Climate Change* (Cambridge University Press, 2014), p. 275.
35 IPCC, "Freshwater resources", in *Climate Change 2014: Impacts, Adaptation, and Vulnerability. Part A: Global and Sectoral Aspects. Contribution of Working Group II to the Fifth Assessment Report of the Intergovernmental Panel on Climate Change* (Cambridge University Press, 2014), p. 232.
36 IPCC, "Ocean systems", in *Climate Change 2014: Impacts, Adaptation, and Vulnerability. Part A: Global and Sectoral Aspects. Contribution of Working Group II to the Fifth*

fisheries productivity and other ecosystem services at risk.[37] These risks to ecosystems caused by climate change are compounded by others associated with human activity, such as over-exploitation of natural resources, pollution of natural systems and land clearing which cause habitat loss and fragmentation.

Various adaptation strategies will be needed to address climate change risks to ecosystems and biodiversity, including measures to protect and restore natural areas, to enhance resilience to changing conditions and extreme climatic events, to create protected area networks between ecosystems and, in some cases, to move populations of species to more suitable areas.[38] The use of policy, law and regulation will clearly be critically important in achieving these outcomes. However, Sterling and colleagues (2010) claim that traditional, linear and compartmentalised thinking continues to be used in this context.[39] Kotzé and Marauhn (2014) suggest that current approaches to biodiversity protection need to be revisited in order to adopt "a more holistic vision that considers global biodiversity governance through the lens of the Earth system".[40] The next section considers whether this way of thinking is evident in EU regulation applicable to the management of biodiversity risks.

III The regulatory regime in the EU

"Biodiversity law" covers the spectrum of rules designed to safeguard the natural world, including the biodiversity by which it is characterised. The regulatory space occupied by biodiversity law is densely populated, comprising a multitude of regulatory instruments at the international, regional, national and local government levels.[41]

In the EU, various regulatory instruments are in place. The 1979 Bern Convention on the Conservation of European Wildlife and Natural Habitats (Bern Convention) applies regionally.[42] It is complemented by various EU

Assessment Report of the Intergovernmental Panel on Climate Change (Cambridge University Press, 2014), p. 415.

37 Ibid. p. 416.

38 A. Trouwborst, "Conserving European biodiversity in a changing climate: The Bern Convention, the European Union Birds and Habitats Directives and the adaptation of nature to climate change" (2011) 20(1) *Review of European Community & International Environmental Law*, pp. 62–77, at 62.

39 E.J. Sterling, A. Gómez, & A.L. Porzecanski, n. 18 above, p. 1094.

40 L.J. Kotzé & T. Marauhn, n. 3 above, pp. 4–5.

41 For example, there are seven international conventions that deal with biodiversity issues, namely, the 1951 International Plant Protection Convention, the 1973 Convention on International Trade in Endangered Species, the 1979 Convention on Conservation of Migratory Species, the 2001 International Treaty on Plant Genetic Resource for Food and Agriculture, the 1971 Ramsar Convention on Wetlands, the 1972 World Heritage Convention (WHC) and the 1993 Convention on Biological Diversity.

42 The Bern Convention has 50 contracting parties, including the EU as a party, EU Member States, other European countries, and African States. There are numerous

directives, which target specific aspects of nature protection, including the 1979 Wild Birds Directive[43] (Birds Directive) and the 1992 Habitats Directive (Habitats Directive).[44] These directives – collectively referred to in this case study as the "EU Directives" – are used to implement the EU's obligations under the Bern Convention. Member States of the EU are bound to give effect to the requirements of the EU Directives, but have flexibility regarding how to do so in practice.[45] The Bern Convention and the EU Directives are the focus of this case study.

I The Bern Convention

The Bern Convention is notable because it was the first international treaty to deal broadly with nature conservation, rather than being focused on specific habitats or species.[46] Indeed, its establishment preceded the CBD, but the Bern Convention now plays an important role in facilitating implementation of the EU's global biodiversity obligations, including under the CBD.[47]

Three core aims underpin the Bern Convention, namely to conserve wild flora and fauna and natural habitats, to promote co-operation between States, and to give particular attention to endangered and vulnerable species, including migratory species.[48] In pursuance of these aims, the Bern Convention imposes general obligations on its Parties to promote national policies for the conservation of wild flora and fauna and their natural habitats;[49] to have regard in the context of regional planning, development and environmental policies to the conservation of wild flora and fauna and to conservation requirements of protected areas;[50] and to take requisite measures to maintain the population of wild flora and fauna at, or adapt it to, a level which corresponds to ecological, scientific and cultural requirements, while taking account of economic and recreational requirements.[51]

other regional conventions of relevance to the conservation of European biodiversity, but the Bern Convention is central to European biodiversity law.

43 2009/147/EC.

44 92/43/EEC. There are other directives of relevance in the EU, particularly the Water Framework Directive (2000/60/EC), Floods Directive (2007/60/EC) and Marine Strategy Framework Directive (2008/56/EC).

45 This case study does not examine the implementation of the Bern Convention and the EU Directives by the EU Member States but focuses instead on the regulatory architecture for biodiversity protection established at the regional level.

46 Council of Europe, "The Bern Convention's contribution to the Strategic Plan for Biodiversity 2011–2020 and the Aichi Targets", in *Standing Committee on the Convention on the Conservation of European Wildlife and Natural Habitats, 39th meeting, Strasbourg, 3–6 December 2019* (T-PVS/Inf(2019)22, 21 November 2019), p. 4.

47 Ibid. p. 5.

48 Article 1, Bern Convention.

49 Article 3.1, Bern Convention.

50 Articles 3.2 and 4.2, Bern Convention.

51 Article 2, Bern Convention.

The Bern Convention deals both with protection of natural habitats as well as species. In relation to habitats, Parties to the Convention (which extend beyond the EU Member States) are obliged to take "appropriate and necessary legislative and administrative measures" to ensure conservation of the habitats of wild flora and fauna species, especially those specified in Appendices I (Strictly protected flora species) and II (Strictly protected fauna species), and the conservation of endangered habitats.[52] They must also give special attention to the protection of areas that are of importance for certain migratory species in Appendix II and in Appendix III (Protected fauna species).[53]

These habitat conservation obligations have been elaborated by a series of interpretative resolutions and non-binding recommendations issued by the Bern Convention's Standing Committee, including Recommendation No. 16 (1989) to establish "Areas of Special Conservation Interest" (ASCIs) to ensure that necessary and appropriate conservation measures are taken.[54] The ASCIs constitute the so-called "Emerald Network" which is an "ecological network" for the conservation of wild animal and plant species and natural habitats of European importance that are listed in resolutions made pursuant to the Bern Convention.[55]

The criteria that apply in identifying an ASCI are set out in Recommendation No. 16 and include requirements that an area contributes substantially to the survival of threatened species, supports significant numbers of species in an area of high species diversity, or otherwise contributes substantially to the achievement of the Bern Convention's objectives. Pursuant to Recommendation No. 16, Parties are requested to designate ASCIs within their territory or under their responsibility in accordance with the requirements of the Bern Convention, and to "take such steps, either by legislation or otherwise to ensure wherever possible that [ASCIs] are the subject of an appropriate regime". Resolution No. 8 (2012) on the national designation of adopted Emerald sites and the implementation of management, monitoring and reporting measures further states that the national designation of Emerald sites "will ensure that they are protected from external threats and subject to an

52 Article 4.1, Bern Convention.
53 Article 4.3, Bern Convention.
54 Recommendation No. 16 (1989) on Areas of Special Conservation Interest is accessible at: https://search.coe.int/bern-convention/Pages/result_details.aspx?ObjectId=0900001680746be2.
55 These habitats and species are listed respectively in Resolution No. 4 (1996) listing endangered natural habitats requiring specific conservation measures (including revised Annex I to Resolution No. 4 (1996), adopted in 2014 by the Standing Committee) and Resolution No. 6 (1998) listing the species requiring specific habitat conservation measures (including revised Annex I to Resolution No. 6 (1998), adopted in 2011 by the Standing Committee).

appropriate regime for achieving conservation status of the relevant species and natural habitats".[56]

In relation to species protection, Parties are obliged to take appropriate and necessary legislative and administrative measures to ensure special protection of the strictly protected wild flora species specified in Appendix I[57] and the strictly protected wild fauna species listed in Appendix II.[58] These obligations are supplemented with specific prohibitions in relation to dealings with these species of wild flora and fauna.[59] An additional obligation also exists to take appropriate and necessary legislative and administrative measures to ensure the protection of the wild fauna species in Appendix III.[60]

2 The EU Directives

The Birds Directive, which was adopted in 1979, aims to protect all wild birds and their most important habitats across the EU.[61] The Habitats Directive was adopted in 1992. It focuses on conservation of natural habitats and habitats of particular species and extends protection to rare, threatened or endemic species of wild animals and plants.[62]

As in the case of the Bern Convention, the EU Directives are characterised by dual sets of obligations regarding both habitat and species protection. In relation to habitat protection, the Habitats Directive and the Birds Directive together establish the "Natura 2000 network" for species and habitat types listed in Annex I (natural habitat types) and Annex II (habitats of particular species) of the Habitats Directive,[63] in Annex I of the Birds Directive (species and sub-species of wild birds), as well as for migratory birds.[64] This network is made up of Sites of Community Importance (SCIs) adopted by the European Commission, which are then designated by EU Member States as Special Areas of Conservation (SACs) under the Habitats Directive and classified by Member States as Special Protection Areas (SPAs) under the Birds Directive.

56 As noted, these habitats and species are listed respectively in Resolution No. 4 (1996) and Resolution No. 6 (1998).

57 Article 5, Bern Convention.

58 Article 6, Bern Convention.

59 Articles 5 & 6, Bern Convention.

60 Article 7, Bern Convention.

61 European Commission, "A starter's guide: Overview on the main provisions of the Water Framework Directive, the Marine Strategy Framework Directive, the Birds and Habitats Directives, and the Floods Directive: similarities and differences", in *Technical Report 103* (2016), p. 5.

62 Ibid. p. 5.

63 Article 3.1, Habitats Directive.

64 Articles 4, Birds Directive.

For each Natura 2000 site, Member States are required to implement necessary conservation measures to maintain and, where relevant, to restore habitats and species.[65] The Habitats Directive aims for "favourable conservation status", which is defined under the Directive as existing for a natural habitat when, among other things, its range is "stable or increasing" and the "specific structure and functions which are necessary for its long-term maintenance exist and are likely to continue to exist for the foreseeable future".[66] The status of a species is taken to be favourable when, among other things, the species "is maintaining itself on a long-term basis as a viable component of its natural habitats", "the natural range of the species is neither being reduced nor is likely to be reduced for the foreseeable future" and "there is, and will probably continue to be, a sufficiently large habitat to maintain its populations on a long-term basis."[67]

Under the Birds Directive, measures must be taken to "maintain the population of the species … at a level which corresponds in particular to ecological, scientific and cultural requirements, while taking account of economic and recreational requirements, or to adapt the population of these species to that level".[68] The Birds Directive elaborates that Members shall take the requisite measures to "preserve, maintain or re-establish a sufficient diversity and areas of habitats" for relevant species of wild birds.[69] The measures must include action for the "creation of protected areas" and "upkeep and management in accordance with the ecological needs of habitats inside and outside the protected zones".[70] The EU Directives require strict protection of species both within and outside Natura 2000 sites.[71]

IV Approach to regulating climate change risks to EU ecosystems

The EU regulatory regime is characterised by multiple regulatory instruments, a dual approach towards the protection of species and habitats, and complex conservation standards. As explained below, the regulatory approach towards regulating climate change risks to ecosystems is not explicit in the regulatory instruments considered for this case study, although relevant supplementary material helps to cast light on this issue.

65 Article 6.1, Habitats Directives. Article 4.1, Birds Directive.
66 Definition of "conservation status of a natural habitat" in Article 1(e), Habitats Directive.
67 Definition of "conservation status of a species" in Article 1(i), Habitats Directive.
68 Article 2, Birds Directive.
69 Article 3.1, Birds Directive.
70 Article 3.2, Birds Directive.
71 Articles 12 and 13, Habitats Directive. Article 5, Birds Directive.

I Regulatory boundary

An initial question that arises in determining whether systemic risks to EU ecosystems caused by climate change are within the boundary of the EU regulatory regime relates to the manner in which ecosystems are conceptualised and described within the regime. The extent to which the regime corresponds to the way ecosystems are structured, function and evolve over time gives an indication of whether the risks to such systems are likely to be effectively addressed under the regime, including the systemic risks caused by climate change.

The Emerald Network and Natura 2000 established under the Bern Convention and the EU Directives respectively convey a sense of the systemic characteristics of ecosystems covered by the EU regulatory regime. The ASCIs, SPAs and SACs are designed to constitute a coherent European ecological network. In theory, such a network can help to address fragmentation of natural areas by linking physically separated habitats to enable species to move between these areas[72] and may consist of core areas, corridors, stepping stones, restoration areas and buffer zones connected by landscape design.[73] The connectivity that ecological networks imply can help facilitate adaptation to climate change by enabling species to disperse and move as their habitats change.[74] However, Verschuuren suggests that Natura 2000 is, in fact, "not a network but a collection of isolated sites".[75]

In keeping with the dynamic nature of ecosystems, and the fact that risks to those systems may cause them to evolve over time, it is important that changes to ecological networks covered under the EU regulatory regime are accounted for under the regime. As noted by Cliquet and colleagues (2009), the identification of protected areas is "not a one-time operation"[76] but, rather, Member States should be continuously reviewing and considering sites for inclusion in Natura 2000, particularly to address the impact of climate change. Ensuring that the Natura 2000 network keeps pace with changes caused by climate change could be supported by, for example, mapping and

72 T. van der Sluis & P. Chardon, "How to define European ecological networks" (2001) 46 *WIT Transactions on Ecology and the Environment*, p. 119.

73 J. Verschuuren, "Climate change: Rethinking restoration in the European Union's Birds and Habitats Directives" (2010) 28(4) *Ecological Restoration*, pp. 431–9, at 435.

74 European Council, "Conserving European biodiversity in a changing climate: The Bern Convention, the EU Birds and Habitats Directives and the adaptation of nature to climate change" prepared by A. Trouwborst, *Group of Experts on Biodiversity and Climate Change, 6th Meeting, Strasbourg, 10–11 October 2011* (T-PVS/Inf (2011)21, 21 September 2011), p. 18.

75 J. Verschuuren, n. 73 above, p. 436.

76 A. Cliquet, C. Backes, J. Harris, & P. Howsam, "Adaptation to climate change: Legal challenges for protected areas" (2009) 5(1) *Utrecht Law Review*, p. 158, at 164.

assessment of ecosystems, such as the EU-wide Ecosystem Assessment which was completed in 2020.[77] These issues raise questions about the capacity of the EU regulatory regime, as a matter of practice (rather than design), to deal with ecosystems in a comprehensive, integrated and coherent way that accounts for their systemic characteristics.

Turning to the question of whether risks to ecosystems arising from climate change are covered by the EU regime, it is notable that the relevant regulatory instruments do not articulate their aims and the manner in which those aims should be achieved in terms of risk. Rather, the focus is on outcomes – specifically, ensuring a favourable conservation status for habitats and species. The regulatory regime does not explicitly address the impact of climate change on achievement of these outcomes, which is unsurprising given that climate change was not an explicit driver for the establishment of the regime.[78] Nonetheless, arguably, the Bern Convention and the EU Directives accommodate an interpretation that climate change should be accounted for in applying these legal instruments and that, therefore, climate change risks fall within the scope of the regime because climate change can adversely impact the achievement of the regime's conservation objectives. This view is supported by the analysis of the Bern Convention and the EU Directives undertaken by Trouwborst (2011) for the Bern Convention's Group of Experts on Biodiversity and Climate Change.[79] He concludes that the instruments comprising the EU regulatory regime impose legal obligations on Member States to take measures needed to facilitate the adaptation of biodiversity in Europe to climate change.[80]

These obligations are informed by various non-binding recommendations issued by the Standing Committee of the Bern Convention.[81] In summary, the recommendations stress "the need to adapt conservation work to the challenges of climate change so as to minimise its impact on the species and natural habitats protected under the Convention".[82] They note that "uncertainties surrounding the precise nature of future climate change and its impacts on biodiversity should not delay practical conservation action".[83] Parties are urged to develop "climate change adaptation activities for biodiversity".[84]

77 European Commission. Joint Research Centre, *Mapping and Assessment of Ecosystems and their Services: An EU Wide Ecosystem Assessment in Support of the EU Biodiversity Strategy* (2020).
78 European Council, n. 74 above.
79 Ibid.
80 Ibid. p. 21.
81 See Recommendations No. 122 (2006), No. 135 (2008), Nos. 142 and 143 (2009) and Nos. 145, 146 and 147 (2010).
82 Recommendations No. 135 (2008) and No. 143 (2009).
83 Ibid.
84 Ibid.

While the above analysis tends to indicate that the EU regulatory regime covers systemic risks to ecosystems caused by climate change, there are remaining questions about the regime's requirements regarding how those risks are to be addressed, including how the risks are assessed and the responses to those risks, which are considered in turn below.

2 Regulatory context

The context for application of the EU regulatory regime assists in determining whether or not a regulatory response will be triggered under the regime to respond to climate change risks to ecosystems because such risks have been assessed as undermining the conservation objectives of the regime. Various aspects of the EU regulatory regime implicitly dictate that context.

The criteria for establishment of Natura 2000 sites, which are protected under the EU regime, are relevant in this regard. The designation of SACs for Natura 2000 under the Habitats Directive is based on ecological criteria and relevant scientific data,[85] whereas under the Birds Directive, SPAs are classified by taking into account the most suitable territories for conservation of protected species based on factors including trends and variations in species population levels.[86]

The conservation standards applicable to the protection of habitats and species under the regulatory regime are also relevant. Specifically, the Bern Convention refers to conservation levels which correspond to "ecological, scientific and cultural requirements, while taking account of economic and recreational requirements and the needs of sub-species, varieties or forms at risk locally".[87] Similar language is used in the Birds Directive.[88] In defining the concept of "favourable conservation status", the Habitats Directive refers to ecological factors including the stability, dynamics, diversity and longevity of ecosystems,[89] but the Directive also makes clear that measures required under the Directive "shall take account of economic, social and cultural requirements and regional and local characteristics".[90]

These criteria for the respective establishment of ecological networks and conservation standards under the EU regulatory regime cover a range of ecological and socio-economic factors. A purely textual interpretation suggests a broad context that could involve assessment of the impact of climate change on ecosystems using scientific information as well as socio-cultural information. Yet, Araújo and colleagues (2011) argue that, in practice, the selection of

85 The criteria are set out in Annex III of the Habitats Directive.
86 Article 4.1, Birds Directive.
87 Article 2, Bern Convention.
88 Article 2, Birds Directive.
89 See definitions of "conservation status of a natural habitat" and "conservation status of a species" in Article 1, Habitats Directive.
90 Article 2.3, Habitats Directive.

protected areas covered by Natura 2000 sites based on criteria contained in the regime fails to take account of the effects of climate change.[91] Further, according to Cliquet and colleagues (2009), definitions of a favourable conservation status are formulated in a "very static way" and, as climate change exerts increasing pressure on species and habitats, "it will become increasingly difficult and will not always make sense to reach the (rather static) conservation goals".[92] Moreno-Mateos and colleagues (2020) further argue that traditional approaches for ecosystem assessment based on "simplified abstractions"[93] need to be replaced with an approach that considers "the full complexity of ecosystems", including the adaptive potential of species and recognition of the fact that restoration of these systems may take very long periods of time.[94]

This commentary indicates that, while a broad analysis appears to be accommodated by textual aspects of the EU regulatory regime, the practical reality of the application of the relevant criteria may be quite different. Sterling and colleagues (2010) explain that "successfully navigating the environmental, social, and financial issues leading to conservation goals will require a basic understanding of the values and motivations of multiple stakeholders".[95] An understanding of these values could be informed through mechanisms that currently exist within the regulatory regime to educate, disseminate information and co-operate,[96] as well as input from NGOs and other members of civil society.[97] Arguably, full account of all the factors referenced in the regulatory regime, including the associated values and perspectives of relevant stakeholders, will result in better conservation outcomes,[98] particularly in the face of the systemic threats posed by climate change.

3 Regulatory content

The content of the EU regulatory regime provides insights as to how threats to the underlying conservation objectives and, more specifically, threats to EU ecosystems are likely to be addressed in practical terms, if at all. The regulatory

91 M.B. Araújo, D. Alagador, M. Cabeza, D. Nogués-Bravo, & W. Thuiller, "Climate change threatens European conservation areas" (2011) 14(5) *Ecology Letters*, pp. 484–92, at 484.
92 A. Cliquet et al., n. 76 above, pp. 168, 174.
93 D. Moreno-Mateos et al., n. 25 above, p. 676.
94 Ibid. p. 683.
95 E.J. Sterling, A. Gómez, & A.L. Porzecanski, n. 18 above, p. 1095.
96 See, for example, Articles 1, 3 and 11, Bern Convention and Articles 17, 18 and 22 Habitats Directive.
97 C.L. Diaz, "The Bern Convention: 30 years of nature conservation in Europe" (2010) 19(2) *Review of European Community & International Environmental Law*, pp. 185–96, at 189–90. Diaz suggests that members of civil society have played an important role in policy development and compliance in the context of the Bern Convention.
98 E.J. Sterling, A. Gómez, & A.L. Porzecanski, n. 18 above, p. 1096.

instruments comprising the regime variously refer to "preservation", "maintenance", "restoration", and "re-establishment" of habitats and species and speak of the need to "avoid or minimise" deterioration and disturbances.[99] These terms to describe what is required under the regulatory regime convey a sense that the framework is geared towards maintaining the status quo rather than facilitating change, even though the latter may be required when climate change causes a permanent ecosystem shift.

Rather than prescribing the manner in which conservation objectives must be achieved, Member States are granted significant discretion and flexibility. The regime requires Member States to take "requisite", "appropriate", and "necessary" legislative and administrative measures,[100] but there are no explicit regulatory requirements, for example, to map and assess ecosystems nor to establish specific action plans to address threats to these systems. The European Court of Justice has interpreted these terms as requiring Member States to simply "do what it takes".[101]

So, what will it take to respond to the ecosystem risks arising from climate change? An approach that accounts for the systemic characteristics of ecosystems and is founded on the concept of resilience is at the heart of many suggestions. For example, Chambers and colleagues (2019) argue that ecological resilience can enable ecosystems to adapt to unpredictable change and facilitate transformation to more desirable states. They note that such an approach involves understanding factors that influence resilience, including the process and patterns of interaction, but operationalising ecological resilience has been difficult in the absence of a systems approach to ecosystem assessment and management.[102] They suggest that, accordingly, new laws may be needed.[103]

V Alignment with the risk governance regulatory ideal

The regime applicable to the regulation of ecosystems in the EU bears limited resemblance to the risk governance paradigm put forward in this book as the basis for regulating systemic risks. Indeed, the EU regime is neither cast in terms of risk, nor is it expressly aimed at managing risks. Rather, the emphasis

99 See Articles 2 and 4.2 of the Bern Convention, Articles 2.2, 3.1, 4.4 and 6.2 of the Habitats Directive, and Articles 2, 3.1 and 4.4 of the Birds Directive.

100 See Articles 4–7 of the Bern Convention, Article 6 of the Habitats Directive and Articles 2–5 of the Birds Directive.

101 European Council, n. 75 above, p. 17; Council of Europe, "Obligations of Bern Convention Parties regarding the conservation of candidate and adopted Emerald network sites: A legal analysis" prepared by A. Trouwborst in *Standing Committee of Convention on the Conservation of European Wildlife and Natural Habitats*, (T-PVS/PA (2020)07, 5 October 2020), pp. 6–12.

102 J.C. Chambers, C.R. Allen, & S.A. Cushman, "Operationalizing ecological resilience concepts for managing species and ecosystems at risk" (2019) 7 *Frontiers in Ecology and Evolution*, Article 241, p. 2.

103 Ibid. pp. 2, 21.

in the regulatory regime is squarely on the achievement on conservation outcomes.

In many ways, this approach makes sense and is commendable for its emphasis on what counts – that is, the preservation, maintenance, restoration, and re-establishment of the EU's habitats and species. Such an approach signals that conservation of species and habitats comprising the EU's ecosystems is the ultimate objective of the regime, regardless of the spectrum of risks to which those ecosystems may be subject. This approach is also practical given the multitude of these risks, which may make it legally and practically challenging, if not impossible, to canvass them all explicitly in the regime.

Nonetheless, the EU regime seems somewhat abstract and lacking in context in light of its silence about the range of risks that could place ecosystems under stress and, potentially, could compromise achievement of the conservation objectives that are central to the regime. In the case of climate change risks, the silence has been addressed through the issuance of guidance under the Bern Convention. Although this guidance is non-binding, it does help to clarify that climate change risks to ecosystems are intended to be covered by the regulatory regime and provides some indication as to how those risks should be addressed.

Questions remain as to whether the approach, objectives and regulatory measures provided for under the regime are well suited to address the systemic risks to ecosystems caused by climate change and, for that matter, other systemic risks that could emerge in the future. At the core of these questions is the fact that, notwithstanding the establishment of an ecological network in the EU in the form of the Emerald Network and Natura 2000, which hint at the systemic characteristics of ecosystems, there is otherwise limited evidence in the regulatory regime of an approach that accounts for the entire natural system, including interactions and feedback, and focuses on system resilience.

By its own admission, the European Commission (2021b) has stated in its "EU Biodiversity Strategy for 2030" (Biodiversity Strategy) that "nature is in crisis".[104] Further, in a 2009 report from the European Commission to the European Council on conservation status of habitat types and species, the Commission acknowledged that the results of the first ever systematic assessment of the conservation status of Europe's most vulnerable habitats at that time showed that "only a small proportion of the habitats and species of Community interest are in a favourable conservation status" and noted the negative impact of climate change on this status.[105]

There is independent evidence indicating that the existing regulatory regime in the EU is unlikely to be effective in addressing climate change risks. For

104 European Commission, *EU Biodiversity Strategy for 2030: Bringing Nature Back into our Lives* (2021), p. 6.

105 European Commission, *Report from the Commission to the Council and the European Parliament on the Conservation Status of Habitat Types and Species as Required under Article 17 of the Habitats Directive, Communication*, COM(2009) 358 (13 July 2009).

example, a study conducted by Araújo and colleagues (2011) to assess the effectiveness of Natura 2000 in conserving European plant and terrestrial vertebrate species in the face of climate change ominously found that, by 2080, around 60% of the species would lose suitable climate in protected areas. Further, the authors found that Natura 2000 areas retain climate suitability for species no better, and sometimes less effectively, than unprotected areas.[106]

Nonetheless, there are heartening signs that the EU is aware of the need to revisit its approach to ecosystem management and, moreover, it appears to be heading in the direction of risk governance. In the 2021 "EU Strategy on Adaptation to Climate Change" to forge a climate-resilient Europe (Adaptation Strategy), the European Commission specifically notes the risks to ecosystems and the need to ensure resilience:

> We need to better understand the interdependencies between climate change, ecosystems, and the services they deliver. Major shifts in terrestrial ecosystems and vegetation types on the European Union's land area are expected during this century, including in protected areas. Water cycle and temperature changes, or sea level rise will put ecosystems under additional stress. Over this century, the ocean is expected to reach unprecedented conditions with increased temperatures, further acidification, and oxygen decline. We need science-based, robust ecosystem restoration and management that helps minimise risks, improves resilience, and ensures the continued delivery of vital ecosystem services and features: food provision, air and water purification, flood protection, biodiversity, and climate mitigation.[107]

The Adaptation Strategy announces that the Commission will, among other things, "close knowledge gaps on climate impacts and resilience" and "improve the state of the art on adaptation modelling, risk assessment and management tools".[108] Moreover, the Biodiversity Strategy announces the establishment of a new governance framework to "steer the implementation of biodiversity commitments agreed at national, European or international level" and notes that specific consideration will be given to whether a legally binding approach to governance is needed.[109] These announcements appear to herald a new era of ecosystem regulation in Europe that is based on a risk governance approach. Time will ultimately tell whether any changes to the design or application of the EU regulatory regime implied by this shift will ensure resilience to climate change risks to the EU's ecosystems in the future.

106 M.B. Araújo et al., n. 91 above.

107 European Commission, "Forging a climate-resilient Europe – the new EU Strategy on Adaptation to Climate Change", *Communication from the Commission to the European Parliament, the Council, the European Economic and Social Committee and the Committee of the Regions*, COM(2021) 82, p. 5.

108 European Commission, n. 105 above, p. 5.

109 European Commission, n. 104 above, p. 23.

Bibliography

Anand, M., Gonzalez, A., Guichard, F., Kolasa, J., & Parrott, L. (2010). "Ecological systems as complex systems: Challenges for an emerging science." *Diversity*, 2, 395–410. https://doi.org/10.3390/d2030395.

Araújo, M.B., Alagador, D., Cabeza, M., Nogués-Bravo, D., & Thuiller, W. (2011). "Climate change threatens European conservation areas." *Ecology Letters*, 14(5), 484–492. https://doi.org/10.1111/j.1461-0248.2011.01610.x.

Baessler, C., Bonn, A., Klotz, S., Schröter, M., & Seppelt, R. (eds.). (2019). *Atlas of Ecosystem Services: Drivers, Risks, and Societal Responses*. Springer International Publishing. https://doi.org/10.1007/978-3-319-96229-0.

Bellard, C., Bertelsmeier, C., Leadley, P., Thuiller, W., & Courchamp, F. (2012). "Impacts of climate change on the future of biodiversity: Biodiversity and climate change." *Ecology Letters*, 15(4), 365–377. https://doi.org/10.1111/j.1461-0248.2011.01736.x.

Brussard, P.F., Reed, J.M., & Tracy, C.R. (1998). "Ecosystem management: What is it really?" *Landscape and Urban Planning*, 40, 9–20. https://doi.org/10.1016/S0169-2046(97)00094–00097.

Cardinale, B.J., Duffy, J.E., Gonzalez, A., Hooper, D.U., Perrings, C., Venail, P., Narwani, A., Mace, G.M., Tilman, D., Wardle, D.A., Kinzig, A.P., Daily, G.C., Loreau, M., Grace, J.B., Larigauderie, A., Srivastava, D.S., & Naeem, S. (2012). "Biodiversity loss and its impact on humanity." *Nature*, 486(7401), 59–67. https://doi.org/10.1038/nature11148.

Ceballos, G., Ehrlich, P.R., & Raven, P.H. (2020). "Vertebrates on the brink as indicators of biological annihilation and the sixth mass extinction." *Proceedings of the National Academy of Sciences*, 117(24), 13596–13602. https://doi.org/10.1073/pnas.1922686117.

Chambers, J.C., Allen, C.R., & Cushman, S.A. (2019). "Operationalizing ecological resilience concepts for managing species and ecosystems at risk." *Frontiers in Ecology and Evolution*, 7, 241. https://doi.org/10.3389/fevo.2019.00241.

Cliquet, A., Backes, C., Harris, J., & Howsam, P. (2009). "Adaptation to climate change: Legal challenges for protected areas." *Utrecht Law Review*, 5(1), 158. https://doi.org/10.18352/ulr.100.

Council of Europe. (2019). "The Bern Convention's Contribution to the Strategic Plan for Biodiversity 2011–2020 and the Aichi Targets." *Standing Committee of the Convention on the Conservation of European Wildlife and Natural Habitats*.

Council of Europe. (2020). "Obligations of Bern Convention Parties regarding the conservation of candidate and adopted Emerald network sites: A legal analysis." *Standing Committee of the Convention on the Conservation of European Wildlife and Natural Habitats*.

Diaz, C.L. (2010). "The Bern Convention: 30 years of nature conservation in Europe." *Review of European Community & International Environmental Law*, 19(2), 185–196. https://doi.org/10.1111/j.1467-9388.2010.00676.x.

Díaz, S., Settele, J., Brondízio, E.S., Ngo, H.T., Agard, J., Arneth, A., Balvanera, P., Brauman, K.A., Butchart, S.H.M., Chan, K.M.A., Garibaldi, L.A., Ichii, K., Liu, J., Subramanian, S.M., Midgley, G.F., Miloslavich, P., Molnár, Z., Obura, D., Pfaff, A., … Zayas, C.N. (2019). "Pervasive human-driven decline of life on Earth points to the need for transformative change." *Science*, 366(1327), eaax3100. https://doi.org/10.1126/science.aax3100.

European Commission. (2021a). "Forging a climate-resilient Europe – the new EU Strategy on Adaptation to Climate Change." *Communication from the Commission to the European Parliament, the Council, the European Economic and Social Committee and the Committee of the Regions.*

European Commission. (2021b). *EU Biodiversity Strategy for 2030: Bringing Nature Back into our Lives.* Publications Office. https://data.europa.eu/doi/10.2779/677548.

European Commission & Directorate-General for Environment. (2016). "A starter's guide: Overview on the main provisions of the Water Framework Directive, the Marine Strategy Framework Directive, the Birds and Habitats Directives, and the Floods Directive: Similarities and differences." In *Technical Report 103.* Publications Office. https://doi.org/10.2779/110229.

European Commission. Joint Research Centre. (2020). *Mapping and Assessment of Ecosystems and their Services: An EU Wide Ecosystem Assessment in Support of the EU Biodiversity Strategy.* Publications Office. https://data.europa.eu/doi/10.2760/757183.

European Council. (2011). "*Conserving European biodiversity in a changing climate: The Bern Convention, the EU Birds and Habitats Directives and the adaptation of nature to climate change.*" Group of Experts on Biodiversity and Climate Change, 6th Meeting, Strasbourg, 10–11 October 2011. https://rm.coe.int/16807469ef.

Goudie, A. (2013). *The Human Impact on the Natural Environment: Past, Present, and Future.* John Wiley & Sons.

Ibrahim, M., Karimi, H., Mustafa, Y., & Hassan, M. (2021). "Mapping ecosystem service: Challenges and solutions." *Journal of Geoinformatics & Environmental Research*, 2(1), 62–77. https://doi.org/10.38094/jgier2130.

IPCC. (2014a). "Summary for policymakers." In *Climate Change 2014: Impacts, Adaptation, and Vulnerability. Part A: Global and Sectoral Aspects. Contribution of Working Group II to the Fifth Assessment Report of the Intergovernmental Panel on Climate Change.* Cambridge University Press.

IPCC. (2014b). "Freshwater resources." In *Climate Change 2014: Impacts, Adaptation, and Vulnerability. Part A: Global and Sectoral Aspects. Contribution of Working Group II to the Fifth Assessment Report of the Intergovernmental Panel on Climate Change.* Cambridge University Press.

IPCC. (2014c). "Ocean systems." In *Climate Change 2014: Impacts, Adaptation, and Vulnerability. Part A: Global and Sectoral Aspects. Contribution of Working Group II to the Fifth Assessment Report of the Intergovernmental Panel on Climate Change.* Cambridge University Press.

IPCC. (2014d). "Terrestrial and Inland Water Systems." In *Climate Change 2014: Impacts, Adaptation, and Vulnerability. Part A: Global and Sectoral Aspects. Contribution of Working Group II to the Fifth Assessment Report of the Intergovernmental Panel on Climate Change.* Cambridge University Press.

Johnson, J.A., Ruta, G., Baldos, U., Cervigni, R., Chonabayashi, S., Corong, E., Gavryliuk, O., Gerber, J., Hertel, T., Nootenboom, C., & Polasky, S. (2021). *The Economic Case for Nature.* World Bank Group. https://openknowledge.worldbank.org/bitstream/handle/10986/35882/A-Global-Earth-Economy-Model-to-Assess-Development-Policy-Pathways.pdf?sequence=1&isAllowed=y.

Kotzé, L.J., & Marauhn, T. (2014). *Transboundary Governance of Biodiversity.* Brill. http://ebookcentral.proquest.com/lib/monash/detail.action?docID=1730307.

Lackey, R.T. (1998). "Seven pillars of ecosystem management." *Landscape and Urban Planning*, 40(1–3), 21–30. https://doi.org/10.1016/S0169-2046(97)00095–00099.

Loreau, M. (2010). *From Populations to Ecosystems: Theoretical Foundations for a New Ecological Synthesis (MPB-46)*. Princeton University Press. http://ebookcentral.proquest.com/lib/monash/detail.action?docID=557151.

Millennium Ecosystem Assessment Board. (2005). *Ecosystems and human well-being: Synthesis*. Island Press. https://www.millenniumassessment.org/documents/document.356.aspx.pdf.

Moreno-Mateos, D., Alberdi, A., Morriën, E., van der Putten, W.H., Roíguez-Uña, A., & Montoya, D. (2020). "The long-term restoration of ecosystem complexity." *Nature Ecology & Evolution*, 4, 676–685. https://doi.org/10.1038/s41559-020-1154-1.

Mori, A.S. (2011). "Ecosystem management based on natural disturbances: Hierarchical context and non-equilibrium paradigm." *Journal of Applied Ecology*, 48, 280–292. https://doi.org/10.1111/j.1365-2664.2010.01956.x.

Naeem, S., Bunker, D.E., Hector, A., Loreau, M., & Perrings, C. (2009). *Biodiversity, Ecosystem Functioning, and Human Wellbeing*. Oxford University Press. https://doi.org/10.1093/acprof:oso/9780199547951.001.0001.

Norris, K. (2012). "Biodiversity in the context of ecosystem services: The applied need for systems approaches." *Philosophical Transactions. Biological Sciences*, 367(1586), 191–199. https://doi.org/10.1098/rstb.2011.0176.

OECD. (2019). *Biodiversity: Finance and the Economic and Business Case for Action*. Report Environment Ministers' Meetings, 5–6 May 2019. https://doi.org/10.1787/a3147942-en.

Prip, C. (2018). "The Convention on Biological Diversity as a legal framework for safeguarding ecosystem services." *Ecosystem Services*, 29, 199–204. https://doi.org/10.1016/j.ecoser.2017.02.015.

Schmitz, O.J. (2010). *Resolving Ecosystem Complexity*. Princeton University Press.

Secretariat of the Convention on Biological Diversity. (2000). *Sustaining Life on Earth: How the Convention on Biological Diversity Promotes Nature and Human Well-Being*. Secretariat of the Convention on Biological Diversity. https://www.cbd.int/doc/publications/cbd-sustain-en.pdf.

Steffen, W., Crutzen, P.J., & McNeill, J.R. (2007). "The Anthropocene: Are humans now overwhelming the great forces of nature?" *Ambio*, 36(8), 614–621.

Sterling, E.J., Gómez, A., & Porzecanski, A. L. (2010). "A systemic view of biodiversity and its conservation: Processes, interrelationships, and human culture." *BioEssays*, 32 (12), 1090–1098. https://doi.org/10.1002/bies.201000049.

Trouwborst, A. (2011). "Conserving European biodiversity in a changing climate: The Bern Convention, the European Union Birds and Habitats Directives and the adaptation of nature to climate change." *Review of European Community & International Environmental Law*, 20(1), 62–77. https://doi.org/10.1111/j.1467-9388.2011.00700.x.

van der Sluis, T. & Chardon, P. (2001). "How to define European ecological networks." *WIT Transactions on Ecology and the Environment*, 46. https://doi.org/10.2495/ECO010121.

Verschuuren, J. (2010). "Climate change: Rethinking restoration in the European Union's Birds and Habitats Directives." *Ecological Restoration*, 28(4), 431–439. https://doi.org/10.3368/er.28.4.431.

Chapter 9

Regulating Climate Change Risks to the UK Health System

Health systems, which comprise the range of institutions, infrastructure, people and resources used to deliver health services to the public, play a critical role in delivering important health outcomes – namely, to improve health while ensuring responsiveness to a population's legitimate expectations regarding their interactions with the health system, and fairness in relation to access to health systems and services.[1] Hsiao (2003) describes a health system as a "means to an end" which "exists and evolves to serve societal needs".[2] The critical functions performed by health systems highlight the importance of ensuring their well-functioning. Strong health systems help to secure positive health outcomes in the face of threats, but weak health systems can put these outcomes at risk.[3]

Health systems and facilities are at the forefront in protecting people from health threats, including those posed by climate change.[4] Yet, given their inherent complexity, the interplay between health systems and external stressors such as climate change can lead to unpredictable consequences that are challenging to plan for. Failure to account for this complexity may undermine effective management of health systems and, in turn, could compromise public health outcomes.[5] As Midgley (2006) puts it "the whole concept of public health is founded on the insight that health and illness have causes or conditions

1 World Health Organization, *Everybody's Business – Strengthening Health Systems to Improve Health Outcomes: WHO's Framework for Action* (World Health Organization, 2007), at p. 2 referring to *The World Health Report 2000 – Health Systems: Improving Performance* (WHO, 2000). C.J.L. Murray, J. Frenk, & World Health Organization. Global Programme on Evidence for Health Policy, "A WHO framework for health system performance assessment" (1999) *Global Programme on Evidence for Health Policy*, p. 6–8.

2 W.C. Hsiao, "What is a Health System? Why should we care?" (2003) *Harvard School of Public Health Working Paper*, p. 3–4.

3 World Health Organization, n. 1 above, p. 1.

4 World Health Organization, *COP26 Special Report on Climate Change and Health: The Health Argument for Climate Action* (World Health Organization, 2021), p. 25.

5 R. Atun, "Health systems, systems thinking and innovation" (2012) 27 *Health Policy and Planning*, pp. iv4–8, at iv4–5, iv6–7.

DOI: 10.4324/9781003253464-9

that go beyond the biology and behavior of the individual" and that responding to public health issues requires a systems approach that accounts for the various interacting variables and involves "looking at things in terms of the bigger picture".[6]

As observed during the global pandemic, health systems are vulnerable to "tipping points".[7] Indeed, the explosion of COVID-19 cases at the height of the pandemic crippled health systems around the world. This raises the question of whether these same systems will be prepared for the next big health threat, such as climate change. A recent survey conducted by the World Health Organization (WHO) highlighted that only half of the world's countries have a national climate change plan or strategy in place to address the impact of climate change on health.[8] This case study considers relevant aspects of the regulation of the UK health system to identify the approach to managing climate change risks to the system and, in particular, to determine whether there is evidence of a risk governance approach reflected in the UK regulatory regime that was outlined earlier in this book.

I The role and importance of health systems

A health system, like any other system, is a set of interconnected parts that must function together to be effective. According to the WHO, a health system consists of all the organisations, people and actions the primary intent of which is to promote, restore or maintain health. This includes aspects of a health system that have a direct impact on health (such as hospitals and practitioners who provide health services) as well as those aspects that have a more indirect health impact by seeking to influence the determinants of health (such as campaigns to encourage behaviour that reduces risks to health).[9] Goodwin (2017) notes that every country can be said to have some form of national health care system, regardless of how unstructured or unsystematic its operation.[10]

The WHO's definition of a health system conveys a sense of their possible scale and complexity and is also suggestive of the enormity of the challenge in ensuring that desirable health outcomes can be delivered. Every health system will comprise a multitude of components and moving parts. Atun (2012)

6 G. Midgley, "Systemic intervention for public health" (2006) 96(3) *American Journal of Public Health*, pp. 466–72, at 466.

7 D. De Savigny, T. Adam, & World Health Organization, *Systems Thinking for Health Systems Strengthening* (World Health Organization, 2009), p. 47.

8 World Health Organization, *2021 WHO Health and Climate Change Global Survey Report* (World Health Organization, 2021), p. x.

9 World Health Organization, n. 1 above p. 2; N. Goodwin, "National health systems: A historical overview", in *International Encyclopaedia of Public Health* (Elsevier, 2017), pp. 201–14, at 201.

10 N. Goodwin, n. 9 above, p. 201.

explains that health systems are open systems consisting of interlinked components that interact with the context within which the health system is situated and that it is this interaction that determines the system's behaviour, rather than any single component of the system.[11] Improvements to one aspect of the system may depend upon contributions from other aspects and, similarly, interventions affecting one part of the system could have repercussions elsewhere within the system.[12] These features highlight the need for a holistic, systems-based approach for the design and management of health systems.

The WHO has identified six common "building blocks" that are essential for the effectiveness of a health system and support the delivery of positive health outcomes. They also serve to simplify and facilitate understanding of the key components of a health system.[13] These are good health services, a well-performing health workforce, a well-functioning health information system, access to essential medical products, vaccines and technologies, a good health financing system, and leadership and governance.[14] Together, these building blocks must work synergistically together to respond to changing health needs so that the system can achieve positive health outcomes.[15]

Strong health systems can also help ensure resilience to negative shocks. According to the WHO (2015), health system resilience refers to "the holistic ability and agility of a system to change and flex – according to circumstances – and continue to function under stress, while undergoing change."[16] Kruk and colleagues (2015) have more specifically explained that this means that the system is able to "maintain core functions when a crisis hits and, informed by lessons learned during the crisis, reorganise if conditions require it".[17] In other words, health system resilience relates to the capacity of the system to cope with and manage health risks in a way that its essential functions, identity and structure are maintained.[18]

Arguably, the building block dealing with leadership and governance of health systems is the most critical in addressing negative shocks because it involves overseeing and guiding the entire system – including its component parts – in a manner that protects the public interest, particularly when the system's operating context dramatically and adversely changes.[19] This building block, which has also been

11 R. Atun, n. 5 above, pp. iv4–5.
12 World Health Organization, n. 1 above, p. 4.
13 Ibid. p. 3.
14 Ibid. p. vi.
15 M. Fridell, S. Edwin, J. von Schreeb, & D.D. Saulnier, "Health system resilience: What are we talking about? A scoping review mapping characteristics and keywords" (2020) 9(1) *International Journal of Health Policy and Management*, pp. 6–16.
16 World Health Organization, *Operational Framework for Building Climate Resilient Health Systems* (World Health Organization, 2015), p. 7.
17 M.E. Kruk, M. Myers, S.T. Varpilah, & B.T. Dahn, "What is a resilient health system? Lessons from Ebola" (2015) 385(9980) *The Lancet*, pp. 1910–12.
18 World Health Organization, n. 16 above, p. 7.
19 World Health Organization, n. 1 above, p. 23.

described as "stewardship", encompasses a broad range of activities, including system design, regulation, policy guidance and fostering collaboration and coalition building.[20] The question arises as to how effective stewardship can be executed in practice – that is, in a manner that accounts for the complexity of health systems and the varied contexts in which health systems are embedded.[21] According to the WHO (2009), "in designing and evaluating system-level interventions or interventions with system-wide effects, a comprehensive assessment of the main effects (intended or not) and the contextual factors that may help explain the success or failure of the intervention are essential".[22] An important contextual consideration of relevance for all health systems is the impact of climate change.

II The impact of climate change on health systems

The WHO has labelled climate change as the biggest global threat to human health of the 21st century.[23] Heat waves, wildfires, floods, droughts and other extreme weather events have already affected the health of millions of people worldwide and resulted in a range of health and socio-economic consequences.[24] The IPCC (2014) has found that, in the future, climate change will intensify existing threats to health and exacerbate health problems that already exist.[25] It is projected to alter the geographical spread and numbers of cases of injuries, illnesses and deaths from climate-related conditions, and to affect the well-functioning of public health and health care systems.[26]

The specific impacts of climate change on health have been well-catalogued.[27] For example, extreme heat can complicate cardiovascular and respiratory disease.[28] Weather-related natural disasters such as extreme storms, floods, and wildfires

20 Ibid. p. 23.
21 J.N. Lavis et al., "Guidance for evidence-informed policies about health systems: Linking guidance development to policy development" (2012) 9(3) *PLoS Medicine*, e1001186–e1001186.
22 D. De Savigny, T. Adam, & World Health Organization, n. 7 above, p. 47.
23 World Health Organization, n. 4 above, p. 15.
24 Ibid. p. 2; World Health Organization, *Climate Change and Health Research: Current trends, gaps and perspectives for the future* (2021), p. 1.
25 IPCC, "Human health: Impacts, adaptation, and co-benefits", in *Climate Change 2014 Impacts, Adaptation, and Vulnerability. Working Group II Contribution to the Fifth Assessment Report. Intergovernmental Panel on Climate Change* (Cambridge University Press, 2014), p. 713.
26 L.F. Wiley, "Climate change adaptation and public health law", in *Research Handbook on Climate Change Adaptation Law* (Edward Elgar Publishing, 2013), pp. 171–213, at 174.
27 See, for example, K.L. Ebi et al., "Human health: Impacts, risks, and adaptation in the United States", in *Impacts Risks, and Adaptation in the United States: Fourth National Climate Assessment, Volume II* (U.S. Global Change Research Program, 2018).
28 L.F. Wiley, "Climate change adaptation and public health law", in *Research Handbook on Climate Change Adaptation Law* (Edward Elgar Publishing, 2013), p. 174–5.

cause physical injuries, mental health issues and deaths.[29] Climate change may also create conditions for the emergence of new infectious pathogens, by altering the ways in which animal species interact with each other and with humans.[30]

In relation to the impact of climate change on health systems, "The Lancet Countdown on health and climate change" (2019) stated that "[l]eft unabated, climate change will define the health profile of current and future generations [and] will challenge already overwhelmed health systems".[31] However, the pressure exerted by climate change on health systems has not been documented as comprehensively nor as clearly as the impacts of climate change on health.[32] Curtis and colleagues (2017) reviewed evidence regarding the impacts of extreme weather on health systems and found that heat waves may impair functionality of hospitals, including medical equipment and storage of medicines.[33] Cold waves can include increased rates of consultation and treatment for respiratory conditions and for injuries resulting from falls in ice and snow.[34] Floods can result in flooded health facilities, interruptions to power and water supplies, safety of and access to patient records and interruption to ambulance services.[35] Extreme weather may lead to increased hospitalisation rates due to event-related injuries.[36]

As awareness of the consequences of climate change both for health and health systems grows, so too does the need to focus on how these consequences should be addressed.[37] However, predicting the likely impact of climate change on health systems is challenging because the location, frequency, scale and impact of many climate-sensitive health risks are uncertain.[38] This necessarily means that there is also uncertainty about the impact of climate change on health systems and raises the important question of how health systems can be prepared for the intensification of climate change risks when the specific impact of those risks on health systems may be far from clear.

29 Ibid. p. 177.
30 Ibid. p. 174.
31 N. Watts et al., "The 2019 report of The Lancet Countdown on health and climate change: Ensuring that the health of a child born today is not defined by a changing climate" (2019) 394(10211) *The Lancet*, pp. 1836–78, at 1838.
32 M. Fridell, S. Edwin, J. von Schreeb, & D.D. Saulnier, n. 15 above, p. 6 referring to A.E. Olafsdottir, P. Allotey, & D.D. Reidpath, "A health system in economic crises: A case study from Iceland" (2013) 41(2) *Scandinavian Journal of Public Health*, pp. 198–205 and S. Thomas, C. Keegan, S. Barry, R. Layte, M. Jowett, & C. Normand, "A framework for assessing health system resilience in an economic crisis: Ireland as a test case" (2013) 13(1) *BMC Health Services Research*.
33 S. Curtis, A. Fair, J. Wistow, D.V. Val, & K. Oven, "Impact of extreme weather events and climate change for health and social care systems" (2017) 16(Suppl 1) *Environmental Health*, p. 128, at 24.
34 Ibid. p. 25.
35 Ibid. p. 25.
36 Ibid. p. 25.
37 N. Watts et al., n. 31 above, p. 1849.
38 World Health Organization, n. 4 above, p. 2.

The WHO advocates the establishment of a "climate-resilient health system" – that is, one that is able to anticipate, respond to, cope with, recover from and adapt to climate-related shocks and stress.[39] The WHO Operational Framework, the goal of which is to enhance the capacity of health systems to protect and improve population health in an unstable and changing climate, indicates that, for an entire health system to become more climate-resilient, its independent building blocks also have to become resilient.[40]

System stewards will clearly be instrumental in this regard.[41] The WHO Operational Framework indicates that leadership and strategic planning will need to address the complex and long-term nature of climate change risks.[42] Among other things, this will involve an assessment of climate-sensitive health risks and the extent to which health systems are capable of responding to such risks.[43] The "vulnerability and adaptation assessments" can then be used to inform the development of national adaptation plans.[44] The Framework also notes that focal points within health ministries can be used to co-ordinate action.[45] Pencheon (2018) further suggests that regulatory frameworks can be used to "accelerate" changes to health systems and "normalise them socially".[46] The role for government, health ministries and public health regulation in driving climate resilience in health systems is clear.

III Public health regulation to tackle climate change risks

In 1920, Winslow (a medical academic) described public health as:

> the science and the art of preventing disease, prolonging life, and promoting physical health and efficiency through organized community efforts for the sanitation of the environment, the control of community infections, the education of the individual in principles of personal hygiene, the organization of medical and nursing service for the early diagnosis and preventative treatment of disease, and the development of the social machinery which will ensure to every individual in the community a standard of living adequate for the maintenance of health.[47]

39 Ibid. p. 8.
40 World Health Organization, n. 16 above, p. 8.
41 D. De Savigny, T. Adam, & World Health Organization, n. 7 above, p. 74.
42 World Health Organization, n. 16 above, pp. 10–11.
43 World Health Organization, n. 4 above, p. 25.
44 Ibid. p. 25.
45 World Health Organization, n. 16 above, p. 15.
46 D. Pencheon, "Developing a sustainable health care system: The United Kingdom experience" (2018) 208(7) *Medical Journal of Australia*, pp. 284–5.
47 C.-E.A. Winslow, "The untilled fields of public health" (1920) 51(1306) *Science*, pp. 23–33, at 30.

While some aspects of this definition may seem outdated today, the comprehensive, holistic and civic-minded conception of public health reflected in this definition still resonates and is consistent with the WHO's definition of the term.[48]

Gostin (2006) suggests that the government has primary responsibility for the public's health[49] and Jochelson (2006) considers that this responsibility means that government is "ultimately responsible for citizens' wellbeing".[50] There is a multitude of ways in which public health could be advanced and protected by government, including by ensuring access to health infrastructure and services, initiatives to promote healthy communities and healthy behaviour by individuals, as well as activities to address sources of health hazards such as food, environmental hazards and natural disasters. Similarly, there is a range of mechanisms that can be used to implement these activities and initiatives, such as through educational campaigns, establishing institutions and developing infrastructure, providing funds and other resources, along with legislating to regulate matters concerning public health.

Gostin and Wiley (2016) define public health regulation as:

> legal powers and duties of the state to assure the conditions for people to be healthy (to identify, prevent, and ameliorate risks to health in the population) and the limitations on the power of the state to constrain the autonomy, privacy, liberty, proprietary, or other legally protected interests of individuals for the common good.[51]

They outline three main "modes" of legal intervention in the domain of public health – namely, direct regulation for the public's health and safety (such as warnings or laws about products and services that could have an adverse impact on health), indirect regulation in the form of tort law to provide redress for adverse health outcomes, and taxation, spending and social support to help establish and maintain public health systems and services.[52] The obvious breadth of each of these modes of legal intervention for public health purposes hints at the spectrum of regulatory instruments that are likely to comprise public health regulation in a particular jurisdiction, each serving a distinct public health purpose and addressing public health issues in different ways.

48 *The World Health Report 2000 – Health Systems: Improving Performance* (WHO, 2000), p. xi.
49 L. Gostin, "Legal foundations of public health law and its role in meeting future challenges" (2006) 120 *Public Health*, pp. 8–14, at 8.
50 K. Jochelson, "Nanny or steward? The role of government in public health" (2006) 120(12) *Public Health*, pp. 1149–55, at 1153.
51 L.O. Gostin & L.F. Wiley, *Public Health Law: Power, Duty, Restraint* (University of California Press, 3rd edition, 2016), p. 4.
52 L.O. Gostin & L.F. Wiley, n. 51 above. See Part 3 – "Modes of Legal Intervention".

In terms of the modes of regulation that have been used so far to address the public health risks caused by climate change, Wiley (2013) suggests that health adaptation efforts to date have focused primarily on threat-specific programs aimed at vector control, disaster preparedness, or heat wave planning.[53] Moreover, in his view, regulatory responses to climate change risks have mostly developed in a "piecemeal, reactionary fashion and do little to meet the need for comprehensive and deliberate health adaptation planning".[54] He advocates more "upstream interventions", which help to prevent health risks from materializing, but he says that such interventions are being neglected under current regulatory frameworks.[55]

Arguably, the "upstream" approach put forward by Wiley constitutes a fourth mode of public health regulation which sits above the other three modes identified by Gostin and Wiley – namely, government's leadership and stewardship of the entire health system, including its various components and aspects, through regulation to mitigate risks to the system. The next section considers whether this approach is evident in the UK public health regulatory regime, particularly in relation to the UK government's management of climate change risks to health systems.

IV The regulatory regime in the UK

The primary legal instrument for the management of the UK health system is the *National Health Service Act 2006* (NHS Act). The origins of the NHS Act date back to 1948 when the UK's National Health Service (NHS) was first established in the wake of World War II.[56] The context for this important social reform included poor standards of health, shortage of hospital facilities, uneven provision of health services and inadequate access to health care for more than half the population.[57] The NHS has undergone significant evolution since its inception, including the devolution of responsibility to the nations comprising the United Kingdom to run the NHS within their respective jurisdictions – namely, in England, Scotland, Wales and Northern Ireland.[58] Nonetheless, the NHS remains the centrepiece of the UK health system and continues to be based on the core principle of universal healthcare.[59]

53 L.F. Wiley, n. 26 above, p. 202.
54 Ibid. p. 184.
55 Ibid. p. 213.
56 "The history of the NHS", accessible at: https://www.nuffieldtrust.org.uk/health-and-social-care-explained/the-history-of-the-nhs/.
57 R. Surender, K. Matsuoka, & P. Ovseiko, "United Kingdom, Health System of", in *International Encyclopedia of Public Health* (Elsevier, 2017), pp. 293–300, at 427.
58 S.L. Greer, "Devolution and health in the UK: policy and its lessons since 1998" (2016) 118(1) *British Medical Bulletin*, pp. 16–24.
59 K. Grosios, P.B. Gahan, & J. Burbidge, "Overview of healthcare in the UK" (2010) 1(4) *EPMA Journal*, pp. 529–34, at 529.

It is clear from the NHS Act, as it applies in England, that the Secretary of State for Health and Social Care (Secretary) sits at the pinnacle of the NHS. Under the Act, the Secretary has ministerial responsibility for the NHS in England.[60] The Secretary's various duties include the promotion of a comprehensive health service,[61] continuous improvement in the quality of services for the protection or improvement of public health,[62] and promotion of research on matters relevant to the health service.[63] In addition, the Secretary has a specific duty in relation to the protection of public health. In particular, section 2A of the NHS Act relevantly provides:

1 The Secretary of State must take such steps as the Secretary of State considers appropriate for the purpose of protecting the public in England from disease or other dangers to health.
2 The steps that may be taken under subsection (1) include –

   a the conduct of research or such other steps as the Secretary of State considers appropriate for advancing knowledge and understanding;
   b providing microbiological or other technical services (whether in laboratories or otherwise);
   c providing vaccination, immunisation or screening services;
   d providing other services or facilities for the prevention, diagnosis or treatment of illness;
   e providing training;
   f providing information and advice;
   g making available the services of any person or any facilities.

There are a few notable aspects of this public health duty. First, the duty is mandatory, but the Secretary has considerable discretion in discharging it. More specifically, under section 2A(1), the Secretary "must" take steps to protect the public health from disease and other dangers to health. The obligatory nature of this duty is clear, although the reference to "steps as the Secretary of State considers appropriate" indicates that the Secretary has flexibility regarding the specific steps to fulfil the duty. Section 2A(2) sets out possible steps, including research and providing information and advice, but the Secretary may take other steps as he or she considers appropriate. Moreover, the Secretary has a general power under the Act to "do anything which is calculated to facilitate, or is conducive or incidental to, the discharge of any function".[64]

The second aspect of the Secretary's public health duty worth highlighting is the reference to "other dangers to health" in section 2A(1). The explanatory notes

60 Section 1(3), NHS Act.
61 Section 1(1), NHS Act.
62 Section 1A(1), NHS Act.
63 Section 1E, NHS Act.
64 Section 2, NHS Act.

accompanying the bill to introduce this duty into the NHS Act explained that this terminology reflects an "all hazards" approach, which does not exhaustively list the dangers to health from which the Secretary of State must protect the public but, rather, helps to ensure that the duty continues to be relevant and effective as new threats to health emerge.[65] The notes also clarify that "'other dangers to health' might include the health effects of climate change (such as flooding and heat waves)". Based on this note, there is a sound legal argument to be made that the Secretary's duty to protect public health includes dealing with the dangers of climate change *to health* and potentially extends to threats caused by climate change *to the health system* because such threats will have inevitable consequences for human health.

Third, the specific manner in which the Secretary may discharge the duty is also notable. Under the Act, the Secretary has power to delegate public health functions to certain bodies,[66] including NHS England, which is an independent statutory body that leads the NHS in England.[67] Each year, the Secretary issues "the mandate" to NHS England, which must set out the objectives of NHS England to be achieved in the exercise of its functions.[68] These objectives are intended to assist NHS England to develop medium- and long-term planning assumptions.[69] Notably, the current mandate[70] and the NHS long-term plan[71] make no explicit reference to the need to adapt the UK health system to climate change impacts.

The Secretary is also supported in the discharge of public health duties by the UK Health Security Agency (UKHSA), which is an executive agency within the Department of Health and Social Care. It is responsible for "protecting every member of every community from the impact of infectious diseases, chemical, biological, radiological and nuclear incidents and other health threats" and for providing "intellectual, scientific and operational leadership at the national and local level."[72] The UKHSA's website refers to its long-term work on climate change and health, including to assess health risks posed by climate change, and stresses that the COVID-19 pandemic reinforced the need for collaboration at global, national and local levels "in order to tackle major threats".[73]

65 Section 2A of the NHS Act was introduced by section 11 of the *Health and Social Care Act 2012* (HSCA). The Explanatory Notes for this Act are accessible at: https://www.legislation.gov.uk/ukpga/2012/7/notes/contents.
66 Section 7A, NHS Act.
67 NHS England is referred to as the National Health Service Commissioning Board or "the Board" in the NHS Act.
68 Section 13A, NHS Act.
69 HSCA Explanatory Notes, n. 65 above.
70 UK Department of Health & Social Care, *The Government's 2021–22 Mandate to NHS England and NHS Improvement* (March 2021).
71 NHS England, *The NHS Long Term Plan* (January 2019).
72 UK Health Security Agency, "About Us" accessible at: https://www.gov.uk/government/organisations/uk-health-security-agency/about.
73 UK Health Security Agency, "Understanding the health effects of climate change" accessible at: https://ukhsa.blog.gov.uk/2021/11/09/understanding-the-health-effects-of-climate-change/.

In fact, the enormous challenges faced by the UK health system during the COVID-19 pandemic led to reflection about how the NHS system could be improved. In a 2021 White Paper entitled "Integration and Innovation: working together to improve health and social care for all" (White Paper), the Department of Health & Social Care remarked that "[w]e are living through the greatest challenge our health and care system has ever faced".[74] The White Paper set out legislative proposals aimed primarily at building on collaboration that occurred across the system as the pandemic unfolded by establishing "Integrated Care Systems" (ICS). These are partnerships between providers and commissioners of NHS services across a geographical area together with local authorities and other local partners to collectively plan health and care services to meet the needs of local populations. The White Paper explains:[75]

> At the heart of the changes being taken forward by the NHS and its partners, and at the heart of our legislative proposals, is the goal of joined up care for everyone in England. Instead of working independently every part of the NHS, public health and social care system should continue to seek out ways to connect, communicate and collaborate so that the health and care needs of people are met. Healthy, fulfilled, independent and longer lives for the people of England will require health and care services, local government, NHS bodies, and others to work ever more closely together. Different professions, organisations, services and sectors will work with common purpose and in partnership. This will be especially important when we seek to focus on the people and communities that are most in need of support.

The White Paper led to the *Health and Care Bill*, which was introduced into Parliament in July 2021 (Bill).[76] Notably, the Bill contains new powers for the Secretary to give directions to NHS England in relation to its functions. The explanatory notes for the Bill explain that the policy intention is to enable the Secretary to ensure that NHS England continues to work effectively with other parts of the system, to support integration and to tackle broader priorities, although these broader priorities are not specifically identified.[77]

74 Department of Health & Social Care, *Integration and Innovation: working together to improve health and social care for all* (2021), p. 3.

75 Ibid. p. 7.

76 At the time of preparing this case study, the Bill had not yet been passed by the UK Parliament.

77 Explanatory Notes, *Health and Care Bill 2021*, accessible at: https://publications.parliament.uk/pa/bills/cbill/58-02/0140/en/210140en.pdf.

V Approach to the management of climate change risks to the UK health system

The NHS Act is a detailed and complex piece of legislation, which covers a spectrum of issues associated with the establishment and operation of the NHS and the delivery of health care services in the UK. While climate change is not explicitly mentioned in the Act, as discussed below, aspects of the regulatory architecture appear to be available to help drive climate resilience of the UK health system.

1 Regulatory boundary

In terms of whether the regulatory boundary of the NHS Act encompasses systemic risks to the UK health system caused by climate change, the scope of the Act is wide-ranging and covers many facets of the NHS, which could indicate that risks to the system as a whole fall within the boundary of the regulatory regime of the Act.[78] Even so, the various elements of the health system and the manner in which they are intended to work together to deliver health services and protect public health are not readily apparent from a perusal of the Act on its own.

The coverage of climate change risks to the health system under the NHS Act is also unclear as the Act does not specifically reference these risks. This is unsurprising. In fact, it would be practically impossible for public health regulation to catalogue each and every risk to public health. Nonetheless, the explanatory notes for the Secretary's public health protection duty under the NHS Act could be relied upon to interpret the duty as extending to dangers to health systems arising from climate change.[79] Therefore, it is at least arguable that the public health protection duty covers the risks to the UK health system arising from climate change and that, therefore, these risks fall within the scope of the regulatory regime.

2 Regulatory context

As there is no explicit recognition of climate change risks to the UK health system in the NHS Act, there is consequently no legislative mechanism to contextualise those risks under that Act either. Nevertheless, the UK *Climate Change Act 2008* (Climate Change Act) helps to fill in that context. The Climate Change Act is the foundation for the UK's approach to reducing greenhouse gas emissions and adapting to the effects of climate change across the country. Among other things, the Act requires the government to publish a

78 This interpretation tends to be supported by the broad powers, duties and functions vested in the Secretary of State and other participants in the NHS under the NHS Act.

79 HCSA Explanatory Notes, n. 65 above.

climate change risk assessment every five years[80] and to develop a National Adaptation Programme to respond to the risk assessment.[81] The UK's progress in relation to its climate change commitments is monitored by the Climate Change Committee (CCC), which is an independent expert body established under the Climate Change Act. In its 2019 Progress Report to Parliament, the CCC was critical of the government's approach to tackling the health impacts of climate change. The CCC said that "little progress" was being made in planning for and addressing climate change risk in key health areas.[82]

In December 2021, the "Third Health and Care Adaptation Report" (HCA Report) was issued by the NHS and UKHSA. The stated aims of the HCA Report are to review progress on adaptation in the health and social care sector and to provide an overview of next steps required at a local, regional and national level to address any identified risks and to build resilience.[83] The report confirms that "[c]limate change is a threat to the significant progress made in health and social care to date, impacting the sector's ability to deliver services and manage population health" and emphasises that "[a]ctions must be taken to build the health and care sector's climate resilience, so it can effectively respond to climate change impacts while maintaining core functions and identifying lessons learned to reorganise if required".[84] The HCA Report highlights the complexity of establishing a climate resilient health and care sector due to factors including the fragmentation between tertiary, primary and social care services.[85]

The HCA Report refers to the Strategic Health Asset Planning and Evaluation (SHAPE) tool, which seeks to support strategic planning for all healthcare services, and ostensibly could be an important practical tool for the contextualisation of climate change risks to the health system. According to the Report, local organisations and health services use the tool to map local risks, to assess the best locations for services, and evaluate any impact of service configuration on population health. It also informs the development of plans, including for natural disasters and emergencies.[86] However, the extent to which climate change risks are specifically reflected in this tool is unclear.[87] Moreover, given that the tool is used independently by various players in the health system and has no statutory basis, it is far from clear whether such a tool as currently designed and applied would be capable of yielding a comprehensive and coherent assessment and evaluation of climate

80 Section 56, *Climate Change Act 2008.*
81 Section 58, *Climate Change Act 2008.*
82 Committee on Climate Change, "Progress in preparing for climate change", *2019 Report to Parliament*, p. 16.
83 NHS and UKHSA, *Third Health and Care Adaptation Report* (2019), p. 7.
84 Ibid., p. 5.
85 Ibid. p. 25.
86 Ibid. p. 12.
87 The SHAPE website does not explicitly refer to climate change risks; accessible at: https://shapeatlas.net/.

change risks to the UK health system as a whole that would lead to effective regulatory intervention.

3 Regulatory content

Regarding the question of whether the content of the UK regulatory regime is appropriate for the regulation of systemic risks caused by climate change, the NHS Act is commendable in its recognition of the critical importance of leadership and governance of the NHS. In particular, the Secretary clearly has a stewardship role under the regulatory regime, which is discharged through core duties, including the duty to protect public health. This duty is coupled with broad discretion to take "appropriate steps" and to "do anything" to facilitate discharge of the Secretary's duty. This formulation provides the Secretary with significant flexibility to use policy and administrative mechanisms to ensure that the climate change risks to the various aspects of the UK health system are effectively managed, although the extent to which this flexibility has been utilised for this purpose so far is unclear.

VI Alignment with the risk governance regulatory ideal

The NHS Act governs the regulation of the NHS. This singularity may suggest that risks to the entire UK health system can be managed in a consistent and coherent manner and that the systemic nature of those risks can be appropriately addressed under the Act. However, the risks to the health system arising from climate change are not explicitly referenced in the regime, which could mean that such risks are not prioritised in practice.

Moreover, consideration of the manner in which these risks are regulated under the NHS Act revealed limited alignment with the ideal for the regulation of systemic risks based on the risk governance paradigm put forward earlier in this book. Indeed, the Act is not couched in terms of risk mitigation, at least not explicitly. In addition to the absence of explicit reference to risks to the UK health system, the NHS Act does not contain a regulatory mechanism to identify, assess and address those risks. In fact, it is very difficult to characterise the regime as a mechanism to mitigate risk at all.

All of this does not necessarily mean that systemic risks to the UK health system, including those caused by climate change, will be neglected notwithstanding the apparent gaps in the regulatory regime. As steward of the UK health system under the NHS Act, the Secretary has the duty, powers and functions to ensure that risks to the system are effectively addressed and, therefore, is well poised – at least from a legal perspective – to take action to ensure that the system is resilient to climate change risks. The emphasis on systems, integration and collaboration in the recent proposed reforms to the NHS and the NHS Act in the wake of the Covid-19 pandemic provides promising evidence that the systemic features of the health system have been identified and, hopefully, will be accounted for in relation to the administration of the NHS moving forward, particularly in the face of the systemic risks posed by climate change.

Bibliography

Atun, R. (2012). "Health systems, systems thinking and innovation." *Health Policy and Planning*, 27, iv4–iv8. https://doi.org/10.1093/heapol/czs088.

Curtis, S., Fair, A., Wistow, J., Val, D.V., & Oven, K. (2017). "Impact of extreme weather events and climate change for health and social care systems." *Environmental Health*, 16(Suppl 1), 128. https://doi.org/10.1186/s12940-017-0324-3.

De Savigny, D., Adam, T., & World Health Organization. (2009). *Systems Thinking for Health Systems Strengthening*. World Health Organization.

Department of Health & Social Care. (2021). *Integration and Innovation: Working together to improve health and social care for all*. DHSC.

Ebi, K.L., Balbus, J., Luber, G., Bole, A., Crimmins, A.R., Glass, G.E., Saha, S., Shimamoto, M.M., Trtanj, J.M., & White-Newsome, J.L. (2018). "Human health. Impacts, risks, and adaptation in the United States." In *Impacts, Risks, and Adaptation in the United States: Fourth National Climate Assessment, Volume* II. U.S. Global Change Research Program. https://doi.org/10.7930/NCA4.2018.CH14.

IPCC. (2014). "Human health: Impacts, adaptation, and co-benefits." In *Climate Change 2014: Impacts, Adaptation, and Vulnerability: Working Group II contribution to the Fifth Assessment Report of the Intergovernmental Panel on Climate Change*. Cambridge University Press.

Fridell, M., Edwin, S., von Schreeb, J., & Saulnier, D.D. (2020). "Health system resilience: What are we talking about? A scoping review mapping characteristics and keywords." *International Journal of Health Policy and Management*, 9(1), 6–16. https://doi.org/10.15171/ijhpm.2019.71.

Goodwin, N. (2017). "National health systems: A historical overview." In *International Encyclopedia of Public Health*. Elsevier. https://doi.org/10.1016/B978-0-12-803678-5.00298-8.

Gostin, L. (2006). "Legal foundations of public health law and its role in meeting future challenges." *Public Health*, 120, 8–14. https://doi.org/10.1016/j.puhe.2006.07.006.

Gostin, L.O., & Wiley, L.F. (2016). *Public Health Law: Power, Duty, Restraint* (Third edition). University of California Press.

Greer, S.L. (2016). "Devolution and health in the UK: Policy and its lessons since 1998." *British Medical Bulletin*, 118(1), 16–24. https://doi.org/10.1093/bmb/ldw013.

Grosios, K., Gahan, P.B., & Burbidge, J. (2010). "Overview of healthcare in the UK." *EPMA Journal*, 1(4), 529–534. https://doi.org/10.1007/s13167-010-0050-1.

Hsiao, W. C. (2003). "What is a health system? Why should we care?"*Harvard School of Public Health Working Paper.*

IPCC. (2014). "Human health: Impacts, adaptation, and co-benefits." In *Climate Change 2014 Impacts, Adaptation, and Vulnerability. Working Group II Contribution to the Fifth Assessment Report. Intergovernmental Panel on Climate Change*. Cambridge University Press.

Jochelson, K. (2006). "Nanny or steward? The role of government in public health." *Public Health*, 120(12), 1149–1155. https://doi.org/10.1016/j.puhe.2006.10.009.

Kruk, M.E., Myers, M., Varpilah, S.T., & Dahn, B.T. (2015). "What is a resilient health system? Lessons from Ebola." *The Lancet*, 385(9980), 1910–1912. https://doi.org/10.1016/S0140-6736(15)60755–60753.

Lavis, J.N., Røttingen, J.-A., Bosch-Capblanch, X., Atun, R., El-Jardali, F., Gilson, L., Lewin, S., Oliver, S., Ongolo-Zogo, P., & Haines, A. (2012). "Guidance for evidence-informed policies about health systems: Linking guidance development to policy development." *PLoS Medicine*, 9(3), e1001186–e1001186. https://doi.org/10.1371/journal.pmed.1001186.

Midgley, G. (2006). "Systemic intervention for public health." *American Journal of Public Health*, 96(3), 466–472. ProQuest One Academic; Social Science Premium Collection.

Murray, C.J.L., Frenk, J., & World Health Organization. Global Programme on Evidence for Health Policy. (1999). "A WHO framework for health system performance assessment." *Global Programme on Evidence for Health Policy*. WHO IRIS. https://apps.who.int/iris/handle/10665/66267.

Olafsdottir, A.E., Allotey, P., & Reidpath, D.D. (2013). "A health system in economic crises: A case study from Iceland." *Scandinavian Journal of Public Health*, 41(2), 198–205.

Pencheon, D. (2018). "Developing a sustainable health care system: The United Kingdom experience." *Medical Journal of Australia*, 208(7), 284–285. https://doi.org/10.5694/mja17.01134.

Surender, R., Matsuoka, K., & Ovseiko, P. (2017). "United Kingdom, health system of." In *International Encyclopedia of Public Health*. Elsevier. https://doi.org/10.1016/B978-0-12-803678-5.00477-X.

The Nuffield Trust. (2019) *The history of the NHS*. https://www.nuffieldtrust.org.uk/health-and-social-care-explained/the-history-of-the-nhs/.

Thomas, S., Keegan, C., Barry, S., Layte, R., Jowett, M., & Normand, C. (2013). "A framework for assessing health system resilience in an economic crisis: Ireland as a test case." *BMC Health Services Research* 13(1), 450.

Watts, N., Amann, M., Arnell, N., Ayeb-Karlsson, S., Belesova, K., Boykoff, M., Byass, P., Cai, W., Campbell-Lendrum, D., Capstick, S., Chambers, J., Dalin, C., Daly, M., Dasandi, N., Davies, M., Drummond, P., Dubrow, R., Ebi, K.L., Eckelman, M., … Montgomery, H. (2019). "The 2019 report of The Lancet Countdown on health and climate change: Ensuring that the health of a child born today is not defined by a changing climate." *The Lancet*, 394(10211), 1836–1878. https://doi.org/10.1016/S0140-6736(19)32596–32596.

Wiley, L.F. (2013). "Climate change adaptation and public health law." In J. Verschuuren, *Research Handbook on Climate Change Adaptation Law*. Edward Elgar Publishing. https://doi.org/10.4337/9781781000083.00012.

Winslow, C.-E.A. (1920). "The untilled fields of public health." *Science*, 51(1306), 23–33. https://doi.org/10.1126/science.51.1306.23.

World Health Organization. (2000). *The World Health Report 2000—Health Systems: Improving Performance*. World Health Organization.

World Health Organization. (2007). *Everybody's Business – Strengthening Health Systems to Improve Health Outcomes: WHO's Framework for Action*. World Health Organization. https://apps.who.int/iris/handle/10665/43918.

World Health Organization. (2015). *Operational Framework for Building Climate Resilient Health Systems*. World Health Organization. https://apps.who.int/iris/handle/10665/189951.

World Health Organization. (2021a). *2021 WHO Health and Climate Change Global Survey Report*. World Health Organization. https://apps.who.int/iris/handle/10665/348068.

World Health Organization. (2021b). "Climate change and health research: Current trends, gaps and perspectives for the future." *Discussion Paper*. https://www.who.int/p

ublications/m/item/climate-change-and-health-research-current-trends-gaps-and-perspectives-for-the-future.

World Health Organization. (2021c). *COP26 Special Report on Climate Change and Health: The Health Argument for Climate Action*. World Health Organization. https://apps.who.int/iris/handle/10665/346168.

Chapter 10

Regulating Climate Change Risks to the US Financial System

Systemic risk is a familiar concept and phenomenon in the financial sector.[1] Indeed, Gohari and Woody (2015) suggest that financial systems are "inherently prone to crises".[2] Systemic risks have been observed in the sector dating back to at least the Middle Ages, when currency was devalued by a deliberate reduction in the silver content of coins to help fund a war.[3] The causes of systemic risks since then have varied over time, including asset bubbles, banking panics and government default on debt.[4]

The Global Financial Crisis – the most significant example of systemic risk to financial systems since the Great Depression[5] – demonstrated the unpredictability and speed of transmission of such risks, as well as the catastrophic consequences they can inflict on a vast scale.[6] Given mounting evidence of and rising concern about the adverse consequences climate change could have for financial systems across the world, this case study considers the approach to the regulation of systemic risks caused by climate change to the US financial system, particularly to determine how these risks are managed under the US regulatory regime and whether the regulatory approach is aligned with the ideal risk governance paradigm for the management of systemic risks described earlier in this book.

1 X. Freixas, L. Laeven, & J.-L. Peydró, *Systemic Risk, Crises, and Macroprudential Regulation* (MIT Press, 2015).
2 B. Gohari & K.E. Woody, "The new global financial regulatory order: Can macroprudential regulation prevent another global financial disaster?" (2015) 40(2) *The Journal of Corporation Law*, p. 403.
3 A. Gottesman & M. Leibrock, *Understanding Systemic Risk in Global Financial Markets: A Professional Guide to Accounting Arbitrations* (Wiley, 2017), pp. 4–5.
4 Ibid. pp. 3–5.
5 B. Gohari & K.E. Woody, n. 2 above.
6 *The Financial Crisis Inquiry Report: Final Report of the National Commission on the Causes of the Financial and Economic Crisis in the United States* (US Financial Crisis Inquiry Commission, 2011); A. Gottesman & M. Leibrock, n. 3 above.

DOI: 10.4324/9781003253464-10

I The role and importance of financial systems

A financial system is the set of institutions, infrastructure, rules and practices that together enable the exchange of funds in a variety of forms between borrowers, lenders and investors. This type of system – which can exist at local, national, regional and global levels – helps to mobilise funds which can then be used for consumption or for productive investment.[7] However, activity that occurs in the context of a financial system is inherently risky, with possible positive or negative outcomes for parties involved in such activity.[8] For example, a lender of funds to a borrower benefits from interest payable on the loan, but also faces the risk that the loan will not be repaid. Similarly, sellers of commodity futures enable them to hedge against the risk of volatile commodity prices, while speculators buy the futures to assume those risks. A critical role of a well-functioning financial system is to intermediate between parties in order to facilitate the management and transformation of risk.[9]

There is empirical evidence to demonstrate the positive impact of a sound financial system on economic growth.[10] More specifically, a financial system can enable parties that may be otherwise financially constrained to access capital, to invest, to innovate and to expand. As noted by the influential political economist Joseph Schumpeter (1961), "[t]he banker … authorizes people in the name of society … to innovate".[11] The activity enabled by financial systems may also increase employment, labour productivity and wages[12] and address inequality and poverty by influencing the allocation of funds to start a business, to pay for education and even to survive.[13]

Stability of a financial system is critical to enable delivery of these benefits. Schinasi (2004) describes financial stability in terms of a financial system's ability to facilitate an efficient allocation of resources and the effectiveness of other economic processes (such as wealth accumulation, economic growth and social prosperity); to assess, price, allocate and manage financial risks; and to maintain the ability to perform these functions when affected by external shocks.[14] He

7 E.H. Neave, *Financial Systems: Principles and Organisation* (Routledge, 1998).
8 Ibid. pp. 14–15.
9 World Bank, *World Development Report 2014: Risk and Opportunity – Managing Risk for Development* (World Bank, 2013).
10 See, for example, S. Nyasha & N.M. Odhiambo, "Do financial systems spur economic growth in the USA? An empirical investigation" (2019) 66(2) *Panoeconomicus*, pp. 165–85.
11 Quoted in R.G. King & R. Levine, "Finance and growth: Schumpeter might be right" (1993) 108(3) *Quarterly Journal of Economics*, p. 717.
12 M. Pagano & G. Pica, "Finance and employment" (2014) 27(69) *Economic Policy*, pp. 5–55.
13 T. Beck, A. Demirgüç-Kunt, & R. Levine, "Finance, inequality and the poor" (2007) 12 *Journal of Economic Growth*, pp. 27–49.
14 G.J. Schinasi, "Defining financial stability" (2004) 04(187) *IMF Working Papers*, p. 8.

proposes the following definition of financial system stability in light of these characteristics:[15]

> A financial system is in a range of stability whenever it is capable of facilitating (rather than impeding) the performance of an economy, and of dissipating financial imbalances that arise endogenously or as a result of significant adverse and unanticipated events.

Conversely, financial instability can undermine the well-functioning of the financial system. Cioffi (2011) explains that "the structural characteristics of modern financial systems drive them to become dangerously unstable and destructively extractive".[16] Banks may be deterred from financing projects, asset prices may deviate from their true value and payments may be delayed. The consequences of these outcomes can be transmitted throughout the financial system, undermining confidence in the system and, at worst, causing bank runs, hyperinflation and stock market crashes.[17] A study conducted by Bardoscia and colleagues (2017) shows that complexity of the financial system can lead to financial market instability because it contributes to creating cyclical structures that tend to amplify financial distress and, thereby, makes crises more likely.[18] In turn, such instability could cause systemic risk, defined by Freixas and colleagues (2015) as "the risk of threats to financial stability that impair the functioning of a large part of the financial system with significant adverse effects on the broader economy".[19]

A resilient financial system is one that is capable of regaining its original shape after a systemic shock.[20] Yet, the sheer complexity of financial systems makes resilience a formidable challenge. In noting the "tangle of interconnections" in the financial system which had a role to play in the transmission and magnitude of the Global Financial Crisis, the US Commission tasked with examining the causes of the crisis commented that:[21]

> The financial system we examined bears little resemblance to that of our parents' generation … The financial markets have become increasingly

15 Ibid. p. 8.

16 J.W. Cioffi, "After the fall: Regulatory lessons from the global financial crisis", in D. Levi-Faur (ed.), *Handbook on the Politics of Regulation* (Edward Elgar Publishing, 2011), p. 644.

17 World Bank, accessible at: https://www.worldbank.org/en/publication/gfdr/gfdr-2016/background/financial-stability.

18 M. Bardoscia, S. Battiston, F. Caccioli, & G. Caldarelli, "Pathways towards instability in financial networks" (2017) 8 *Nature Communications*, 14416.

19 X. Freixas, L. Laeven, & J.-L. Peydró, n. 1 above, p. 13.

20 C.-Y. Tsang, "Balancing the governance of the modern financial ecosystem: A new governance perspective and implications for market discipline" (2018) 40(2) *Houston Journal of International Law*, pp. 531–612, at 537.

21 US Financial Crisis Inquiry Commission, n. 6 above.

> globalized. Technology has transformed the efficiency, speed, and complexity of financial instruments and transactions. There is broader access to and lower costs of financing than ever before. And the financial sector itself has become a much more dominant force in our economy.

A 2016 briefing paper prepared for the US Office of Financial Research (OFR) maps the "network of networks",[22] which characterises the US financial system and the numerous potential "channels of contagion".[23] The briefing paper further notes that in the case of the Global Financial Crisis, "[r]isks emerged and spread in unanticipated ways".[24] The uncertainty associated with the occurrence and progression of systemic risks through the financial system due to its inherent complexity is likely to be compounded by the additional complexity associated with the stressors to which the financial system may be subject, including climate change, which is considered in the next section.

II The impact of climate change on financial systems

It is generally accepted that climate change causes risks to the stability of financial systems, including the US financial system.[25] Climate-related risks to financial systems have been grouped into two broad categories – namely, physical risks and transition risks, which manifest in financial systems as traditional financial risks, such as credit risk, liquidity risk, market risk, and operational risk.[26]

Physical risks refer to the harm to people, physical assets and natural capital arising from the impacts of climate change, including extreme weather events (such as floods, bushfires and severe storms) as well as longer-term phenomena (such as increasing temperatures, sea level rise and changes in rainfall patterns).[27] For example, low-lying coastal infrastructure may be damaged due to higher sea levels and extreme weather events may destroy crops. As a consequence of these types of impacts, the value of land, infrastructure and other physical assets may fall, affecting profitability, investments, employment and, eventually,

22 R. Bookstaber & D.Y. Kenett, "Looking deeper, seeing more: A multilayer map of the financial system" (2016) *Office of Financial Research Brief Series*, p. 2.

23 Ibid. p. 1.

24 Ibid. p. 1.

25 Financial Stability Board, *The Implications of Climate Change for Financial Stability* (Financial Stability Board, 2020); Financial Stability Oversight Council, *Report on Climate-Related Financial Risk* (2021); US Commodity Futures Trading Commission, *Managing Climate Risk in the U.S. Financial System* (US Commodity Futures Trading Commission, 2020).

26 Financial Stability Oversight Council, n. 25 above, p. 24.

27 Ibid. p. 12.

GDP.[28] Default risk on loans could increase, banks could tighten lending conditions and insurance claims could increase.[29]

Transition risks are those that could arise in sectors adversely impacted by an adjustment towards a low-carbon economy as part of the global response to climate change.[30] Such risks could arise from technological innovations (for example, a break-through in battery technology), changes in consumer sentiment and behaviour (such as a decline in the purchase of plastics) and changes in policy and regulation (including the introduction of a carbon tax).[31] Decarbonisation could result in business disruption, investor pressure and declines in asset prices, income and profitability in sectors that are emissions-intensive.[32]

Physical and transition risks could also combine. For example, a major flood event linked to climate change that causes extensive physical damage could lead to dramatic changes in regulation, such as a prohibition on infrastructure development in areas that are particularly prone to flooding. The combination of physical and transition risks in such cases could amplify the destabilising effects on financial systems.[33] In summary, the physical and transition risks caused by climate change may result in credit and market risks associated with loss of income, default risk could increase as asset values decline, liquidity risks may arise with changing demand for liquidity, and operational risks may eventuate from damage or destruction of infrastructure and assets.[34]

Lael Brainard, Member of the Board of Governors of the Federal Reserve (2020) suggests that the physical and transition risks caused by climate change could compromise financial stability.[35] More recently, Brainard (2021) has also noted that "[u]nlike episodic or transitory shocks, climate change is an ongoing, cumulative process, which is expected to produce a series of shocks".[36] When comparing climate change to the systemic risk that caused the Global Financial Crisis, Rudebusch (2021) states:[37]

28 I. Monasterolo, "Climate change and the financial system" (2020) 12(1) *Annual Review of Resource Economics*, pp. 299–320, at 303.

29 P. Grippa, J. Schmittmann & F. Suntheim, "Climate change and financial risk" (2019) *Finance & Development*, pp. 26–9, at 27.

30 Financial Stability Oversight Council, n. 25 above, p. 13.

31 J. Ryan-Collins, "Beyond voluntary disclosure: Why a 'market-shaping' approach to financial regulation is needed to meet the challenge of climate change" (2019) *UCL Institute for Innovation and Public Purpose Policy Brief*, p. 3; I. Monasterolo, n. 28 above, p. 304.

32 G.D. Rudebusch, "Climate change is a source of financial risk" (2021) 2021–03 *Federal Reserve Bank of San Francisco Economic Letter*, at 2; P. Grippa, J. Schmittmann & F. Suntheim, n. 29 above, p. 27.

33 Financial Stability Board, n. 25 above, pp. 1 and 4.

34 Financial Stability Oversight Council, n. 25 above, p. 13.

35 L. Brainard, "Strengthening the financial system to meet the challenge of climate change" (2020) *The Financial System & Climate Change: A Regulatory Imperative*, p. 3.

36 L. Brainard, "Financial stability implications of climate change", (2021) *Transform Tomorrow Today Ceres 2021 Conference* (2021), p. 4.

37 G.D. Rudebusch, n. 32 above, p. 4.

> [T]he financial impacts of climate change appear even more complex, uncertain, opaque, and persistent [compared to the Global Financial Crisis]. As a result, there is considerable scope for abrupt shifts in perceived climate risks and a rapid repricing of assets threatened by climate change or reliant on carbon emissions. The greater likelihoods of such a disruptive, cascading asset repricing could increase financial vulnerabilities.

While the existence of risks to financial systems caused by climate change appears clear, it is nevertheless difficult to quantify these risks so that they can be appropriately accounted for by regulators and regulated entities alike because of compounding uncertainty associated with such risks. In particular, there is uncertainty about climate change impacts, including their specific nature, magnitude, timing, frequency and location.[38] Uncertainty also exists because of the "tight interconnectedness" within the financial system,[39] which can lead to unexpected and unpredictable transmission of adverse shocks through the system. There is additional uncertainty about the particular measures that might be taken at the local, national and global levels to respond to climate change and the impact of those measures on individual participants within the financial system as well as the financial system as a whole.[40] In summary, the future path of climate change and its impact on the financial system are highly uncertain and could be non-linear over time.[41]

The challenges associated with quantifying risks to financial stability from climate change are compounded by analytical complexity, data gaps, and inadequate financial models. Specifically, estimation of the impact of climate change on financial exposures is likely to be complex.[42] There is inadequate data to measure financial institutions' exposure to climate-related risks.[43] In addition, climate models typically rely upon long time horizons, whereas shorter-term data may be of better use to the financial sector.[44] Moreover, standard financial risk analysis is typically backward-looking, and uses linear pricing techniques,[45] whereas climate change is likely to cause complex, uncertain, non-linear dynamics within the financial system for which a forward-looking approach is required.[46]

38 Financial Stability Oversight Council, n. 25 above, p. 23.
39 I. Anabtawi & S.L. Schwarcz, "Regulating systemic risk: Towards an analytical framework" (2011) 86(4) *Notre Dame Law Review*, p. 1349, at 1360.
40 Financial Stability Board, n. 25 above, pp. 3–4; J. Ryan-Collins, n. 31 above, p. 4.
41 Financial Stability Board, n. 25 above, p. 3.
42 Financial Stability Board, *Stocktake of Financial Authorities' Experience in Including Physical and Transition Climate Risks as Part of Their Financial Stability Monitoring* (2020), p. 11.
43 Financial Stability Board, n. 25 above, p. 3.
44 N.J. Michel, D.R. Burton, & N.D. Loris, "Using financial regulation to fight climate change: A losing battle", *Backgrounder* (2021) (3634) *The Heritage Foundation*, p. 4.
45 J. Ryan-Collins, n. 31 above, p. 4.
46 I. Monasterolo, n. 28 above, p. 315.

Monasterolo (2020) suggests that climate-related risks are not currently being priced into financial transactions[47] and Ryan-Collins (2019) asserts that it is "widely accepted that climate risks are not accurately reflected in the valuations of financial assets, nor in the lending criteria of banks and nonbank financial institutions".[48] Further, based on a stock-take of the various approaches adopted by financial authorities, the Financial Stability Board (2020a) concluded that "[n]o approach to quantification [of climate change risks to financial stability] provides a holistic assessment of climate-related risks to the global financial system".[49]

In summary, attempting to model future financial risks caused by climate change is fraught with difficulty. Given the inherent complexity and profound uncertainty associated with this task, there is a strong case in favour of a precautionary approach which involves steering the financial system in a direction that will help to minimise the impact of future shocks. The next section considers whether the regulatory regime governing the US financial system is capable of achieving this outcome.

III The regulatory regime in the US

The US financial regulatory regime is complex, with a multitude of regulators that oversee a variety of markets and transactions, and a broad range of regulated entities that undertake a diverse spectrum of financial activities.[50] The regime has been described as "fragmented" and the product of "piecemeal" evolution in response to various financial crises.[51]

The Dodd-Frank Wall Street Reform and Consumer Protection Act (Dodd-Frank Act)[52] was enacted in 2010 in the wake of the Global Financial Crisis and made some significant changes to the regime, including in relation to the management of systemic risk in the US financial system to promote financial stability. The preamble of the Dodd-Frank Act explains the objectives of the Act:

> To promote the financial stability of the United States by improving accountability and transparency in the financial system, to end "too big to fail," to protect the American taxpayer by ending bailouts, to protect consumers from abusive financial services practices, and for other purposes.

47 Ibid. p. 300.

48 J. Ryan-Collins, n. 31 above, p. 2.

49 Financial Stability Board, n. 42 above, p. 1.

50 M. Labonte, "Who regulates whom? An overview of the U.S. financial regulatory framework" (2020) *Congressional Research Service*.

51 Ibid. p. 1.

52 *Dodd-Frank Wall Street Reform and Consumer Protection Act*, Public Law 111–203, 124 Stat. 1376 (2010).

The emphasis on financial stability is notable and signals a shift away from a microprudential to a macroprudential approach to regulation. Such a shift is a response to criticism that the pre-crisis regulatory regime failed to provide sufficient regulatory oversight to identify and manage systemic risk.[53] The former approach primarily focuses on risks faced by individual participants in the system and is based on the assumption that the management of risk at this micro level helps to address risk at the systemic level.[54] In contrast, a macroprudential approach targets risks to the financial system as a whole with the objective of preventing or, at least, mitigating systemic shocks.[55]

To address the fragmented nature of the regulatory regime and to establish an institutional structure for a macroprudential approach, the Dodd-Frank Act created the Financial Stability Oversight Council (FSOC) – a council of regulators and experts, including the Federal Reserve Board of Governors, chaired by the Secretary of the Treasury.[56] The FSOC is the overarching body charged with monitoring the financial system, including identifying potential threats to the country's financial stability. Given its prominence in the architecture for the regulation of systemic risks, the FSOC and its powers and functions are the focus of discussion in this case study.

The purposes of the FSOC are set out in section 112 of the Dodd-Frank Act:[57]

> (A) *to identify risks to the financial stability* of the United States that could arise from the material financial distress or failure, or ongoing activities, of large, interconnected bank holding companies or non-bank financial companies or that *could arise outside the financial services marketplace*;
>
> (B) to promote market discipline, by eliminating expectations on the part of shareholders, creditors, and counterparties of such companies that the Government will shield them from losses in the event of failure; and
>
> (C) to respond to *emerging threats to the stability of the United States financial system.*
> (emphasis added)

These purposes are notable because they provide a sense of the FSOC's place in relation to the management of risks to the US financial system under the regulatory regime. First, the FSOC clearly has an important role to play in identifying and responding to risks to financial stability. Second, the FSOC's purview includes emerging threats to the stability of the US financial system – that is, systemic risks. Finally, the FSOC purposes make reference to risks associated with banks and non-bank financial enterprises as well as risks that arise "outside the financial

53 M. Labonte, "Introduction to financial services: Systemic risk" (2021) *Congressional Research Service.*
54 B. Gohari & K.E. Woody, n. 2 above, pp. 406–7; X. Freixas et al., n. 1 above, p. 199.
55 B. Gohari & K.E. Woody, n. 2 above, p. 410; X. Freixas et al., n. 1 above, p. 254.
56 Dodd-Frank Act, n. 52 above, section 111.
57 Ibid. section 112(a)(1).

services marketplace", including both endogenous and exogenous risks to the financial system.

As indicated by its name, the FSOC's role is one of oversight over the member agencies that comprise the Council, rather than one involving direct regulatory intervention in the US financial system. The FSOC's specific duties are to "monitor the financial services marketplace in order to identify potential threats to the financial stability of the United States",[58] to "assess risks to the United States financial system,"[59] and to "identify gaps in regulation that could pose risks to the financial stability of the United States".[60] Among other things, the FSOC provides "a forum for discussion and analysis of emerging market developments and financial regulatory issues"[61] and to "facilitate information sharing and coordination among the member agencies and other Federal and State agencies".[62] The FSOC also has authority to recommend to member agencies "general supervisory priorities and principles reflecting the outcome of discussions among member agencies".[63] In addition, the FSOC may require the Board of Governors to supervise certain financial institutions that may pose risks to financial stability[64] and to make recommendations concerning heightened prudential standards.[65] The FSOC can advise and make recommendations to Congress in areas that will enhance stability.[66]

The OFR supports the FSOC and helps it to fulfil its purposes and discharge its duties, including in relation to the management of systemic risks causing financial instability. In particular, the OFR is tasked with collecting relevant data, developing tools for risk measurement and monitoring, and conducting research.[67] The management of systemic risks is further facilitated by the FSOC's "Systemic Risk Committee" (SRC), which comprises relevant staff from the member agencies, and serves as a forum to identify and analyse potential risks, including risks that extend beyond the jurisdiction of any one agency.[68]

The FSOC is required to produce an annual report to Congress outlining "potential emerging threats to the financial stability of the United States,"[69] including recommendations to "enhance … the stability of United States financial markets".[70] The latest annual report, published in 2021, notes that, as part of its responsibility to identify risks to financial stability in the US, the

58 Ibid. section 112(a)(2)(C).
59 Ibid. section 112(a)(2)(A).
60 Ibid. section 112(a)(2)(G).
61 Ibid. section 112(a)(2)(M)(i).
62 Ibid. section 112(a)(2)(E).
63 Ibid. section 112(a)(2)(F).
64 Ibid. section 112(a)(2)(H).
65 Ibid. section 112(a)(2)(I).
66 Ibid. section 112(a)(2)(D).
67 Ibid. section 153(a).
68 *2021 Financial Stability Oversight Council Annual Report* (2021), p. 137.
69 Dodd-Frank Act, n. 52 above, section 112(a)(2)(N)(iii).
70 Ibid. section 112(a)(2)(N)(vi)(I).

FSOC considered climate-related financial risks during the course of 2021.[71] The FSOC's Report on Climate-Related Financial Risk (FSOC Climate Risk Report)[72] provides important insights into the FSOC's approach to the management of systemic risks – particularly, those caused by climate change – and is considered in the next section.

IV Approach to the management of climate change risks to the US financial system

By "Executive Order on Climate-Related Financial Risk" (Order), in May 2021, the US President directed the FSOC together with other government bodies and agencies to engage with their respective members to consider "assessing, in a detailed and comprehensive manner, the climate-related financial risk, including both physical and transition risks, to the financial stability of the Federal Government and the stability of the US financial system".[73] The preface to the Order explains the policy behind this direction:[74]

> The intensifying impacts of climate change present physical risk to assets, publicly traded securities, private investments, and companies – such as increased extreme weather risk leading to supply chain disruptions. In addition, the global shift away from carbon-intensive energy sources and industrial processes presents transition risk to many companies, communities, and workers. At the same time, this global shift presents generational opportunities to enhance U.S. competitiveness and economic growth, while also creating well-paying job opportunities for workers. *The failure of financial institutions to appropriately and adequately account for and measure these physical and transition risks threatens the competitiveness of U.S. companies and markets, the life savings and pensions of U.S. workers and families, and the ability of U.S. financial institutions to serve communities.* In this effort, the Federal Government should lead by example by appropriately prioritizing Federal investments and conducting prudent fiscal management.
>
> It is therefore the policy of my Administration to advance consistent, clear, intelligible, comparable, and accurate disclosure of climate-related financial risk … including both physical and transition risks; *act to mitigate that risk and its drivers*, while accounting for and addressing disparate impacts on disadvantaged communities and communities of color … and spurring the creation of well-paying jobs; and achieve our target of a net-zero emissions economy by no later than 2050. This policy will marshal the creativity,

71 2021 FSOC Annual Report, n. 68 above, p. 137.

72 Financial Stability Oversight Council, n. 25 above.

73 Executive Order 14030 on Climate-Related Financial Risk (2021) *The White House*, accessible at: https://www.federalregister.gov/documents/2021/05/25/2021-11168/climate-related-financial-risk, at Section 3.

74 Ibid. Section 3.

> courage, and capital of the United States necessary to bolster the resilience of our rural and urban communities, States, Tribes, territories, and financial institutions in the face of the climate crisis, rather than exacerbate its causes, and position the United States to *lead the global economy to a more prosperous and sustainable future.*
>
> (emphasis added)

The FSOC Climate Risk Report (2021), relevant aspects of which are discussed below, was produced in response to the Order.[75]

I Regulatory boundary

Described as the "lengthiest piece of banking legislation in US history",[76] the Dodd-Frank Act is expansive in terms of the matters that are regulated. This is evident by simply casting an eye down the Act's extensive table of contents.[77] It is obvious from the preamble and other relevant provisions of the Act that it has systemic dimensions and, further, that the regulatory regime covers systemic risks posed to the US financial system.

In particular, the financial system is explicitly referenced in the preamble and in the FSOC's purposes and duties. Further, the mitigation of threats to that system is a consistent theme throughout the regulatory regime. The multi-layer map of the system prepared for the OFR (referred to earlier in this chapter) helps to concretise the concept of the US financial system governed by the regulatory regime, but the map also provides a sense of the immense complexity and associated challenges for the FSOC and its member agencies to address vulnerability and exposure of the system to stressors that could result in financial instability.

Risk and, specifically, systemic risk that could compromise financial stability are clearly within the Act's purview. While the Act does not make explicit reference to climate change, nor the risks to the US financial system caused by climate change, the FSOC Climate Risk Report (2021) states that climate-related financial risks are regarded as an "emerging threat to the financial stability of the United States",[78] thereby placing such risks clearly within the ambit of the Act and the FSOC's remit. The combination of the FSOC's ongoing power to monitor threats to financial stability[79] and the obligation to report on these threats[80] provides a useful mechanism to ensure that new,

75 Financial Stability Oversight Council, n. 25 above, p. 2.
76 J.R. Barth, A.P. Prabha, & C. Wihlborg, "The Dodd-Frank Act: Key features, implementation progress, and, financial system impact", in *The First Great Financial Crisis of the 21st Century: A Retrospective* (World Scientific, 2015), pp. 337–76, at 341.
77 Dodd-Frank Act, n. 52 above, p. 1.
78 Financial Stability Oversight Council, n. 25 above, pp. 1–2.
79 Dodd-Frank Act, n. 52 above, section 112(a)(2)(C).
80 Ibid. section 112(a)(2)(N)(iii).

emerging threats – like those caused by climate change – are captured with the scope of the FSOC's powers and functions and, therefore, are also governed by the regulatory regime.

2 Regulatory context

The regulatory regime includes various mechanisms for establishing context for the regulatory response to systemic risk, including systemic risks caused by climate change. For example, the FSOC is specifically empowered to assess risks to the US financial system.[81] However, the Act does not prescribe how the risk assessment is to be undertaken. This is potentially an important omission because the scope and criteria for the risk assessment – including the hazards, exposure and vulnerability of those entities and individuals affected by risk – will likely have a significant impact on the outcome of the risk assessment and, consequently, on how the relevant risks are evaluated and eventually dealt with under the regulatory regime.

The complexity of the assessment of climate-related financial risks is specifically noted in the FSOC Climate Risk Report. The FSOC has observed that integrating climate-related risks into financial assessments is a "new area of focus for which tools are under development" and that "policymakers need to develop new approaches as part of the process of measuring, managing, and mitigating climate-related financial risks".[82] The FSOC has indicated that scenario analysis is likely to be an emerging tool to assess climate-related financial risk, particularly given that such an approach is forward-looking, rather than being rooted in historical experience. However, the FSOC has also conceded that scenario analysis will likely be complex, involving multiple scenarios and models.[83]

While the FSOC has authority to recommend "supervisory priorities",[84] the Dodd-Frank Act does not specify how evaluation of risks is to be undertaken so that they can be prioritised. Nevertheless, the FSOC Climate Risk Report notes that a Climate-Related Financial Risk Advisory Committee (CFRAC) will be established and tasked with identifying priority areas for assessing and mitigating climate-related risks for the financial system.[85] It may well be that the critically important task of risk evaluation will be undertaken through this body.

The FSOC Climate Risk Report includes a range of proposed measures to enhance communication about risks to financial stability, but also to engage relevant stakeholders. In relation to risk communication, the FSOC has recommended member agencies to include descriptions of their activities

81 Ibid. section 112(a)(2)(A).
82 Financial Stability Oversight Council, n. 25 above, p. 89.
83 Ibid. pp. 89–90.
84 Dodd-Frank Act, n. 52 above, section 112(a)(2)(F).
85 Financial Stability Oversight Council, n. 25 above, p. 5, Recommendation 1.1.

related to climate-related financial risks in their annual reports and consider incorporating climate-related financial risks in relevant risk reports that they publish to help keep the public informed.[86] In addition, FSOC members have been encouraged to make climate-related data for which they are the custodians freely available to the public, as appropriate.[87] The FSOC has also announced that the CFRAC will engage with a broad array of stakeholders, including climate science experts, representatives from non-governmental research institutions, academia, the financial services industry, as well as consumers and investors.[88] The specific purposes, nature and scope for such engagement in the context of the management of systemic risks to financial stability are, as yet, unclear. Nonetheless, stakeholder engagement could be critically important in the assessment and evaluation of systemic risks, including those caused by climate change, and may thereby provide important context for the management of systemic risks to the US financial system caused by climate change.

3 Regulatory content

The content of the Dodd-Frank Act indicates in clear terms that the FSOC's role regarding the identification and management of systemic risks to the US financial system is confined to monitoring, overseeing, advising and informing. The tools available to the FSOC under the Act to execute its role are "soft" in the sense that they are designed to facilitate rather than enforce action to ensure that systemic risks are effectively managed. Ultimately, the effectiveness of the regulatory regime in mitigating such risks will be determined by the actions of downstream regulators over whom the FSOC has oversight.

In this regard, the FSOC Climate Risk Report refers to the FSOC's expectations of these regulators in relation to the management of systemic risks caused by climate change. In particular, the FSOC has recommended that member agencies should continue their efforts to consider the incorporation of climate-related risks into their regulatory and supervisory programs and update those programs as necessary. As part of this work, the members have been encouraged by the FSOC to review regulated entities' efforts to address climate-related risks and clarify or enhance risk management requirements for regulated entities where necessary to promote appropriate consideration of climate-related financial risks.[89] In addition, FSOC members have been advised to review existing regulations, guidance and regulatory reporting relevant to climate-related risks, including credit risks, market risks, counterparty risks, and other financial and operational risks, to assess whether updates are necessary to

86 Ibid. p. 6, Recommendation 1.4.
87 Ibid. p. 6, Recommendation 1.5.
88 Ibid. p. 5, Recommendation 1.2.
89 Ibid. p. 9, Recommendation 4.6.

appropriately address climate-related financial risks.[90] Further, the FSOC has recommended these agencies to evaluate whether additional regulations or guidance specific to climate-related risks are necessary to clarify expectations for regulated or supervised institutions regarding management of climate risks, taking into account an institution's size, complexity, risk profile and existing enterprise risk management processes.[91]

V Alignment with the risk governance regulatory ideal

From a systems standpoint, there is a lot to recommend the US regulatory regime. The systemic dimensions of the US financial system are apparent from the text of the Dodd-Frank Act. Moreover, risk is front and centre of the regime, and systemic risks that could cause financial instability are explicitly referenced. Further, governance arrangements for oversight of the management of risks to the US financial system are embedded in the regime. In fact, there appears to be relatively good alignment between the US regulatory regime and the risk governance paradigm to regulate systemic risks outlined earlier in this book. At least theoretically, this alignment could help ensure that risks to the US financial system are managed in a manner that is in keeping with the systemic characteristics of those risks. However, this outcome will essentially depend upon the outcome of a number of inter-related challenges that are inherent in the regime.

First, there are practical limits as to what a single regulatory regime and those overseeing it can achieve, particularly given that it covers the spectrum of risks to the stability of the US financial system. There is always the risk that an ambitious regulatory regime may under-achieve the intended objectives because it has been stretched too far. Nonetheless, there is evidence that the FSOC is taking climate change risks to US financial system seriously – climate-related financial risks are currently a key priority area for the Council and its members.[92]

Second, the capacity of the regulatory regime to effectively respond to systemic risks caused by climate change, will hinge on a sound understanding of climate change and its impact on US financial system, notwithstanding the underlying uncertainty. In relation to physical risks that could have an impact on the US financial system, the FSOC Climate Risk Report notes that, while a large amount of potentially relevant data currently exists, more work is needed to improve access to and usability of this data, particularly to standardise it, so that it can be incorporated into financial risk assessments.[93] Regarding transition risks, the FSOC has observed that the forward-looking nature of transition

90 Ibid. p. 9, Recommendation 4.7.
91 Ibid. p. 9, Recommendation 4.8.
92 Ibid. p. 26.
93 Ibid. p. 49.

risks requires analysis of different climate-related scenarios but that it may be challenging to measure and project the impact of such risks given the uncertainty of future developments.[94] The FSOC has also noted that disclosure requirements applicable to relevant companies and financial entities with respect to climate-related risks do not result in disclosures that are consistent, comparable, and "decision-useful"[95] but that enhanced climate-related disclosures could help address this gap.[96] Moving forward, the CFRAC will support the FSOC by gathering information on climate-related financial risks from a broad array of stakeholders.[97]

Third, even with comprehensive and useful information about climate-related risks to the US financial system, a key challenge will be to make sense of this information and, more specifically, to use it to map a pathway forward for the US financial system to help ensure its resilience to these risks. Gottesman and Leibrock (2017) suggest that systemic risk analysis is a nascent area and that there is a need for "new and enhanced tools to assist the industry in its efforts to understand, quantify, monitor and mitigate systemic risks" coupled with a "roadmap" to guide utilisation of these tools.[98]

Finally, the success of the US regulatory regime in mitigating systemic risks caused by climate change is likely to depend critically on the FSOC's ability to ensure collective, coordinated and coherent regulatory responses to such risks across the various agencies for which it has oversight. In this regard, it is notable that the same agencies presided over the Global Financial Crisis. Barth and colleagues (2015a) argue that these agencies were informed about the growing risks that led to that crisis, had the regulatory power to respond, but failed to do so in a "timely and appropriate manner".[99] Similarly, in its recent report about managing climate risk in the US financial system, the US Commodity Futures Trading Commission (2020) noted that, in general, existing legislation provides the FSOC's member agencies with wide-ranging and flexible powers to respond to climate-related risk to the US financial system, but that these tools are not being fully utilised. The Commission has indicated that legislation may be necessary to ensure a coordinated national response.[100]

Presumably, this type of coordination is the very intent of the Dodd-Frank Act and the purpose of the FSOC. Rather than enacting more legislation, effective governance of the systemic risks caused by climate change is likely to come down to how existing legislation, and the institutional mechanisms to

94 Ibid. pp. 51 and 52.
95 Ibid. p. 84.
96 Ibid. p. 88.
97 Ibid. p. 5, Recommendation 1.2.
98 A. Gottesman & M. Leibrock, n. 3 above, pp. 6–7.
99 J.R. Barth, G. Caprio, & R. Levine, "Misdiagnosis: Incomplete cures of financial regulatory failures", in J.R. Barth & G.G. Kaufman (eds.), *The First Great Financial Crisis of the 21st Century: A Retrospective* (World Scientific, 2015), pp. 399–431, at 425.
100 US Commodity Futures Trading Commission, n. 25 above, p. iii.

support it, are employed in practice. There is a swathe of important recommendations of relevance to financial regulators set out in the FSOC Climate Risk Report. It is incumbent upon member agencies to follow the guidance provided by the FSOC and to work together to ensure that the stability of the US financial system is maintained as the risks associated with climate change continue to materialise and intensify over time.

Bibliography

Anabtawi, I., & Schwarcz, S.L. (2011). "Regulating systemic risk: Towards an analytical framework." *Notre Dame Law Review*, 86(4), 1349.

Bardoscia, M., Battiston, S., Caccioli, F., & Caldarelli, G. (2017). "Pathways towards instability in financial networks." *Nature Communications*, 8, 14416. https://doi.org/10.1038/ncomms14416.

Barth, J.R., Caprio, G., & Levine, R. (2015a). "Misdiagnosis: Incomplete Cures of Financial Regulatory Failures." In J.R. Barth & G.G. Kaufman (eds.), *The First Great Financial Crisis of the 21st Century: A Retrospective(Vol.*9). World Scientific. https://doi.org/10.1142/9789814651257_0015.

Barth, J.R., Prabha, A.P., & Wihlborg, C. (2015b). "The Dodd-Frank Act: Key features, implementation progress, and financial system impact." In J.R. Barth & G.G. Kaufman (eds.), *The First Great Financial Crisis of the 21st Century: A Retrospective (Vol.*9). World Scientific. https://doi.org/10.1142/9789814651257_0013.

Beck, T., Demirgüç-Kunt, A., & Levine, R. (2007). "Finance, inequality and the poor." *Journal of Economic Growth*, 12, 27–49. https://doi.org/10.1007/s10887-007-9010-6.

Bookstaber, R., & Kenett, D.Y. (2016). *Looking Deeper, Seeing More: A Multilayer Map of the Financial System.* Office of Financial Research Brief Series.

Brainard, L. (2020). "Strengthening the financial system to meet the challenge of climate change." *The Financial System & Climate Change: A Regulatory Imperative.* Federal Reserve, Washington, DC. https://www.federalreserve.gov/newsevents/speech/files/brainard20201218a.pdf.

Brainard, L. (2021). "Financial stability implications of climate change." *Transform Tomorrow Today, Ceres 2021 Conference.* https://www.federalreserve.gov/newsevents/speech/brainard20210323a.htm.

Cioffi, J.W. (2011). "After the fall: regulatory lessons from the global financial crisis." In D. Levi-Faur (ed.), *Handbook on the Politics of Regulation.* Edward Elgar Publishing.

Financial Stability Board. (2020a). *Stocktake of Financial Authorities' Experience in Including Physical and Transition Climate Risks as Part of Their Financial Stability Monitoring.* https://www.fsb.org/wp-content/uploads/P220720.pdf.

Financial Stability Board. (2020b). *The Implications of Climate Change for Financial Stability.* Financial Stability Board. https://www.fsb.org/wp-content/uploads/P231120.pdf.

Financial Stability Oversight Council. (2021a). *Report on Climate-Related Financial Risk.* https://home.treasury.gov/system/files/261/FSOC-Climate-Report.pdf.

Financial Stability Oversight Council. (2021b). *Annual Report.* https://home.treasury.gov/system/files/261/FSOC2021AnnualReport.pdf.

Freixas, X., Laeven, L., & Peydró, J.-L. (2015). *Systemic Risk, Crises, and Macroprudential Regulation.* MIT Press. https://doi.org/10.7551/mitpress/9780262028691.001.0001.

Gohari, B., & Woody, K.E. (2015). "The new global financial regulatory order: Can macroprudential regulation prevent another global financial disaster?" *The Journal of Corporation Law*, 40(2), 403.

Gottesman, A., & Leibrock, M. (2017). *Understanding Systemic Risk in Global Financial Markets: A Professional Guide to Accounting Arbitrations*. Wiley.

Grippa, P., Schmittmann, J. & Suntheim, F. (2019). "Climate change and financial risk." *Finance & Development*, 56(4), 26–29.

King, R.G., & Levine, R. (1993). "Finance and growth: Schumpeter might be right." *Quarterly Journal of Economics*, 108(3), 717.

Labonte, M. (2020). *Who Regulates Whom? An Overview of the U.S. Financial Regulatory Framework*. Congressional Research Service. https://sgp.fas.org/crs/misc/R44918.pdf.

Labonte, M. (2021). *Introduction to Financial Services: Systemic Risk*. Congressional Research Service. https://sgp.fas.org/crs/misc/IF10700.pdf.

Levi-Faur, D. (2011). *Handbook on the Politics of Regulation*. Edward Elgar Publishing. https://doi.org/10.4337/9780857936110.

Michel, N.J., Burton, D.R., & Loris, N.D. (2021). "Using financial regulation to fight climate change: A losing battle." *Backgrounder 3634*. The Heritage Foundation.

Monasterolo, I. (2020). "Climate change and the financial system." *Annual Review of Resource Economics*, 12(1), 299–320. https://doi.org/10.1146/annurev-resource-110119-031134.

Neave, E. H. (1998). *Financial Systems: Principles and Organisation*. Routledge.

Nyasha, S., & Odhiambo, N.M. (2019). "Do financial systems spur economic growth in the USA? An empirical investigation." *Panoeconomicus*, 66(2), 165–185. https://doi.org/10.2298/PAN160517012N.

Pagano, M., & Pica, G. (2014). "Finance and employment." *Economic Policy*, 27(69), 5–55. https://doi.org/10.1111/j.1468-0327.2011.00276.x.

Rudebusch, G.D. (2021). "Climate change is a source of financial risk." Federal Reserve Bank of San Francisco Economic Letter, 2021–03. https://www.frbsf.org/economic-research/files/el2021-03.pdf.

Ryan-Collins, J. (2019). "Beyond voluntary disclosure: Why a 'market-shaping' approach to financial regulation is needed to meet the challenge of climate change." UCL Institute for Innovation and Public Purpose Policy Brief, April 2019. https://www.ucl.ac.uk/bartlett/public-purpose/publications/2019/apr/beyond-voluntary-disclosure.

Schinasi, G.J. (2004). "Defining financial stability." *IMF Working Papers*, 04(187). https://doi.org/10.5089/9781451859546.001.

The White House. (2021, May 25). *Climate-Related Financial Risk*. Federal Register. https://www.federalregister.gov/documents/2021/05/25/2021-11168/climate-related-financial-risk.

Tsang, C.-Y. (2018). "Balancing the governance of the modern financial ecosystem: A new governance perspective and implications for market discipline." *Houston Journal of International Law*, 40(2), 531–612.

US Commodity Futures Trading Commission. (2020). *Managing Climate Risk in the U.S. Financial System. Report of the Climate-Related Market Risk Subcommittee, Market Risk Advisory Committee of the U.S. Commodity Futures Trading Commission*.

US Financial Crisis Inquiry Commission. (2011). *The Financial Crisis Inquiry Report: Final Report of the National Commission on the Causes of the Financial and Economic Crisis in the United States*. Financial Crisis Inquiry Commission.

World Bank. (2013). *World Development Report 2014: Risk and Opportunity - Managing Risk for Development*. World Bank. https://doi.org/10.1596/978-0-8213-9903-3.

Chapter 11

Regulating Climate Change Risks to Australian Infrastructure Systems

Infrastructure comprises the range of structures, facilities and buildings that comprise the built environment. It includes privately-owned infrastructure such as residential homes and commercial buildings, as well as public, critical infrastructure like roads, railways, telecommunication towers, pipes for the delivery of water and power lines. Infrastructure plays an important role in supporting economic activity, linking people to services, improving productivity and competitiveness, and generally enhancing people's lives.[1] Indeed, infrastructure has been labelled by the Australian Climate Institute (2012) as a "critical enabler for activity across all sectors of the economy";[2] it can have a significant impact on settlement patterns, and may also lead to long-term changes in the economy and society as a whole.[3]

Yet, the bulk of Australia's infrastructure has been and continues to be built and maintained to standards based on historic – not future – climate patterns.[4] Meanwhile, infrastructure increasingly faces a variety of external forces, including hazards associated with climate change.[5] Based on modelling undertaken by the Climate Council of Australia (2019), the property market is expected to lose hundreds of billions of dollars by 2030 due to the impact of climate change and extreme weather on infrastructure, and will continue to lose value in the following decades while carbon emissions causing climate change remain high.[6] This case study considers relevant aspects of the regulatory regime in the State of Victoria (Australia) applicable to the

1 E.J. Oughton, W. Usher, P. Tyler, & J.W. Hall, "Infrastructure as a complex adaptive system" (2018) *Complexity*, pp. 1–11, at 6. Australian Government, *The Role of Regulation in Facilitating or Constraining Adaptation to Climate Change for Australian Infrastructure*, Maddocks Report for the Department of Climate Change and Energy Efficiency (Department of Climate Change and Energy Efficiency, 2011), p. xiii. Climate Institute, "Coming ready or not: Managing climate risks to Australia's infrastructure" (2012), p. 15.

2 Climate Institute, n. 1 above, p. 15. S. Hallegatte, "Strategies to adapt to an uncertain climate change" (2009) 19(2) *Global Environmental Change*, pp. 240–7, at 241.

3 E.J. Oughton & P. Tyler, "Infrastructure as a complex adaptive system" (2013) *ITRC Working Paper Series*, at 1.

4 Climate Institute, n. 1 above, p. 15; S. Hallegatte, n. 2 above, p. 241.

5 E.J. Oughton & P. Tyler, n. 3 above, p. 1.

6 Climate Council, *Compound Costs: How Climate Change Is Damaging Australia's Economy* (2019), p. II.

DOI: 10.4324/9781003253464-11

management of one type of hazard linked to climate change – namely, bushfires. The aim of the case study is to determine the extent to which the Victorian regulatory regime is aligned with an ideal regime for the regulation of systemic risks based on risk governance, which has been set out earlier in this book.

I Systems of infrastructure

"Infrastructure" has been defined by Beeferman and Wain (2016) as:

> facilities, structures, equipment, or similar physical assets … that are vitally important, if not absolutely essential, to people having the capabilities to thrive as individuals and participate in social, economic, political, civic or communal, household or familial, and other roles in ways critical to their own well-being and that of their society, and the material and other conditions which enable them to exercise those capabilities to the fullest.[7]

Defined as such, it is clear that infrastructure includes a very broad spectrum of buildings and structures, ranging from small-scale infrastructure such as residential homes, schools and libraries to large critical energy, water and communications infrastructure.

These physical assets, which together support the well-functioning of human settlements, can together be conceived as a "system".[8] In fact, the Victorian *Climate Change Act 2017* (which, among other things, establishes a regulatory framework for the management of climate change risks in the State) refers to the "built environment system" comprising the built environment and how people use and interact with the built environment. Latora and Marchiori (2005) suggest that "infrastructure systems are a key ingredient of modern society"[9] and Saidi and colleagues (2018) explain that such systems help facilitate and manage flows, movement and exchange.[10] In particular, infrastructure is typically organised around patterns of existing settlements. Interconnections exist between components of the infrastructure system (e.g., between homes, roads and schools, as well as between energy and water facilities) as well as between users of the system (including inhabitants, infrastructure users and service providers). Components of urban infrastructure systems are becoming increasingly interdependent because, among other things, they are subject to common drivers and risks, including population growth and climate change.[11]

7 L. Beeferman & A. Wain, "Infrastructure: Defining matters" (2016) *SSRN Electronic Journal*, p. 15.
8 D.S. Thorpe, "A systems approach for the development and management of physical infrastructure", *1st International Conference on Systems Thinking in Management* (2000).
9 V. Latora & M. Marchiori, "Vulnerability and protection of infrastructure networks" (2005) 71(1) *Physical Review E*, 015103, p. 1.
10 S. Saidi, L. Kattan, P. Jayasinghe, P. Hettiaratchi, & J. Taron, "Integrated infrastructure systems – A review" (2018) 36 *Sustainable Cities and Society*, pp. 1–11, at 1.
11 Ibid. p. 1.

In fact, according to Saidi and colleagues, there are important benefits associated with recognising interdependent infrastructure as a system, including to help mitigate the impact of extreme events.[12] In fact, viewing infrastructure as a system is essential if risks to the system are to be effectively managed given the central role played by infrastructure in society. Systems thinking enables critical nodes, interconnections, vulnerabilities and weaknesses to be identified within the system. Among other things, such an approach helps to "understand the dynamics and complexity of the system and avoid ineffective responses."[13]

II The impact of bushfires on infrastructure

A study undertaken by Neumann and colleagues (2015) in the United States to determine the impact of climate change on US infrastructure found that the "impacts of climate change in this sector could be large, especially in the second half of the 21st century as sea level rises, temperature increases, and precipitation patterns become more extreme and affect the sustainability of long-lived infrastructure".[14] Similarly, a 2018 parliamentary report by the Australian Senate on the current and future effects of climate change on Australia's infrastructure noted that:[15]

> Australia has a significant amount of residential dwellings, other buildings and infrastructure. …
>
> Other essential infrastructure networks are also extensive and of high value, such as roads, railways and energy infrastructure.
>
> These buildings and other infrastructure assets need to withstand the Australian climate. Australians are already familiar with extreme events such as heatwaves, cyclones, bushfires and floods that have caused extensive damage to cities and infrastructure and resulted in deaths and injuries. Due to climate change, however, many types of extreme events are expected to become more frequent or more intense.

"Critical infrastructure" is at particular risk because of the heavy dependence of settlements on this type of infrastructure. Infrastructure of this kind has been defined as "the basic facilities, services, and installations needed for the functioning of a community or society, such as transportation and communications

12 Ibid. p. 1.
13 Ibid. p. 2.
14 J.E. Neumann, J. Price, P. Chinowsky, L. Wright, L. Ludwig, R. Streeter, R. Jones, J.B. Smith, W. Perkins, L. Jantarasami, & J. Martinich, "Climate change risks to US infrastructure: Impacts on roads, bridges, coastal development, and urban drainage" (2015) 131(1) *Climatic Change*, pp. 97–109, at 97.
15 Commonwealth of Australia, *Current and Future Impacts of Climate Change on Housing, Buildings and Infrastructure* (Senate Environment and Communications References Committee, 2018), p. 5; E.J. Oughton & P. Tyler, n. 3 above, p. 1.

systems, water and power lines, and public institutions including schools, post offices and prisons".[16] Jennings and Lodge (2010) clarify that, while definitions of critical infrastructure may be diverse, they share a common focus upon "highly interdependent supply chains, complex organisational decision structures, communication networks and transport linkages that enable mass social and economic activities".[17] According to the Climate Council of Australia (2019), around $17 billion (in present value terms) will need to be spent between 2015 and 2050 on rebuilding critical infrastructure in Australia following natural disasters.[18]

Some types of critical infrastructure are themselves interdependent in that one type may affect the well-functioning of another – such as, the interdependence between energy infrastructure and many other types of critical infrastructure, including transport, communications and water.[19] For example, an extreme heatwave in Melbourne, Australia in early 2009 crippled the system for the generation and transportation of electricity and resulted in rolling blackouts across the city. There were also effects across the broader "urban system" which depended upon electricity for its well-functioning, particularly for rail and road transport.[20]

In fact, failure of critical infrastructure can lead to cascading consequences across multiple types of critical infrastructure and other lower-level systems of infrastructure.[21] Florin and Trump (2016) suggest that the "cascading effect is perhaps one of the more complex and unpredictable components of systemic risk and is particularly dangerous in an environment of increasing system interdependency and connectivity".[22] This domino effect is likely to augment

16 W. Steele, K. Hussey, & S. Dovers, "What's critical about critical infrastructure?" (2017) 35(1) *Urban Policy and Research*, pp. 74–86, at 76 referring to J. Moteff and P. Paformak, *Critical Infrastructure and Key Assets: Definition and Identification,* CRS Report for Congress (Congressional Research Service, The Library of Congress, 2004).

17 W. Jennings & M. Lodge, "Critical infrastructures, resilience and organisation of mega-projects: The Olympic Games", in B.M. Hutter (ed.), *Anticipating Risks and Organising Risk Regulation*, (Cambridge University Press, 2010), pp. 161–84, at 161.

18 Climate Council, n. 6 above, p. 17 citing Deloitte Access Economics, *Building Resilience to Natural Disasters in our States and Territories,* Australian Business Roundtable for Disaster Resilience & Safer Communities (2017).

19 G.H. Kjølle, I.B. Utne, & O. Gjerde, "Risk analysis of critical infrastructures emphasizing electricity supply and interdependencies" (2012) 105 *Reliability Engineering & System Safety*, pp. 80–9.

20 D. McEvoy, I. Ahmed, & J. Mullett, "The impact of the 2009 heat wave on Melbourne's critical infrastructure" (2012) 17(8) *Local Environment*, pp. 783–96.

21 Climate Institute, n. 1 above, p. 3; M.A. Centeno, M. Nag, T.S. Patterson, A. Shaver, & A.J. Windawi, "The emergence of global systemic risk" (2015) 41(1) *Annual Review of Sociology*, pp. 65–85, at 75.

22 M.-V. Florin and B.D. Trump, "Resilience in the context of systemic risks: Perspectives from IRGC's Guidelines for the Governance of Systemic Risks" in B.D. Trump, M.-V. Florin and I. Linkov (eds.) *IRGC Resource Guide on Resilience (Volume 2)* (EPFL International Risk Governance Center, 2016), p. 60.

the impact of each individual failure of infrastructure and, consequently, increase the overall effort and resources required to recover from a particular adverse event.[23]

A risk to infrastructure that is of particular concern in Australia arises from bushfires. A "bushfire" (known in countries other than Australia as a wildfire) is a fire that burns in grass, bush or woodland. Although there are differences in the detailed weather patterns across the country, Australia has a characteristically hot, dry climate linked to low and unreliable rainfall. This renders many parts of Australia vulnerable to bushfires.[24] Climate change is predicted to lead to an increase in the frequency of very high temperature days, especially in inland areas,[25] an increase in the number and severity of high fire risk days, and an earlier start and later end to the fire season.[26]

Bushfires pose a range of risks to people, to the environment and to the economy. The economic impacts are typically associated with the destruction of property and businesses, including the spectrum of infrastructure upon which communities depend.[27] In 2010, the Australian Building Codes Board found that, on average, 118 houses had been burnt in Australia per year during the preceding 50 years, mostly in the south-east of Australia.[28] Unfortunately, historical averages have been far exceeded, including during the devastating "Black Summer" bushfires along the east coast of Australia in the summer of 2019/2020.[29]

A decade earlier, in February 2009, the deadly "Black Saturday" bushfires in Victoria occurred and were a turning point for the regulation of bushfire risks in the State. Those bushfires followed more than a week of record or near

23 M.A. Centeno, M. Nag, T.S. Patterson, A. Shaver, & A.J. Windawi, n. 21 above, p. 75.

24 S. Ellis, P. Kanowski, & R.J. Whelan, *National Inquiry on Bushfire Mitigation and Management* (Council of Australian Governments, 2005), p. 11.

25 K. Hennessy, C. Lucas, N. Nichols, J. Bathols, R. Suppiah and J. Ricketts, *Climate Change Impacts on Fire-weather in South-east Australia* (CSIRO, Bushfire CRC and Australian Bureau of Meteorology, 2005).

26 C. Lucas, K. Hennessy, G. Mills and J. Bathols, *Bushfire Weather in Southeast Australia: Trends and Projected Climate Change Impacts*, Consultancy Report prepared for The Climate Institute of Australia (Bushfire CRC and CSIRO, 2007).

27 C. Stephenson, *A Literature Review on the Economic, Social and Environmental Impacts of Severe Bushfires in South-eastern Australia: Fire and Adaptive Management*, Report No. 87 (Victorian Department of Sustainability and Environment, 2010).

28 Australian Building Codes Board, *An Investigation of Possible Building Code of Australia (BCA) Adaptation Measures for Climate Change* (ABCB, 2010), p. 72.

29 Climate Council, *Summer of Crisis* (Climate Council of Australia, 2020). The Climate Council stated at p. II that "Australia's Black Summer was unprecedented in scale and harm. The bushfire season was the worst on record for New South Wales in terms of the scale of the bushfires, the number of properties lost and the amount of area burned". See also S.M. Davey & A. Sarre, "Editorial: The 2019/20 Black Summer bushfires" (2020) 83(2) *Australian Forestry*, pp. 47–51 indicating that 3,100 homes were lost during the Black Summer bushfires.

record temperatures across much of south-eastern Australia and demonstrated the frightening implications of bushfires for the country. At the time, they were the worst bushfires in Australia's history, killing 173 people and having a direct impact on 78 towns throughout the State.[30] The fires destroyed over 2,000 houses with thousands more damaged.[31] The economic cost associated with the disaster, due in large part to the loss of infrastructure, was estimated to be in excess of $4 billion.[32] The Black Saturday bushfires illustrate the particular risks posed by bushfires when they encroach upon human settlements, where extensive systems of infrastructure are likely to be located. Whereas in the past, bushfire risks may have been considered to be largely confined to rural areas, events in recent years both in Australia and abroad have illuminated the real and significant threats associated with bushfires in built-up urban areas. Indeed, it is in relation to the interface between urban areas and vegetated non-urban areas that responding to bushfire risks is arguably most challenging.[33]

The manner in which bushfire risks are regulated differs somewhat between jurisdictions in Australia. This case study examines the regulatory regime to address climate change risks to infrastructure associated with bushfire risks in Victoria because Australia's south-east (particularly, Victoria) is likely to be significantly affected by the heightened bushfire risk associated with climate change.[34] Indeed, while the average annual area affected by bushfire in northern Australia exceeds that in the south,[35] the risks are greater in the south, where human settlement is more expansive and dense compared to the north.[36]

III The regulatory regime in Victoria, Australia

Despite the longstanding historical awareness of bushfires and associated risks in Australia, the aftermath of the Black Saturday bushfires provides stark and bitter

30 Victorian Government, *Final report* (2009 Victorian Bushfires Royal Commission, 2010), p. 1; C. Stephenson, J. Handmer and A. Haywood, "Estimating the net cost of the 2009 Black Saturday Fires to the affected regions" (2012), accessible at: https://www.bushfirecrc.com/sites/default/files/news/estimating_the_net_cost_of_the_2009_black_saturday_fires_to_the_affected_regions.pdf.

31 C. Stephenson et al., ibid.

32 Victorian Government, n. 30 above, p. 1.

33 March, A. and Rijal, Y. (2015) "Reducing bushfire risk by planning and design: A professional focus", *Planning Practice & Research* Vol. 30, No. 1, 33–53, pp. 34–35.

34 H. Clarke, C. Lucas, & P. Smith, "Changes in Australian fire weather between 1973 and 2010" (2013) 33(4) *International Journal of Climatology*, pp. 931–44, at 939, 940, 941 and 942; A. Foerster, A. Macintosh, & J. McDonald, "Transferable lessons for climate change adaptation planning? Managing bushfire and coastal climate hazards in Australia" (2013) 30(6) *Environmental and Planning Law Journal*, pp. 469–90, at 471.

35 S. Ellis, P. Kanowski, & R.J. Whelan, n. 24 above, p. 5.

36 Commonwealth of Australia, n. 15 above, pp. 15–16, referring to CSIRO Submission 45, p. 10 and Climate Council of Australia, Submission 40, pp. 6–7. See also A. Foerster, A. Macintosh, & J. McDonald, n. 34 above, p. 470.

evidence that the regulatory regime in place in Victoria at the time of those bushfires was not effective to protect humans and human settlements from these risks. In fact, Black Saturday was the tragic catalyst for major regulatory reform, including reform of the regime designed to address bushfire risks to infrastructure.

The 2009 Victorian Bushfires Royal Commission (VBRC) was established following Black Saturday to investigate the causes and responses to the bushfires. Amongst other things, the VBRC evaluated the regulatory regime applicable to the development of infrastructure – particularly, the regulatory frameworks that applied to land-use planning and building in bushfire-prone areas. The VBRC found that a well-informed, holistic and strategic approach to hazard mapping is essential for planning and building decisions.[37] The VBRC observed that such mapping provides baseline information that can inform bushfire-specific planning and building controls in high-risk areas. Yet, it found significant deficiencies in the mapping of bushfire risk throughout Victoria at the time of the Black Saturday bushfires.[38]

In relation to land-use planning, the general purpose of which is to establish the conditions regarding the use of land that are needed to protect and create efficient and attractive urban environments,[39] the VBRC concluded that the regulatory framework failed to attach sufficient importance to bushfire risk and the potential threat to people's lives in bushfire-prone areas.[40] It noted, more specifically, that the pre-existing approach to planning did not take account of the fact that there are some areas where the risk to life from bushfire is so high that new settlements should not be established.[41]

Regarding the regulation of buildings, the VBRC found that the standard for construction in bushfire-prone areas failed to adequately cover all the important components of bushfire risk.[42] The VBRC also considered that maintaining a defendable space around buildings is crucial to establishing a line of defence against bushfires, particularly by reducing the risk of radiant heat and direct flame contact and that the land-use planning system needed to facilitate this.[43]

An "Integrated Planning and Building Framework" for management of bushfire risks was established in response to the VBRC's findings and associated recommendations. It was designed to strengthen consideration of bushfire risks at different stages of the planning process and better integrate the planning and building systems.[44] In summary, the framework involves three key elements.

37 Victorian Government, n. 30 above, p. 215.
38 Ibid. p. 215.
39 N. Gurran, *Australian Urban Land Use Planning: Principles, Systems and Practice* (Sydney University Press, 2nd ed., 2011), p. 7.
40 Victorian Government, n. 30 above, p. 215.
41 Ibid. p. 215.
42 Ibid. p. 215.
43 Ibid. p. 215.
44 L. Groenhart, A. March, & M. Holland, "Shifting Victoria's emphasis in land-use planning for bushfire: Towards a place-based approach" (2012) 27(4) *Australian Journal of Emergency Management*, pp. 33–7, at 36.

First, bushfire planning provisions were incorporated into each local land-use planning scheme in Victoria to minimise exposure to bushfire risks and maximise resilience for new development. Second, building regulation was amended to require that buildings in bushfire-prone areas must be assessed at the building permit stage to determine the construction requirements to address the local bushfire hazard. Finally, the rules regarding vegetation management were altered to ensure that vegetation around homes at risk of bushfires is adequately managed to mitigate bushfire risk. The integrated approach means that a combined consideration of planning, building and vegetation responses can be made to determine whether or not risk has been reduced to an acceptable level. In addition, as explained below, a discrete regulatory framework applies to the management of risks to critical infrastructure, which is supplemented by overarching obligations under the *Climate Change Act 2017* to have regard to guiding principles and policy objectives "to build the resilience of the State's infrastructure, built environment and communities through effective adaptation and disaster preparedness action".

I Land-use planning

Land-use planning in Victoria is governed by local planning schemes, which are made pursuant to the *Planning and Environment Act 1987 (Vic)*. A planning scheme is a statutory document that sets out the rules regarding land use in each local municipality within the State. Every planning scheme is constructed using the Victoria Planning Provisions (VPPs), which help to ensure that the construction and content of planning schemes is consistent across Victoria, regardless of the particular municipality in which they are applied. The VPPs, contain planning policies, strategies and specific planning controls to address diverse risks, including bushfire risks.

Every local planning scheme must include the State Planning Policy Framework (SPPF), which is set out in the Victoria Planning Provisions and describes planning policy on various issues. Among other things, the SPPF sets out the planning policies applicable to environmental risks, including the risks associated with climate change, which are contained in clause 13 of the VPPs.[45] Clause 13 explicitly states that planning should prepare for and respond to the impacts of climate change.[46] More specifically, the VPPs provide that planning should strengthen the resilience and safety of communities by adopting a "best practice environmental management and risk management approach".[47]

A stated objective of the VPPs is to minimise the impacts of natural hazards and adapt to the impacts of climate change through "risk-based planning".[48] Various tiers of strategies are contained within the VPPs outlining how its objectives are to be achieved. They emphasise the need to reduce vulnerability

45 Clause 13 VPPs ("Environmental Risks and Amenity").

46 Ibid.

47 Ibid.

48 Clause 13.01–1S VPPs ("Natural hazards and climate change").

and exposure to risk. The VPPs contain general strategies in relation to adaptation to climate change. These strategies include the requirement to consider risks associated with climate change in planning and management decision-making processes, to identify risk areas using the "best available data" and climate change science, and to adopt planning controls to implement risk mitigation or risk adaptation strategies.[49] Specific climate change adaptation strategies are identified in relation to new development and existing settlements. For new development, the VPPs state that development should be directed to low-risk locations and that risk to life, property, the natural environment and community infrastructure from natural hazards should be minimised.[50] Regarding existing settlements, the VPPs provide that adaptation response strategies should be developed for settlements located in risk areas to accommodate change over time.[51]

The VPPs also contain strategies specifically for bushfire planning. These include the strategy of reducing the vulnerability of communities to bushfire through the consideration of bushfire risk in decision-making at all stages of the planning process.[52] The VPPs specifically refer to the need for bushfire risk to be properly assessed applying the "best available science" to identify vegetation, topographic and climatic conditions that create a bushfire hazard. Decision-makers must also consider the "best available information" about bushfire hazard, including the map of designated Bushfire Prone Areas (BPAs).[53] Finally, a specific planning control to manage bushfire risks – the so-called Bushfire Management Overlay (BMO) – must be applied to areas where the extent of vegetation can create an extreme bushfire hazard.[54]

Clause 44.06 of VPPs contains the BMO. An overlay such as this one applies special planning controls to areas that have special features or are prone to certain risks. The primary purpose of the BMO is to ensure that development in high-risk bushfire areas only takes place after full consideration of a range of bushfire safety issues. The BMO's objectives and standards provide a benchmark for complying with State policy and determining "acceptable risk".[55] More specifically, development is only permitted where the risk to life, property and community infrastructure from bushfire can be reduced to an acceptable level.[56] The BMO requires consideration of the location, design and construction of development and the implementation of bushfire protection measures in making this assessment and a schedule to the BMO can be used to ensure that the particularities of each municipality are properly accounted for.[57]

49 Clause 13.01 VPPs ("Natural hazards and climate change").
50 Ibid.
51 Ibid.
52 Clause 13.02–1S VPPs ("Bushfire planning").
53 Ibid.
54 Ibid.
55 L. Groenhart, A. March, & M. Holland, n. 44 above, p. 36.
56 Clause 53.02 VPPs ("Bushfire Planning").
57 Clause 44.06–4 VPPs ("Requirements of Clause 53.02").

A Bushfire Prone Area is an area identified by the Victorian Minister for Planning as subject to, or likely to be subject to, bushfires.[58] This determination is based on mapping that uses a detailed set of criteria, including landscape conditions, vegetation types and sizes, to determine the overall bushfire hazard level.[59] Land within BPAs that involve the highest risks is subject to the BMO.

The VPPs provide that a risk assessment is required under the BMO as a precondition to the grant of a permit to authorise development of infrastructure on land.[60] Amongst other things, applications for planning permits in relation to land subject to the BMO must be accompanied by three critical risk management documents, which are submitted by the permit applicant. A bushfire hazard site assessment must include a plan that describes the bushfire hazard within 150 metres of the proposed development. A bushfire hazard landscape assessment must include a plan that describes the bushfire hazard of the general locality more than 150 metres from the site. A bushfire management statement must describe how the proposed development responds to the requirements in the VPPs.[61]

2 Building regulatory framework

The regulatory framework for buildings is centred on the Building Code of Australia (BCA), which forms part of the National Construction Code (NCC).[62] The BCA is a national technical code which applies to all new buildings, upgrades of existing buildings, buildings undergoing a change of use, but does not apply to existing buildings that are not being changed or altered.[63] The BCA is called up by applicable state legislation – in the case of Victoria, this is the *Building Act 1993 (Vic)*.

The BCA is a performance-based code.[64] For each aspect of construction, the BCA sets out performance requirements, which are mandatory, minimum requirements that must be met regarding the design and construction of buildings. Compliance with the performance requirements may be achieved by complying with prescriptive "deemed-to-satisfy" provisions which, as the name suggests, are deemed to satisfy the performance requirements. Alternatively, the performance requirements may be met through an "alternative solution",

58 A formal determination to designate Bushfire Prone Areas is made by the relevant Minister under section 192A of the *Building Act 1993 (Vic)* ("Designated bushfire prone areas").
59 Victorian Government, *Bushfire Mapping Methodology and Criteria*, Fact Sheet (Department of Environment, Land, Water and Planning, 2019).
60 Clause 44.06–2 VPPs ("Permit requirement").
61 Clause 44.06–3 VPPs ("Application requirements").
62 The NCC comprises the BCA (Volumes One and Two) and the Plumbing Code of Australia (Volume 3).
63 Introduction to the National Construction Code, BCA, Volume 1.
64 See the ABCB website accessible at https://ncc.abcb.gov.au/ncc-online/How-it-works.

which must undergo an independent verification to ensure that it meets the relevant performance requirements. The performance-based approach provides building sector participants with considerable flexibility to respond to the effects of climate change including bushfire risks and, in theory, the approach accommodates innovative building solutions that address climate change impacts as they become available without the need for regulatory change.

Building construction and design can minimise the impact of ember attack and radiant heat on a building. Accordingly, the BCA contains overarching objectives and performance requirements in relation to achieving an appropriate construction standard to improve the ability of a building to withstand bushfire attack[65] – in other words, to ensure a building's resilience. In particular, the BCA contains a performance requirement which states that, for a building that is constructed in a designated BPA, to the degree necessary, it must be designed and constructed to reduce bushfire risk.[66] The verification method for compliance with this performance requirement essentially involves a risk assessment, which accounts for the relative "importance level" of the building or structure based on the type of building or structure and the number of people it is designed to contain, as well as the annual probability of exceedance (AEP) for bushfire exposure or for particular weather conditions.[67]

The deemed-to-satisfy provision requires that residential buildings in designated BPAs must comply with *Australian Standard 3959:2018 Construction of buildings in bushfire prone areas* (AS 3959).[68] This standard was updated in 2009 following the Black Saturday bushfires to improve the ability of buildings in bushfire prone areas to withstand bushfire attack. The Victorian building regulator explains that the standard reflects a more "finely tuned risk assessment" compared to the previous standard.[69] A methodology to determine building construction requirements in BPAs is set out in AS 3959. The key mechanism to determine applicable construction requirements is the Bushfire Attack Level (BAL), which is a way of measuring the ability of a building to withstand attack from bushfire and is based on the modelled severity of a building's potential exposure to ember attack, radiant heat and direct flame contact. As the bushfire threat increases, so do the construction requirements.

65 Part G5, BCA, Volume 1 ("Construction in bushfire prone areas").
66 GP5.1, BCA, Volume 1 ("Bushfire resistance"). More specifically, the performance requirement requires that the design and construction of a building in a BPA must reduce the risk of ignition from a bushfire appropriate to the potential for ignition caused by burning embers, radiant heat or flame generated by a bushfire and intensity of the bushfire attack on the building.
67 Verification method GV5 (Verification methods for building in bushfire prone areas).
68 G5.2 BCA, Volume 1 ("Deemed-to-Satisfy Provisions – Protection").
69 Victorian Building Authority, *A Guide to Building in Victoria after Bushfires*, p. 1.

3 Regulation of the built environment system and critical infrastructure

In addition to the regulatory frameworks dealing with land use and building described above, Victoria's regulatory regime also includes the *Climate Change Act 2017*, which provides in section 34 for the preparation of a climate change adaptation plan in respect of a range of systems, including the built environment system as well as emergency management legislation, which requires certain responsible entities for the State's most critical infrastructure to demonstrate their assets are resilient to risks in the event of an emergency.[70] Under the *Climate Change Act 2017*, the preparation of an adaptation action plan must involve consideration of guiding principles, including the principle of risk management to carefully evaluate the best practicably available information about the potential impacts of climate change and the principle of community engagement in order to provide appropriate information and to facilitate engagement with members of the community. Under emergency management legislation, each responsible entity must identify the risks to the relevant critical infrastructure and specify the emergency risk management actions or activities that the responsible entity proposes to take to mitigate those risks. Subordinate legislation prescribes the minimum standards that must be applied by responsible entities in complying with their risk management obligations.[71]

IV Approach to regulating bushfire risks caused by climate change risks to Australian infrastructure

Systemic risks are complex. There will seldom be a single point of regulatory intervention that will be effective in controlling such risks. Rather, a multi-pronged approach, which accounts for the interactions and interdependencies of each constituent part of the system, will often be needed. The regulation of bushfire risks in Victoria is a case in point, where multiple regulatory frameworks to manage these risks comprise the regulatory regime. However, as explained below, there are also likely to be challenges associated with this approach.

1 Regulatory boundary

The approach to the identification of risks covered by each of the relevant regulatory frameworks comprising the Victorian regulatory regime differs, which makes the extent of the boundary for the regulation of bushfire risks to infrastructure systems as a whole in Victoria somewhat unclear.

The risks to built environment system regulated under the *Climate Change Act 2017* relate to the potential impacts of climate change whereas the risks to

70 Part 7A of the *Emergency Management Act 2013 (Vic)*.

71 *Emergency Management (Critical Infrastructure Resilience) Regulations 2015 (Vic)*.

critical infrastructure that must be accounted for by relevant responsible entities in managing such infrastructure have not been specifically identified under the emergency management regulatory framework. Instead, responsible entities are tasked with identifying those risks themselves, which may or may not include bushfire risks. In the case of the building regulatory framework, the stated goal of the BCA is to "enable the achievement of nationally consistent, minimum necessary standards of relevant health, safety (including structural safety and safety from fire), amenity, accessibility and sustainability objectives efficiently".[72] These objectives indicate that the explicit requirement to reduce bushfire risk in the BCA should be interpreted by reference to considerations, including personal health and safety. Bushfire risks are also specifically identified in the land-use planning framework. The VPPs expressly refer to the need to manage risks to life, property, the natural environment and infrastructure caused by natural hazards, including bushfire risks.[73]

2 *Regulatory context*

The regulatory regime includes a number of mechanisms to provide context for the regulatory response to bushfire risks. Requirements to undertake risk assessments exist under the *Climate Change Act 2017* as well as under each of the land-use planning, building, and emergency management frameworks. In the case of land-use planning, these are the bushfire hazard site assessment and the bushfire hazard landscape assessment. For building, the BCA's performance requirements require a risk assessment which accounts for the relative "importance level" of the building or structure.[74] In the case of critical infrastructure, each responsible entity must prepare an emergency risk management plan for vital critical infrastructure to prepare for an emergency.[75]

In addition, BPAs, which are relevant for both land-use planning and building, provide broad context for responding to bushfire risk by identifying areas that are, or are likely to be, exposed to bushfires. This contextual information is supplemented by schedules to the BMO, which can be used to ensure that the particularities of each municipality, including any vulnerability to bushfire risk, are properly accounted for in determining the appropriate planning response. Similarly, the determination of the BAL under the building regulatory framework involves a site-specific assessment of a variety of fire risk factors associated with the particular location where new construction is proposed, including the likely exposure to bushfires.

72 Introduction to the National Construction Code, NCC 2019, Volume 1.
73 Clauses 13.01 VPPs ("Natural hazards and climate change") and 44.06 VPPs ("Bushfire Management Overlay").
74 Verification method GV5 (Verification methods for building in bushfire prone areas).
75 Section 74P(1), *Emergency Management Act 2013 (Vic)*.

In terms of the evaluation of risk, the underlying premise of the regulatory frameworks comprising the Victorian regime is that bushfire risks must be reduced to an "acceptable level". Under the VPPs, development is only permitted where bushfire risk can be reduced to such a level.[76] While the land-use planning framework does not prescribe what constitutes this level, it does include factors that decision-makers must take into consideration, including the location, layout and siting of the proposed development, the extent of defendable space and the proposed design of the construction.[77] In relation to critical infrastructure, the risk management standards that must be complied with in establishing emergency risk management plans for critical infrastructure by responsible entities also embed a requirement for risk evaluation to be undertaken.[78] Building requirements in the deemed-to-satisfy provisions of the BCA remove the need for an explicit judgement to be made about relative acceptability of bushfire risk because risk tolerances are already embedded in the specific construction standards that apply for each BAL level. The higher the BAL, the more demanding the construction requirements for the relevant building under AS 3959.

3 Regulatory content

The content of the various regulatory frameworks comprising the Victorian regime is characterised by a flexible regulatory approach. In the case of land-use planning, the framework includes broad objectives, with considerable flexibility as to how those objectives are to be achieved. Similarly, the building regulatory framework establishes general performance standards, although these can be met by complying with the deemed-to-satisfy requirements – in the case of bushfire risks, these requirements include compliance with AS 3959. The regulatory requirements applicable to the management of critical infrastructure under emergency management legislation also impose broad risk management obligations on responsible entities to plan for emergencies, but do not specify in detail how those obligations should be discharged. The *Climate Change Act 2017* specifies the matters that must be taken into consideration when preparing a climate change adaptation plan, including an adaptation plan for the built environment system, but there is also some flexibility regarding the methodology for and contents of such plans to enable them to be tailored to the particular contexts in which they will be applied.

V Alignment with the risk governance regulatory ideal

As has been explained in this chapter, the regulatory regime to address bushfire risks to infrastructure in Victoria comprises multiple regulatory frameworks

76 Clause 53.02 VPPs ("Bushfire Planning").

77 Ibid.

78 Regulation 5, *Emergency Management (Critical Infrastructure Resilience) Regulations 2015 (Vic)*, which requires compliance with AS/NZS ISO 31000:2009 Risk Management – Principles and guidelines, as published from time to time.

dealing respectively with land-use planning, building and critical infrastructure. This multi-pronged, approach helps to ensure that bushfire risk to infrastructure is addressed in a comprehensive manner, rather than through a single regulatory instrument with an unduly narrow focus and limited scope of application.

Risk is conspicuous throughout the regulatory regime, and helps to guide regulatory decisions and actions made under the regime. Moreover, bushfire risks have been singled out for special attention. Specifically, management of bushfire risk is a core policy objective of the land-use planning regulatory framework and is provided for in construction requirements under the BCA.

Even though the regulatory regime employs pre-existing and longstanding regulatory frameworks to address bushfire risks it, nevertheless, bears indicia of a risk governance approach. In particular, the regime provides for these risks to be contextualised, including through assessment and evaluation, and enables a flexible approach in responding to risk. This apparent alignment with the ideal for the regulation of systemic risks based on the risk governance paradigm outlined earlier in this book could be attributed to the origins of the regime. In particular, aspects of the regime were re-designed following the Black Saturday disaster, which resulted in loss not only of infrastructure, but also of many lives. The disaster provided the impetus to initiate a well-resourced and powerful formal review of the previous regulatory arrangements. This, in turn, facilitated a comprehensive re-consideration of the pre-existing regulatory regime in an effort to avoid a repeat of the 2009 bushfire disaster.

While the Victorian regime is commendable from a risk governance perspective, an explicit recognition of the "infrastructure system" and the systemic implications of bushfire risks is absent from the land-use planning and building regulatory frameworks. In particular, the land-use and building regulatory frameworks apply to proposed individual uses and development of land on a case-by-case basis and largely concern new infrastructure, rather than existing infrastructure. In addition, the risk management obligations under emergency management legislation only apply to certain types of critical infrastructure that have been assessed as "vital".

Indeed, the design of the regulatory regime as a whole does not obviously respond to the possibility that a major bushfire event could result in destruction of whole systems of infrastructure across multiple municipalities and, potentially, across State borders. The recent Black Summer bushfires provide potent evidence of these apparent gaps in a regime that has already been overhauled following a similar catastrophe in the not-too-distant past. In the context of the federal 2020 Bushfires Royal Commission, note was taken of the emphasis on new infrastructure,[79] as well as "deficiencies with integrating critical infrastructure into planning processes" and "inconsistencies in the extent to which the vulnerability of essential infrastructure is accounted for in government … risk

79 Commonwealth of Australia, *Royal Commission into National Natural Disaster Arrangements*, Interim Observations (2020), p. 17

management".[80] The Victorian Built Environment Climate Change Adaptation Action Plan for 2022 - 2026, made pursuant to the *Climate Change Act 2017*, helps to address these deficiencies by, among other things, adopting a systems approach to the management of risks to the built environment and contemplates further changes to planning schemes and building standards.

A further question that arises in the context of the Victorian regime is its capacity to respond to contextual change, particularly as bushfire risks intensify over time. The regime, as a whole, embeds flexibility to modulate regulatory responses based on the context, which suggests that the regime will be able to keep pace with these contextual changes. The public participation mechanism to engage with stakeholders in relation to such responses at a systemic level under the *Climate Change Act 2017* is notable; such a mechanism could help to ensure that regulatory responses across the entire system are truly tailored and responsive to the context in which they will be applied and acceptable to those directly affected by those responses. Given the public's sensitivity to bushfire risk – particularly, in Victoria which has suffered through the Black Saturday bushfires and, more recently, the Black Summer bushfires – this public participation mechanism could go a long way in building public trust and confidence in the way bushfire risks to infrastructure systems are being regulated in practice. Such trust is of utmost importance for the effective regulation of systemic risks caused by climate change.

Bibliography

Australian Building Codes Board. (2010). *An Investigation of Possible Building Code of Australia (BCA) Adaptation Measures for Climate Change*. Australian Building Codes Board.

Beeferman, L., & Wain, A. (2016). "Infrastructure: Defining matters." *SSRN Electronic Journal*. https://doi.org/10.2139/ssrn.2714308.

Centeno, M.A., Nag, M., Patterson, T.S., Shaver, A., & Windawi, A.J. (2015). "The emergence of global systemic risk." *Annual Review of Sociology*, 41(1), 65–85. https://doi.org/10.1146/annurev-soc-073014-112317.

Clarke, H., Lucas, C., & Smith, P. (2013). "Changes in Australian fire weather between 1973 and 2010." *International Journal of Climatology*, 33(4), 931–944. https://doi.org/10.1002/joc.3480.

Climate Council. (2019). *Compound Costs: How Climate Change Is Damaging Australia's Economy*. Climate Council.

Climate Council. (2020). *Summer of Crisis*. Climate Council.

Climate Institute. (2012). *Coming Ready or Not: Managing Climate Risks to Australia's Infrastructure*. Climate Institute.

Commonwealth of Australia. (2018). *Current and Future Impacts of Climate Change on Housing, Buildings and Infrastructure*. Senate Environment and Communications References Committee.

80 Ibid. p. 15.

Davey, S.M., & Sarre, A. (2020). "Editorial: The 2019/20 Black Summer bushfires." *Australian Forestry*, 83(2), 47–51. https://doi.org/10.1080/00049158.2020.1769899.

Deloitte Access Economics. (2017). *Building Resilience to Natural Disasters in our States and Territories*. Australian Business Roundtable for Disaster Resilience & Safer Communities.

Ellis, S., Kanowski, P., & Whelan, R.J. (2005). *National Inquiry on Bushfire Mitigation and Management*. Council of Australian Governments.

Florin, M.-V., & Trump, B.D. (2016). "Resilience in the context of systemic risks: Perspectives from IRGC's Guidelines for the Governance of Systemic Risks." In B. D. Trump, M.-V. Florin and I. Linkov (eds.), *IRGC Resource Guide on Resilience* (Volume 2). EPFL International Risk Governance Center.

Foerster, A., Macintosh, A., & McDonald, J. (2013). "Transferable lessons for climate change adaptation planning? Managing bushfire and coastal climate hazards in Australia." *Environmental and Planning Law Journal*, 30(6), 469–490.

Groenhart, L., March, A., & Holland, M. (2012). "Shifting Victoria's emphasis in land-use planning for bushfire: Towards a place-based approach." *Australian Journal of Emergency Management*, 27(4), 33–37.

Gurran, N. (2011). *Australian Urban Land Use Planning: Principles, Systems and Practice*. Sydney University Press (2nd edition).

Hallegatte, S. (2009). "Strategies to adapt to an uncertain climate change." *Global Environmental Change*, 19(2), 240–247. https://doi.org/10.1016/j.gloenvcha.2008.12.003.

Hennessy, K., Lucas, C., Nichols, N., Bathols, J., Suppiah, R. & Ricketts, J. (2005). *Climate Change Impacts on Fire-weather in South-east Australia*. CSIRO, Bushfire CRC and Australian Bureau of Meteorology.

Jennings, W., & Lodge, M. (2010). "Critical infrastructures, resilience and organisation of mega-projects: The Olympic Games." In B.M. Hutter (ed.), *Anticipating Risks and Organising Risk Regulation*. Cambridge University Press. https://doi.org/10.1017/CBO9780511761553.009.

Kjølle, G.H., Utne, I.B., & Gjerde, O. (2012). "Risk analysis of critical infrastructures emphasizing electricity supply and interdependencies." *Reliability Engineering & System Safety*, 105, 80–89. https://doi.org/10.1016/j.ress.2012.02.006.

Latora, V., & Marchiori, M. (2005). "Vulnerability and protection of infrastructure networks." *Physical Review E*, 71(1), 015103. https://doi.org/10.1103/PhysRevE.71.015103.

Lucas, C., Hennessy, K., Mills, G. & Bathols, J. (2007). *Bushfire Weather in Southeast Australia: Trends and Projected Climate Change Impacts*, Consultancy Report prepared for The Climate Institute of Australia. Bushfire CRC and CSIRO.

McEvoy, D., Ahmed, I., & Mullett, J. (2012). "The impact of the 2009 heat wave on Melbourne's critical infrastructure." *Local Environment*, 17(8), 783–796. https://doi.org/10.1080/13549839.2012.678320.

Moteff, J., & Parfomak, P. (2004). *Critical Infrastructure and Key Assets: Definition and Identification*, CRS Report for Congress. Congressional Research Service, The Library of Congress.

Neumann, J.E., Price, J., Chinowsky, P., Wright, L., Ludwig, L., Streeter, R., Jones, R., Smith, J.B., Perkins, W., Jantarasami, L., & Martinich, J. (2015). "Climate change risks to US infrastructure: Impacts on roads, bridges, coastal development, and urban drainage." *Climatic Change*, 131(1), 97–109. https://doi.org/10.1007/s10584-013-1037-4.

Oughton, E.J., & Tyler, P. (2013). "Infrastructure as a complex adaptive system." *ITRC Working Paper Series*. https://doi.org/10.1155/2018/3427826.

Oughton, E.J., Usher, W., Tyler, P., & Hall, J.W. (2018). "Infrastructure as a complex adaptive system." *Complexity*, 1–11. https://doi.org/10.1155/2018/3427826.

Saidi, S., Kattan, L., Jayasinghe, P., Hettiaratchi, P., & Taron, J. (2018). "Integrated infrastructure systems – A review." *Sustainable Cities and Society*, 36, 1–11. https://doi.org/10.1016/j.scs.2017.09.022.

Steele, W., Hussey, K., & Dovers, S. (2017). "What's critical about critical infrastructure?" *Urban Policy and Research*, 35(1), 74–86. https://doi.org/10.1080/08111146.2017.1282857.

Stephenson, C. (2010). *A Literature Review on the Economic, Social and Environmental Impacts of Severe Bushfires in South-Eastern Australia: Fire and Adaptive Management.* Victorian Department of Sustainability and Environment.

Thorpe, D.S. (2000). "A systems approach for the development and management of physical infrastructure." 1st International Conference on Systems Thinking in Management.

Victorian Government. (2010). *Final report. 2009.* Victorian Bushfires Royal Commission.

Chapter 12

Conclusions and the Way Forward for Regulating Systemic Risks

This book has explored the intersection between risk and regulation. It considered various facets of risk and examined how regulation can be employed to manage systemic risks by responding to the key characteristics of this unique type of risk. The particular context for this exploration was systemic risks caused by climate change. The primary objective of the book was to determine how systemic risks can be regulated, if at all. Various important observations surfaced in pursuing this objective, which are outlined below.

1 Overview

The book has revealed the utility of using risk as the primary lens through which to design and evaluate regulatory regimes. Regulation is conventionally viewed as rules with which compliance is expected. Once regulation is reconceptualised as a vehicle to manage risks, the potential offered by regulation to contend with society's most intractable problems, not least of which is climate change, becomes clearer. At least in theory, regulation could be mobilised to manage risks in a targeted, contextualised and strategic manner in the same way that risk management tools are used in other contexts.

The book also deepened understanding of systemic risks, particularly the risks to human and natural systems caused by climate change. Systemic risks were defined, the particularities of modern systemic risks were discussed, and the implications for risk management were considered. As explained in the book, the context, characteristics and consequences of such risks are distinct from other types of risks; they are typically complex, uncertain and ambiguous. Characterising a risk as systemic implies a shift in focus away from individual *causes and consequences of risks to the system* and, instead, a greater emphasis is required on the *system* itself. A systemic perspective avoids a compartmentalised and fragmented approach to risk management. Rather, it demands consideration of the entire system when assessing risks and determining how best to manage them in the context of the system as a whole. In clearly distinguishing systemic risks from other types of risks, the challenges that the key features of systemic risks pose for their regulation were revealed and articulated.

DOI: 10.4324/9781003253464-12

The book additionally took on the challenge of examining the appropriate paradigm to regulate systemic risks. In so doing, a distinction was drawn between a traditional approach to the regulation of risk on the one hand, and a risk governance approach on the other. By identifying and analysing the features of a traditional approach to the regulation of risk, an understanding of the capacity of this traditional approach to address systemic risk was consolidated and elaborated.

The analysis indicated that a traditional approach to the regulation of risk is unlikely to be adequate to effectively manage systemic risks. The limitations of a traditional approach relate to its emphasis on anticipating risks (because it is challenging to anticipate systemic risks, which are characterised by deep uncertainty), the linear process it implies when managing risks (as systemic risk may be too complex and uncertain to be accommodated within a linear model) and the lack of interdisciplinary interaction between regulatory context and content (which is necessary to help address the spectrum of issues that could arise in relation to systemic risks). In addition, a traditional approach to the regulation of risk focuses on stabilising outcomes to avoid deviation from the accepted status quo, whereas the nature of systemic risks is such that regulation may instead be needed to guide systems to change in line with change to the system's context.

Risk governance was examined in the book as an alternative basis for the regulation of systemic risks. Compared to a traditional approach to the regulation of risk, a risk governance approach is likely to be more appropriate to address the challenges associated with systemic risks. Risk governance is a broader, more flexible, inclusive and responsive approach to the regulation of risk, compared to a traditional approach. Risk governance anticipates a process where a broad range of voices will be heard in determining the response to risk, it embeds a holistic, interdisciplinary assessment of risk, and is designed to deliver a response to risk that will be fit-for-purpose and capable of evolution over time. Risk governance, therefore, holds promise for the regulation of systemic risks.

The book went further and considered how a regulatory regime based on a risk governance approach could be designed. By positing an ideal to regulate systemic risks, which was developed by translating into regulatory form an existing risk governance framework to address systemic risks (namely, the IRGC Framework), the book delivered a concrete blueprint for the regulation of systemic risks that did not previously exist. This ideal could form the basis for the design and evaluation of regimes that regulate systemic risks, including those caused by climate change.

What follows is a discussion of the central themes that emerge from the book, both from the theoretical analysis and the case studies where actual regulatory regimes applicable to the management of systemic risks caused by climate change were examined.

II The rise of risk as an organising principle in regulation

Risk as an organising principle has undeniably come of age. This is evident from the extensive academic discourse on risk in a broad variety of disciplines as well as in many different natural and social contexts, including in a regulatory context.

This book has demonstrated that there is a close association between risk and regulation. Kessler (2008) suggests that the "turn" to risk reflects an effort to connect the present with the future and implies that the objective of the increasing reference to risk as an organising principle is to pre-empt risk before it materialises.[1] The question arises as to whether this turn to risk is merely a semantic shift, both in terms of the narrative about the purpose and function of regulation, as well as the way regulation is articulated in the rules and standards. Or, rather, is it the case that a more focused concentration on risk by policy- and law-makers is leading to a fundamental change in the way regulation is designed and applied in practice? Both explanations are plausible and, moreover, they are not mutually exclusive.

In a regulatory context, this transformational orientation towards risk implies that regulation can facilitate different futures through the effective management of risk. In other words, regulation could dramatically change societal outcomes by effectively mitigating risks that would otherwise arise. However, it may not be possible to anticipate systemic risks – at least, not through a traditional approach to the regulation of risk – because systemic risks are typically characterised by inherent complexity, profound uncertainty and, at times, overwhelming ambiguity. As a consequence, it may not be possible to predict the future for this type of risk and this, in turn, may hamper the ability of policy- and law-makers to determine whether and how to employ regulation to intervene. Hence the conundrum: given their challenging characteristics, does regulation have a role to play in addressing systemic risks?

III The regulatory conundrum arising from our dependence on systems

That we are profoundly dependent on a vast array of complex human and natural systems is indisputable. Our lives are inextricably intertwined with many systems that provide a broad range of services which support our way of life and, moreover, keep us alive. This reliance leaves society extremely vulnerable, particularly in the face of a range of potentially catastrophic systemic risks, including those caused by climate change. In the words of Wiener (2016, 2017), these risks give rise to the "tragedy of the uncommons"[2] involving the misperception and mismanagement of rare yet catastrophic risks.

1 O. Kessler, "Is risk changing the politics of legal argumentation?" (2008) 21(4) *Leiden Journal of International Law*, pp. 863–84, at 864.

2 J.B. Wiener, "The tragedy of the uncommons: On the politics of apocalypse" (2016) 7 *Global Policy*, pp. 67–80; J.B. Wiener, "Risk regulation and future learning" (2017) 8(1) *European Journal of Risk Regulation*, pp. 4–9.

Regulation is, at its essence, a mechanism to manage risk. As such, regulation would seem to be an important tool to facilitate adaptation to the systemic risks associated with climate change. However, while there has been much attention paid by academics, policy and law-makers to the design of regulatory mechanisms for climate change mitigation, climate change adaptation has been largely neglected in comparison. In many ways, this neglect is unsurprising. Regulating systemic risks poses considerable challenges for policy- and law-makers because they are the product of a combination of endogenous and exogenous factors, which collectively lead to complexity, uncertainty and ambiguity.

Systems – like ecosystems, health systems, financial systems and infrastructure systems – are usually intrinsically dynamic and constantly changing, are often characterised by tightly coupled and interdependent components, their behaviour cannot be predicted or explained purely on the basis of an analysis of the system's individual components, and adverse change resulting from the impact of various stressors can cause the system to move from one state to another, which is normally less favourable. Recognising and accounting for the *complexity* of systems and the stressors to which systems may be subject, such as climate change, is critically important to ensure that systemic risks are effectively regulated. As explained in the book, systems thinking could play an important role in developing more of a systems-based approach to determining how to regulate systemic risks. Moreover, there are a number of mechanisms that could be employed in a regulatory context to map a system and stressors to that system, which help deconstruct complexity and, thereby, assist in determining the most appropriate point or points of regulatory intervention. Yet, based on the case studies considered in this book, the extent to which systems thinking has been embraced in designing regulatory regimes to address systemic risks caused by climate change appears limited.

There is an important distinction between risk and *uncertainty* in the context of the regulation of systemic risks, like those caused by climate change. Whereas risk refers to events subject to a known or knowable probability distribution, uncertainty relates to events for which it is not possible to specify numerical probabilities. This distinction between risk and uncertainty is critical in the context of climate change, where the past may not be a good predictor of the future for many climate change hazards. The conclusion that certain aspects of climate change are better characterised as involving uncertainty rather than risk leads to a regulatory dilemma. In particular, it is difficult to know whether and how to regulate systemic risks like those associated with climate change because the risks are so complex as well as deeply uncertain and, potentially, characterised by high degrees of ambiguity. However, failing to regulate may result in materialisation of the risks, which could compromise the operation, and, potentially, the very existence of the system at risk.

It becomes apparent from the discussion about how systemic risk should be defined that regulation of climate change risks is less about regulating known risks but, rather, is concerned with navigating uncertainty, which is a very different endeavour. Various conceptions and theories have been identified in the book that are aimed at coming to terms with uncertainty, particularly in relation to the definition, deconstruction and assessment of systemic risk. They reveal a dichotomy and inherent tension between, on the one hand, knowledge which is based on fact and established information and, on the other, beliefs, values, emotions and social norms. This dichotomy can be simplified as a tension between objectivity and subjectivity.

Objective approaches may be more suitable for known risks, whereas subjective approaches may be relied upon for aspects of risk that are not known, nor knowable. Policy- and law-makers cannot avoid the dissonance that this tension creates but, rather, need to be able to reconcile the competing forces at play if there is any hope of effectively managing systemic risks. It may, therefore, be necessary to embed both objective and subjective assessments of risk in regulatory responses to highly uncertain risks, like many risks associated with climate change. Additionally, regulatory mechanisms to mediate between the two approaches may be needed depending upon the relative uncertainty associated with the risks at issue, although this is likely to be novel for policy- and law-makers to design, as well as for regulators required to apply these mechanisms in practice. Indeed, there was limited explicit evidence that the challenges of assessing uncertain systemic risks have been effectively tackled in the various regulatory regimes considered in the case studies.

Ambiguity arises when determining whether and how to regulate systemic risks. Given the spectrum of modern systems and our heavy dependence on them, the consequences of systemic risks will be widely felt. Those affected will include diverse groups of people with differing perceptions of risk and expectations regarding the way in which risks should be responded to, if at all. Diversity of risk perception is particularly pronounced in the context of climate change, which highlights the need for an inclusive approach in responding to systemic risks and the need to embed public participation in regulatory responses to systemic risks. However, the case studies considered in this book indicate that the potential offered by public participation as a means to enhance responses to the systemic risks caused by climate change has not been fully explored.

It is clear that regulating systems, and risks to systems, requires a unique approach because there will seldom be a single point of regulatory control to influence system outcomes. A multi-faceted regulatory regime may be required, including an array of regulatory tools to manage the spectrum of risks that are likely to be faced by the system. Moreover, regulation may be called upon to perform an unfamiliar role – namely, to guide systems to change in response to the stressors to which they might be subject. This book suggests that risk governance could be a good basis for the regulation of such risks.

IV A new era of regulation to govern systemic risks

Risks in modern society are evolving. Systemic risks, like those caused by climate change, pose an existential threat to society and compel a new era of regulation. Risk governance in a regulatory form offers a means to rise to the challenge. Risk governance fuses theories, approaches and perspectives from varied disciplines and domains. It calls on regulators to transcend a traditional approach to the regulation of risk – which is typically rational yet rigid, linear and static – by, as Corbett puts it, effectively "dancing with the system".[3] This approach helps to ensure that risks to the system are coherently and dynamically managed, thereby avoiding the system being catapulted to a state that could inflict catastrophic and widespread harm.

A key feature of risk governance is the flexible, contextualised, inclusive and responsive approach to managing risk. Such an approach involves defining the regulatory problem as one of influencing systems by guiding them towards desired states, rather than managing risks to a system in a rigid, static and linear way. Risk governance prescribes a decision-making process, but not a specific solution or outcome. Among other things, such an approach promises a better understanding and appreciation of underlying risk by decision-makers that is appropriately contextualised and is more likely to yield a response to manage risk that is acceptable to relevant stakeholders, including the public at large.

The prominent role of public participation in a risk governance approach through communication and engagement is significant and signals a decentring and democratisation of regulation. Despite the political, governance and practical challenges associated with balancing technocratic and democratic approaches that a public-centric model of regulation implies, public participation would appear to be indispensable to address the complexity of problems underlying systemic risks, where "post-normal science" necessarily applies.

Risk governance effectively vests decision-makers tasked to manage systemic risks with discretion and flexibility to arrive at an appropriate regulatory response. At least in theory, such flexibility enables decision-makers to account for any trade-offs that may be required to effectively manage risks to a system, ensure an appropriate balance between anticipation and resilience, and tailor the regulatory response to the specific context in which those risks have arisen, noting that this response may change over time as the relevant system's context changes.

V An ideal for the regulation of systemic risks

In bemoaning the limitations of what has been characterised in this book as a traditional approach to the regulation of risk, Fisher (2013) makes a plea for

3 A. Corbett, "A systems approach to regulatory excellence", in C. Coglianese (ed.), *Achieving Regulatory Excellence* (Brookings Institution Press, 2017), citing D. Meadows, *Thinking in Systems: A Primer* (Chelsea Green, 2008), p. 188.

greater "imagination".[4] The book has responded to this call to action by identifying the building blocks for the regulation of systemic risk based on a risk governance approach. These building blocks are then used as the foundation for the development of an ideal for the regulation of systemic risk, which goes beyond high-level suggestions that have been made to date about the general legal approach and principles to address climate change risks.

Choosing a basis from which to establish a vision of how systemic risks should be regulated was a challenging exercise. Indeed, where should one start in establishing a paradigm for risks of a nature, scale and consequence beyond those that have been seen before? While systemic risks are not new, the global context in which they are now arising indicates that a traditional approach to the regulation of risk will not be adequate. Ultimately, the IRGC Framework was selected as the appropriate starting point, given its promise to effectively manage modern systemic risks. The Framework proved useful in several important respects. First, it helped identify the most important aspects of a process to understand and address systemic risks – particularly, the need for a structured and holistic approach to risk identification, assessment and management. Second, it highlighted the importance and benefits of an interdisciplinary perspective for the management of systemic risks. Finally, it crystallised the critical importance of providing for effective public participation for the management of systemic risks. However, the IRGC Framework has not, to date, been expressed in a way that can be readily applied in a regulatory context.

Accordingly, building blocks were established by interpreting relevant aspects of the IRGC Framework in a regulatory context. These building blocks were identified and elaborated to ensure that the essential elements of an ideal regulatory regime to manage systemic risks were explicit and clear and were then "assembled" within the architecture of a regulatory regime. The first set of building blocks included mechanisms to establish the boundary of an ideal regime for the regulation of systemic risks. The second set of building blocks specified the regulatory mechanisms to provide context for responding to systemic risks. The third set of building blocks concerned establishment of the content of regulatory regimes to manage systemic risks – particularly, the regulatory approach, objectives, principles and tools.

The ideal yielded from this building block process gives life to the concept of risk governance – a concept that is increasingly gaining traction but, to date, has been more abstract than concrete in a regulatory setting. The ideal provides a novel yet rational and structured basis upon which to analyse and compare existing regulatory regimes and suggests a potential way to design new regulatory regimes in the future. This is not to say that a standardised regulatory response to systemic risks is being advocated. Indeed, the variety of systems and the stressors to which a system may be subject clearly militates against a one-size-fits-all

4 E. Fisher, "Framing risk regulation: A critical reflection" (2013) 4(2) *European Journal of Risk Regulation*, pp. 125–32, at 132.

approach. On the contrary, the regulatory ideal seeks to avoid a deterministic approach to responding to risk but, rather, attempts to establish a structured process that can deliver a response to risk that is contextualised, tailored and fit-for-purpose. It dispels the notion that there is no room for idealism in the context of regulating risk.

VI Comparison of the regulatory ideal with reality

The risk governance ideal for the regulation of systemic risks outlined in this book was used to evaluate existing regulatory regimes to determine whether idealism is borne out in practice. In particular, regulatory regimes were considered in relation to four different types of systems in various jurisdictions around the world – namely, ecosystems in the EU, the health system in the UK, the financial system in the US, and infrastructure systems in Australia – to determine the extent to which any of the regimes were aligned with the ideal.

The case studies related to very different systems. On the surface, there appears to be little in common between them. However, as revealed in the case studies, there are similarities in their essential characteristics. In particular, each of the systems performs important societal functions, they are each incredibly complex and they are all subject to a variety of uncertain stressors that may affect their stability and, therefore, their capacity to operate efficiently and effectively. These similarities are important because they suggest that there may be lessons to be learned regarding the way in which systemic risks are regulated in each of these contexts.

It would be reasonable to expect that the key systems upon which society depends, like those covered by the case studies, would be obviously identifiable and represented in the applicable regulatory regimes. However, the case studies suggested a different reality. There were varying degrees to which the relevant systems were represented within the corresponding regulatory regimes, ranging from explicit reference to the financial system in the US regime, to no conspicuous representation of an interconnected infrastructure system within aspects of the Australian regulatory regime. A lack of clear representation of a system within a regulatory regime may be due to the fact that the regime does not directly regulate the entire system or does so only partially (an inference that could be drawn about the UK regime), the system may have evolved since establishment of the regime (as in the case of the EU regime), or policy- and law-makers who originally designed the regime may not have recognised a system as such (this is apparent in relation to aspects of the Australian regime). The absence of explicit recognition and representation of a system in the relevant regulatory regime is problematic because it means that the capacity of the regime to comprehensively and effectively address risks to the system may be undermined.

Another striking observation arising from the case studies is the fact that, for the most part, the regulatory regimes were not couched in terms of risk. For example, the UK regime focuses on the establishment, operation and management of the National Health Service, rather than the management of risks associated with the health system. Indeed, based on the terms of the regulatory framework, this regime does not appear to be an obvious vehicle to manage risks at all. Therefore, an analysis of the manner in which risks are regulated under such regimes is somewhat artificial. Nonetheless, it is important to look behind the detailed regulatory provisions to ascertain whether risks to the relevant systems are implicitly recognised and that, in turn, helps to determine how those risks are likely to be managed in practice under the regulatory regime, if at all.

Notably, the Australian regime is the only regime considered in the case studies where climate change risks are explicitly referenced in statutory, legally binding instruments that directly regulate the systems in question. The silence about climate change risks in other regimes may be due to the fact that the regime predates concerns about climate change (as in the case of the EU regime), the regime takes an "all hazards approach", obviating the need to specifically identify all relevant risks (this approach is implicit in the public health protection duty in the UK regime), or there are other ways in which relevant risks may be identified and accounted for (such as the characterisation of climate change risks to the US financial system as "an emerging threat to financial stability", thereby bringing these risks within the purview of the regulatory regime).

In fact, various mechanisms may be used in regulatory regimes where express reference to climate change risks is absent to recognise these risks as falling within the scope of the regime. For example, non-binding recommendations have been issued in the context of the EU regime to stress the importance of accounting for climate change for the purposes of ensuring conservation of ecosystems. In the case of the UK regime, explanatory notes that accompanied the introduction of a public health protection duty into the regime speak of the need to ensure that the duty continues to be relevant and effective as new health threats emerge and specific reference is made in this regard to climate change. As for the US regime, a recent Presidential executive order has helped to ensure consideration of climate change risks in the context of the regulatory regime governing the US financial system. In addition, various bodies have been tasked to assist with the assessment of those risks, such as a "Climate-Related Financial Risk Committee" in the context of the US regime. The use of these types of mechanisms to fill in what could be considered as regulatory gaps regarding the management of climate change risks is heartening. They indicate that pre-existing regulatory frameworks may have the capacity to evolve over time to address emerging risks and, therefore, avoid obsolescence as the context within which those regimes operate changes.

In order to establish how climate change risks are managed under each regime, the regimes were "unpacked" in the case studies and scrutinised using the lens of risk. The core questions that were posed while undertaking this exercise included: Is risk explicitly identified? If so, how is risk managed? Is there evidence that the systemic nature of risk has been accounted for? Does the approach reflect a risk governance approach? Two of the regimes considered in the case studies displayed some alignment with the regulatory ideal – namely, the US regime concerning the US financial system and the Australian regime dealing with bushfire risks to infrastructure in the State of Victoria. These regimes were reformed in the wake of major catastrophes that threw a spotlight on the inadequacies of the pre-existing regimes – namely, the 2008 Global Financial Crisis in the case of the US regime, and the 2009 Black Saturday bushfires in relation to the Australian regime. This context may explain, at least partly, why the post-reform versions of the respective regimes are more clearly focused on the management of risk compared to the other regimes considered in this book. While the degree of alignment of the EU regime with the ideal was found to be limited, a recent EU initiative in the context of its strategy to adapt to climate change identifies that a new governance framework will be established to, among other things, help steer EU ecosystems in the right direction. As for the UK regime, the learnings from the impact of the COVID-19 pandemic on health systems have led to proposed reforms to the regime that appear to better recognise the systemic features of the health system and, if passed, will hopefully pave the way for an integrated and coherent response to systemic shocks, such as those that could be caused by climate change.

In the case of all regimes, there are key areas where risk governance, and the regulatory ideal based on this approach, could enhance outcomes in relation to the management of systemic risks. The first relates to applying systems thinking to better understand systems, including their strengths but particularly their weaknesses, so that regulatory intervention can be better targeted. The second concerns clarifying how systemic risks will be contextualised, which will help navigate the uncertainty associated with such risks. The third involves the categorisation of systemic risks based on the relative degree of complexity, uncertainty and ambiguity so that appropriate regulatory strategies can be deployed that maximise the chance of effectively mitigating risk. The final way in which risk governance could enhance outcomes in the context of the regulation of systemic risks is public participation, which can help to build trust, expand understanding of risk and overcome obstacles that might otherwise prevent the effective management of such risks.

VII The way forward

As climate change risks intensify, society faces the prospect of an onslaught of systemic risks given our dependence on a vast array of natural and human systems that are vulnerable to climate change. If we are to have any real chance of effectively managing these risks, all available tools will need to be mobilised. Law, policy and regulation at the international, national and local levels will be critically important ammunition among the arsenal needed to combat these risks.

This book indicates that we are really only just coming to grips with what it might take to effectively manage systemic risks caused by climate change through regulation. What is clear is that it is time to part company with the past and a traditional approach to the regulation of risk. Rather, we need to use regulation to navigate the uncertainty that will no doubt characterise our future. Risk governance, which enables systemic risks to be managed in a structured, holistic, contextualised and interdisciplinary manner, where all relevant voices are heard, offers a way forward.

Bibliography

Corbett, A. (2017). "A systems approach to regulatory excellence." In C. Coglianese (ed.), *Achieving Regulatory Excellence*. Brookings Institution Press. https://doi.org/10.7864/j.ctt1hfr226.19.

Fisher, E. (2013). "Framing risk regulation: A critical reflection." *European Journal of Risk Regulation*, 4(2), 125–132. https://doi.org/10.1017/S1867299X00003299.

Kessler, O. (2008). "Is risk changing the politics of legal argumentation?" *Leiden Journal of International Law*, 21(4), 863–884. https://doi.org/10.1017/S0922156508005426.

Wiener, J.B. (2016). "The tragedy of the uncommons: On the politics of apocalypse." *Global Policy*, 7, 67–80. https://doi.org/10.1111/1758-5899.12319.

Wiener, J.B. (2017). "Risk regulation and future learning." *European Journal of Risk Regulation*, 8(1), 4–9. https://doi.org/10.1017/err.2017.2.

Index

Note: page numbers in italic type refer to Figures; those in bold type refer to Tables.

For Product Safety Concerns and Information please contact our EU
representative GPSR@taylorandfrancis.com
Taylor & Francis Verlag GmbH, Kaufingerstraße 24, 80331 München, Germany

www.ingramcontent.com/pod-product-compliance
Lightning Source LLC
LaVergne TN
LVHW021121300926
849160LV00010B/805

9781032182155